A RIVER IN THE CITY OF FOUNTAINS

A RIVER IN THE CITY OF FOUNTAINS

AN ENVIRONMENTAL HISTORY OF KANSAS CITY AND THE MISSOURI RIVER

AMAHIA MALLEA

UNIVERSITY PRESS OF KANSAS

© 2018 by the University Press of Kansas
All rights reserved

Published by the University Press of Kansas (Lawrence, Kansas 66045), which was organized by the Kansas Board of Regents and is operated and funded by Emporia State University, Fort Hays State University, Kansas State University, Pittsburg State University, the University of Kansas, and Wichita State University.

Library of Congress Cataloging-in-Publication Data

Names: Mallea, Amahia, author.

Title: A river in the city of fountains : an environmental history of Kansas City and the Missouri River / Amahia Mallea.

Description: Lawrence, Kansas : University Press of Kansas, [2018] | Includes bibliographical references and index.

Identifiers: LCCN 2018027421

ISBN 9780700627103 (cloth:alk. paper)

ISBN 9780700627110 (pbk.:alk. paper)

ISBN 9780700627127 (ebook)

Subjects: LCSH: Kansas City (Mo.)—Environmental conditions. | Missouri River—Environmental conditions. | Water quality management—Missouri—Kansas City—History. | Drinking water—Missouri—Kansas City—History.

Classification: LCC GE155.M85 M35 2018 | DDC 333.91/621609778411—dc23.

LC record available at https://lccn.loc.gov/2018027421.

British Library Cataloguing-in-Publication Data is available.

Printed in the United States of America

10 9 8 7 6 5 4 3 2

The paper used in this publication is acid free and meets the minimum requirements of the American National Standard for Permanence of Paper for Printed Library Materials Z39.48–1992.

For Family/Sendikidea

CONTENTS

Acknowledgments ix

1 An Introduction to Health and Wealth 1

Part I City: Urban Innards 15

2 A View from the Bluffs, 1903 17

3 Drinking the Water 33

4 Sewers: The Waste Stream 70

Part II Region: Borders, Bonds, and Bodies 103

5 Sister Cities 105

6 Social Innards 132

Part III Basin: Health and Wealth 165

7 A Broader Vision 167

8 A View from the Bluffs, 1951 198

9 Downstreamers and Postwar Pollution in a Federal Era 211

10 Concluding with a View from the River 234

Notes 253

Bibliography 313

Index 333

ACKNOWLEDGMENTS

Because the book you are holding represents work spread over eighteen years, it is impossible to thank everyone. I will try to retrace my steps and highlight the people who helped shape this book and make it possible—directly or indirectly. Many of them, I'm happy to say, are still part of my life. No one is forgotten; you are in this book.

Many fine archives, collections, and librarians have helped along the way. I would like to single out the fine staff of the once-named Western Historical Manuscripts Collection in Columbia (thanks, Bill, David, and John) and the Kansas City Public Library, notably Mary Beveridge and, recently, Michael Wells.

From the start I had wonderful teachers and mentors, especially my graduate advisor and hero Susan Flader. Gordon Dodds, Bill Lang, Patricia Schechter, and Tim Garrison inspired and nurtured me at Portland State University, as did my classmates Rose Murdock and Mary K. Gayne. I received excellent training from professors like Mary Neth, Catherine Rymph, and LeeAnn Whites at the University of Missouri.

Good friends at Mizzou like Kristine Stilwell, Bob Faust, Jon Taylor (thanks for the last-minute editorial advice), and other basement pals made graduate school fun and intellectually stimulating. I was fed by the food and camaraderie of people (thanks, Amanda and Pat) associated with Main Squeeze. My friendship with Kris Lawson has lasted through TAing, dissertations, and into manuscripts and jobs, and I thank her for her humor and buoyancy, keen insights, and writing retreats. The Lawson family provided me omelets, wine, and quiet space. One day, we will open a coffee shop and pie counter with full-spectrum lighting and a big sandbox so that professors can work on the beach. I've also made other friends along the way who have supported my growth as a citizen and thinker. Jeff Durbin, Xiao Wang, Amy Dona, and Bob Kauffman come to mind. In fact, I'm in debt to all bike mechanics and river cleanup organizers.

Thanks to the good cheer of my UND officemate Bret Weber, who saw me through one winter in Grand Forks as I shed the ABD for a PhD. Over the years I've made many acquaintances through the American Society for Environmental History and I thank them for their contributions to my growth as a scholar. These include Andrew Hurley, Martin Melosi, Joel Tarr, Sarah Elkind, Craig Colten, Mara Drogan, John Herron, and Robert Kelley Schneiders. The NEH summer institute at the University of Arizona resulted in eye-opening experiences and fresh perspectives about environment and borders. Thank you to Katherine Morrissey, Samer Alatout, Kirsten Gardner, Heather Theissen-Riley, Neil Prendergast, Paul Hirt, and Marsha Weisiger, among others.

I am lucky to inhabit a sane and friendly History Department at Drake. Karen Leroux, Matt Esposito, Glenn McKnight, and the other members have been supportive and wonderful people to work alongside. They understand what it's like to be torn between teaching and writing. My time at Drake has been enhanced by collaborative friends (Michael Haedicke, Carlyn Crowe, and Carol Spaulding-Kruse, to name but a few). Through the Anderson Gallery I met the artist Scott Robert Hudson, who, to me, is an environmental historian who paints bones and carves wood instead of writing books. His friendship and work add a valuable dimension to my understanding of the field.

At the press, I thank my first editor, Ranjit Arab, as well as the readers (I'm indebted). My patient editor Kim Hogeland supported me at a critical time and Amy Sherman was a fantastic copy editor. Sara Vélez Mallea has shared her sage advice all the way through. What I've molded is subject to the pressures one feels after spending so many years on a manuscript. If I were starting over again today, I might write a different book. I am grateful to the Drake University Center for the Humanities for financially supporting the publication of this book.

Family is my backbone. My parents, Karen and Joxe, gave me a love of the outdoors, books, and humanity. I never saw myself as trying to follow in their footsteps, but I realize now that I did. Nikane and Erik have been raucously funny and inspiring siblings. The newest member of my family, my partner, Mark Kende, gets the final, loving mention. Thank you, family.

AN INTRODUCTION TO HEALTH AND WEALTH

I

> Kansas City, Missouri, is the preeminent city on the longest river in America, but you would never know it from talking to the inhabitants, not because of their usual modesty but rather their forgetfulness of the Missouri. In the self-proclaimed City of Fountains there is no spiritual link between them and it and only a distant awareness of its connection to their iced tea, potted geraniums, and baptized babies.
>
> William Least Heat-Moon, River-Horse (1999)

Only blocks away from the confluence of the Missouri and Kaw (Kansas) Rivers stands the *Muse of the Missouri*, one of over 150 fountains in the Kansas Cities. For perspective, only Rome, Italy, has more. The Kansas Cities' construction of fountains dates back to the Progressive Era, when the city made water more accessible to horses and residents without plumbing. Over the twentieth century, as public water became ubiquitous, the fountains became artwork and status symbols. The rivers literally ran through the cities—fountains included—because urbanites were reliant on the rivers for drinking water and sewerage. The *Muse of the Missouri* depicts a fisherwoman, her net abundant with fish, symbolizing the river's service to the city. It was installed in the 1960s—a time when the Kansas Cities were abandoning the river districts. The rivers became invisible to Kansas Citians, even though they hide in plain view—in fountains, as drinking water, and under expressways. In the 1990s, a period of renewal, Kansas City, Missouri, appropriately branded itself the "City of Fountains," each fountain a celebration of the river that has given the city life.

This book is about how intertwined the city and river are and how, despite the difficulty of seeing the Missouri, the river and city are actually in a long-term, intimate relationship. This book aims to make the river more visible and to explore the reasons the city came to ignore its most important asset. If you aren't from Kansas City, or you've never visited, then know that this story is replicated in many cities and on many rivers. What makes Kansas City unique is that it's an extreme example of a river city. I hope that, by the end of this book, readers will want to *see*—literally and symbolically—the Missouri River, or whichever waterway is at the heart of their lives.

I use the term *urban environment* to describe a physical and cultural middle ground that avoids the human/nature dichotomy. It is easy to conjure the difference between the proper nouns *Kansas City* and *Missouri River*, but I wish to draw your attention to the built environment that mediates.[1] The river-city relationship can be explored through infrastructure, fountains, a water purification process, a well-plumbed suburban neighborhood, and a concrete company owned by a political boss—all of which have acted as arbiters between people and the river. A sewer, for example, required technology and human labor to convert a clay deposit into a pipe that blends the city and river and lies at the heart of urban function. The location and service of that sewer was influenced by political power.

The river has been (and is) ubiquitous; its waters have been found in pipes, toilets, public bathing pools, breweries, blocks of ice, and firehoses—even the streets were paved with sand dredged from the river. The city has been present in the river as human and industrial waste and as engineering. What we have is a complex relationship in which the social matrix is constantly and inseparably interacting with the ecological system, and together they make the urban environment.

The Kansas Cities consist of Kansas City, Missouri, and Kansas City, Kansas, as well as dozens of other smaller cities that make up the metropolitan area. Sometimes I generically refer to the collective "Kansas City" or cities, but when the evidence refers to only one of the cities, I am specific. Kansas Citians used to fondly refer to their particular place with

Muse of the Missouri. Courtesy of Bradley Cramer and the City of Fountains Foundation

the phrase "at this bend, atop these bluffs." The physical characteristics of the landscape provided the base for a river city. We know that earlier heartland cities grew along rivers—like Cahokia on the Mississippi, Etzanoa on the Arkansas, and the Mandan villages on the Missouri. William Gilpin, a nineteenth-century booster, believed that the American West would develop a "Centropolis" and that Kansas City, due to its "natural advantages," would be it.[2] (Spoiler: Chicago won.) Early historians rooted the origins of Kansas City's greatness in its geography, topography, and natural resources.[3] Later in the twentieth century, historians pointed to cultural explanations for Kansas City. The best historical understanding of this river city melds environmental and human agency.

Through the nineteenth century, the Missouri River was a liquid freeway for tribes, traders, and explorers, and Kansas City served as the "jumping-off" point to overland trails. The Kansas Cities have been a regional hub in the continent's breadbasket since the Civil War. Despite the decline of steamboats and the rise of railroads and then automobiles, political and economic interests in the Kansas Cities continued to promote river transport. It amounted to a century-long obsession with managing the river for navigation and flood control. During the twentieth century, these leaders led the campaign to enlist the federal government and billions of public dollars to engineer the Missouri River in support of the industrial economy. These powerful Kansas City interests are the entity most responsible for reshaping the Missouri River, especially in the lower basin.

The Missouri River has been too narrowly managed for the last one hundred years; the current shape of the river is not working for the majority nor to the benefit of human and environmental health. The river is controlled for barge navigation, which is not essential to the economy, whereas the river *is* essential to the region's health and to the daily function of the urbanized Missouri basin. Millions of people rely on the river for drinking water. This fact was the basis of the 2016 protests by the Standing Rock Sioux Tribe, who worried that the oil in the Dakota Access Pipeline would put the tribe's water supply at risk as it passed under the Missouri. Although urban and environmental concerns now receive more attention in river management, even today, public health is not the priority but a competing interest.

Cities of the early twentieth century figured little in the shaping of broader resource policies although they shaped landscapes and health delivery systems (like municipal services) within their own jurisdictions. The Missouri and Kaw Rivers were the most important elements of Kansas City health, yet local public health officials had little power over river management or water quality. Ultimately, the way local, state, and federal governments did (or did not) work together was decisive in determining human and environmental health. Because they lacked control upstream, individual cities and states could not guarantee high water quality, nor would they sacrifice and cooperate with each other. This bind caused Progressive Era public health officials (then called sanitarians) to call for federal aid to coordinate and oversee environmental and public health issues. Not until later in the century would federal policies (like the Clean Water Act) and agencies (like the Environmental Protection Agency) play an increasingly important and positive role in health.

This urban environmental history adds the complicating layers of multiple political jurisdictions and expands beyond city limits to better see the river as a force. Political boundaries complicated the watershed because the river respected no city, county, or state lines in its run from the Rockies to the Gulf. Like many rivers in the United States and world, the lower Missouri forms the border of several states and in so doing reinforces the idea that rivers bound, not bind. I show that the river is not a border—it is the connective matter between places and people.

Today, residents, researchers, and visitors undoubtedly notice the socioeconomic segregation in the metropolitan region. This stark urban geography has historical roots.[4] The racial and class inequalities of the Kansas Cities have been reinforced by an urban system that unevenly distributes public resources, infrastructure, and risk, all of which relate to the rivers. The cities' urban innards—like drinking water and sewerage—have always been tied to the river. Political boundaries were a reinforcement tool for the Kansas Cities. In general, the city and river are organized to keep power flowing to the powerful. Environmental justice activists connect social and environmental issues and make the power flow visible, seeking to interrupt it. By pointing out the high incidence of environmental risks in poor neighborhoods and those with people of color, environmental justice activists argue that urban resources and good health should be more equitably distributed. They are in agreement with those officials who thought that public health was the most important attribute of the river, and who were concerned with the equitable distribution of the essential resource of water.

The Missouri River is a captured resource, managed in order to serve an industrial economy, but the river's most important function is health. A healthy river sustains wetlands that reduce flooding, enrich bottomland soils, and provide wildlife habitat for species that are commonly hunted or are uncommon and protected by the Endangered Species Act. A healthy river protects the millions who rely on it for drinking water. Water quality matters more than whether or not the Corps of Engineers can barge gravel to maintain its levee system or the occasional barge loaded with corn can get to New Orleans. Kansas Citians, midwesterners, and all urbanites design their economies through planning, policy, regulation, and daily choices. If we want a healthy river city, then we need to design and develop an economy that supports good human and ecological health.

* Flood events have been far too influential in resource management and policy. This has been detrimental to both city and river. Take note that a flood does *not* best characterize the relationship Kansas Citians have with the Missouri River, though that may be the first thing that comes to mind. Floods showcase exceptional ways the river affected the city, but ordinary uses—like slaking thirst or flushing a toilet—best represent the river-city relationship. These daily uses are so central and mundane as to be invisible until disrupted. The river and city did more than meet at the banks—they mingled. This makes the invisibility of the river, the historical amnesia that leaves the region's residents alienated from its innards, all the more fascinating. Looking beyond the river's identity as a flood threat, I wish to reframe the river-city relationship as one of intimate intertwining.

The primary way this book evaluates river-city intermingling is through health. An (un)healthy body is a tangible way to examine urban environmental interaction. The majority of Americans are urban; the city is the environment we know best. Yet there has been an American cultural tendency to see cities as the antithesis of nature—as unnatural. That is not a very accurate or useful way of understanding our urban surroundings.[5] The river, as it flowed through infrastructure and entered and exited bodies, has always been at the core of public health in Kansas City.

Not all bodies in Kansas City are equal. Because social structure is a determinate in health, those Kansas Citians who are advantaged (for example, white middle-class professionals moving into newer neighborhoods far south of the river) and those who are disadvantaged (working classes and minorities who work in industry and live in older homes nearer the river) experience public health differently. Race, class, and gender influence exposure to risk, sanitation, healthcare, and access to city services. Therefore, disparate socioeconomic groups have had differing relationships to the Missouri River. The intersection between people and their environment is historically evident to us through public health reports and typhoid fever cases.[6]

Similar to the human body, the city has a circulatory system of underground arteries that nourish and cleanse the city. A city's infrastructure must deliver resources to a dense population within a limited space. Urban historian Martin Melosi relays the wonder of the city's body when he writes about the "pipes, conduits, and wires creating a hydraulic, pneumatic, and electrical maze below the streets" that are "integral components in a dynamic environmental system."[7] Lewis Mumford called this physical and technological system the "invisible city"; I call it the "urban innards."[8] Drinking-water flow lines and concrete sewer pipes are extensions of both residents and the river and constitute the intimate river-city relationship. Next time you walk around your city, look for the iron manhole covers. They are evidence of the network of urban innards that connect you to upstreamers and downstreamers. In Kansas City, manhole covers are emblazoned with the fountain logo of the waterworks. Ultimately, I hope that readers—wherever they may live—will see the river in the city, and the city in the river.

The late nineteenth-century Kansas Cities underwent explosive population growth and began a formative era of city building. This era gave the city its shape and established the indissoluble connection of the two cities while revealing the different political characters of their respective states. The urban innards built at the turn of the twentieth century would form the basis of urban environmental interactions for the coming decades, and in that the river was critical. While this work is not meant to be a comprehensive urban history, it lends a new angle to our understanding of the cities. It reassesses urban politics and city growth by integrating insights from recent social histories. Unlike other works, this one integrates the river and recognizes the importance of relationships—from neighborhoods to the basin.[9] The resource of water was a powerful tool to enable wealth and health for some, and deny it to others.

Just as the city cannot be understood without the river, the city can't be understood without both sides of the border. Kansas City, Missouri, has received a fair amount of historical attention, but there is little on Kansas City, Kansas, or the Kansas Cities—a problem for understanding the region historically.[10] Though politically divided, the Kansas Cities have been an economic, social, and environmental unit, and their history cannot be dissected neatly along the state line. The bias toward Missouri obscures economic and social issues. In the early twentieth century, the means of production—the industries and diverse workforce—were more apt to reside on the Kansas side of the river bottoms, whereas the owners and wealth tended to concentrate on the Missouri side, above and away from the river districts. Slighting the Kansas side historically has been a convenient way to ignore socioeconomic disparity and the costs of industrialization. Daily, people have traversed boundaries for jobs, recreation, and shopping, making the border less relevant. This bistate social geography has been inextricably tied to the Missouri and Kaw Rivers. Lastly, the environment has not adhered to political boundaries; the contour of the bluffs and bottoms, the flow of water, and movement of microbes have all permeated the state line.

The state boundary that bisects the city has had a far-reaching and sometimes negative impact. While the political boundaries have been stable, the social geography has shifted. One hundred years ago, Missouri residents held most of the wealth. By the late twentieth century, city, county, and state lines further accentuated a socioeconomically fractured city. Today, the wealth has shifted to the Kansas side—particularly suburban Johnson County, which is among the nation's wealthiest counties. Whereas Kansas City, Missouri, boomed in the early twentieth century, today the city proper is losing population compared to its outlying suburbs. This "hollowing out" has contributed to segregation, sprawl, a loss of economic base, an educational crisis, and other costs associated with a loss of density.[11] Though socially divided throughout the century, the Kansas Cities have functioned as an environmental whole—the best evidence is the common water supply. Rather than continue as an exaggerated example of the way political borders undermine public health and resource management, this study suggests that a focus on commonalities, like human and environmental health, can transcend boundaries and aid metropolitan governance. The Kansas Cities showcase federalism's challenge to protect health. States have eagerly sought federal aid for economic development but have played the states rights' card when it comes to protecting human and ecological health.

At the turn of the century, the Kansas Cities were a politically progressive place—the most progressive in Missouri and the least progressive in Kansas. The cities contributed to the regional and national flavor of progressivism. During an era of reform, from the 1890s to the 1930s, Kansas City, Missouri's urban growth was shaped by the presence of one of America's longest-running urban political machines. The Pendergast brothers, first Jim and then Tom, controlled voting blocs and influenced the city's purse strings. Municipal services and city building set the machine's power in concrete—literally. City departments like streets and water became infamous nests for political favors and corruption, and Pendergast's own investments included local construction companies. Though never elected to office, "Boss" Tom Pendergast would go on to hold a commanding seat of power in the state of Missouri during the New Deal.

Social reformers were another powerful influence in Kansas City, usually counter to the machine. Reformers displayed their white, middle-class values as they sought to steer the populace, notably through public health campaigns and institutions. Economic interests and boosters also focused on aspects of urban reform. Thus, at the center of political struggles were resources like river water. At times, these disparate interests aligned to "Make Kansas City a good place to live in," as the Kansas City Commercial Club advocated. This phrase was the progressive watchword of twentieth-century Kansas City. Opinions differed about how to achieve a "good place"; nevertheless, most could agree upon the end result: a city that was "up to date." The 1943 Rodgers and Hammerstein Broadway musical *Oklahoma!* features a song sung by Will, a young rural man who traveled to Kansas City:

Ev'rythin's up to date in Kansas City.
They've gone about as fur as they c'n go!
They went and built a skyscraper seven stories high—
About as high as a buildin' orta grow.
Ev'rythin's like a dream in Kansas City.
It's better than a magic-lantern show!
Y' c'n turn the radiator on whenever you want some heat.
With ev'ry kind o' comfort ev'ry house is all complete.
You c'n walk to privies in the rain an' never wet yer feet![12]

This popular description of Kansas City reveals its reputation for being well planned and progressive, a definition tied to the built environment, municipal services, and indoor plumbing.

✶ The Missouri River has a rich environmental and social history, but it has been overshadowed. Although it received attention during the bicentennial years of the Corps of Discovery, compared to the storied Mississippi, the Missouri continues to surprise the unsuspecting. John Steinbeck described with surprise his first encounter: "Someone must have told me about the Missouri River . . . or I must have read about it. In either case, I hadn't paid attention. I came on it in amazement."[13] Fans of the Missouri are fond of pointing out that it is North America's longest river, it is the world's fourth-largest river (it used to be third—engineering shortened it), its basin comprises 40 percent of the continental United States, and the Mississippi River incorrectly assumes the title at the confluence near Saint Louis. Technically, it should be the Missouri River from the Continental Divide in the Rocky Mountains, all the way to the Gulf of Mexico.

The Missouri River of today little resembles the wide, shallow, meandering, and unpredictable river that Lewis and Clark wrote about in their journals. State and federal governments have tried to put the river in a straitjacket. During World War II, a multiple-use plan was adopted, commonly referred to as the Pick-Sloan Plan. Combining the US Army Corps of Engineers and the Bureau of Reclamation, Pick-Sloan technocrats dramatically altered the landscape and how residents interacted with the river. Six large mainstem dams were built in the arid upper basin to hold back spring flow for irrigation and downstream flood control. In the lower basin, from Sioux City to Saint Louis, a series of levees and wing dikes straightened, narrowed, and deepened the channel to make it ideal for barge navigation. This engineering organized the river to produce profits but not to protect health. The river ceased to function as it had historically and a loss of habitat and wildlife followed. Most astounding is that, despite the costly straitjacket, navigation is negligible and flood risk is exacerbated.

Two visions of the Missouri River have competed throughout the twentieth century. I call one the Economic River. Its principles, policies, and management were organized around a capitalist model of economic development and focused on interstate commerce. I call the other vision the Healthy River. It was held by sanitarians and, later, environmentalists. This vision was organized around society and public health; later in the century the vision embraced environmentalism and ecosystem restoration. Readers will note that, similar to these dueling visions of the river, *health* and *wealth* could be competing ideologies. This stripped-down dichotomy pits the Economic and Healthy Rivers against one another. The profit of economic development came at the risk and cost of health, whereas prioritizing environmental and public health dampened economic growth and profits. In this history, though we see that the Economic River had an advantage, the river's history runs nearer the middle; neither vision of the Missouri fully materialized, nor was health inimical to wealth.

Most histories about the Missouri River are political and economic works interested in engineering and the Economic River. There is a paucity of consideration for social and urban aspects or the mundane uses of the river, meaning that people have been missing from the river's history.[14] Notable exceptions are literary travel writing and nineteenth-century history, when heroic exploration dominates.[15] The river appears in Kansas City's past but is absent in its present and future, as if nature no longer influences modern humans in their urban endeavors. But all this time, the Missouri is there, vital to the city and its citizens. "We've turned our backs on it, but we're looking over our shoulders at it now," as a modern river rat put it; "maybe we should take another look."[16]

✷ Since the late twentieth century, the river has seen a resurgence of interest. Worldwide, we see how rusty industrial cities like Glasgow, Bilbao, and Portland, Oregon, have worked to revitalize their waterfronts. These vital urban landscapes reorganize the relationships residents have with rivers and watersheds. Kansas City is amid this process of revival. Despite the continued importance of the rivers in their daily lives, most postwar Kansas Citians had not been near the Kaw or Missouri and knew very little about them. They associated the river bottoms with floods, poverty, and pollution. The river districts were once inaccessible to all but the hardiest souls who didn't mind crossing the tracks, traversing the industrial

"Imagining a river past, and a city future." This spliced image, made in the 1950s, depicts an old and new Kansas City along the Missouri River. Steamboats and bankside community gave way to skyscrapers high above the river. Missouri Valley Special Collections, Kansas City Public Library, Kansas City, Missouri

district, and putting in on a mud bank. Slowly, with public support, federal legislation forced cities and states to address water quality and abate pollution. Urban redevelopment has recaptured valuable real estate and brought people nearer the river. Recreationists have also made demands for access and young people are thirsty for knowledge about their surroundings.

From the bluffs today, we can observe the basic elements of this study. With this bird's-eye view, the three historic locations of the Kansas City waterworks are visible along the rivers. Surveying a wide swath, the observer sees watersheds creasing the urban landscape. Most streams have been buried underground as sewers, but topography and gravity reveal where the main lines are laid. The raw sewage outfalls no longer exist, but the round basins of one of the city's seven wastewater treatment facilities is visible in the West Bottoms, as is the roaring outfall from the plant. The

Blue River's treatment plant has a white stack standing as a beacon of the technology and infrastructure that mediates the river-city relationship. In sum, from the bluffs, we can still see that the river runs through the city.

To investigate these topics, I have organized the book in concentric circles of relationships, from the local to the regional. The chapters broaden in time and place, from the city to the bistate metropolitan region and related boundary-crossing social issues, and finally to the lower Missouri River basin. In short chapters that punctuate—the way floods interrupt—I examine the 1903 and 1951 floods and then will conclude with the 1993 flood. I hope to illustrate how the lens of environment can help us see the river and city not as distinct entities but as a system with feedback loops. Concentric circles, from the neighborhood to the watershed, show everyone has lived upstream and downstream.

To illustrate the importance of the river-city relationship, and to present the main themes in this book, I begin locally, with a view from the Kansas City bluffs during flood time. The flood of 1903 is a cultural benchmark—it punctuates public memory and was a common way to perceive river-city interaction. A flood illuminates both the well-known dramatic and the lesser-considered daily uses of the river, all in a span of weeks. By looking at this cycle of flooding in 1903, we will see that the function and physical layout of the city was predicated by waterways, that the river was central to public health, how social geography correlated to environment, and how the river-city relationship was mediated by power and social status. City dwellers need to know that urban environments have sometimes been managed to reinforce socioeconomic inequity.

The following chapters discuss the history of the city and river and, to help tell the story, they highlight a few people who left behind a paper trail. By following the lives and careers of engineers like Robert E. McDonnell and public sanitarians like Dr. Samuel Crumbine, we can better understand formation of the urban innards and the process of the river running through. These were well-educated Anglo-Saxon men who successfully shaped their world. Driven by civic-mindedness, McDonnell became wealthy building water and waste infrastructure in the region and he wielded his political power to achieve progressive goals. Crumbine was a medical doctor who became a Kansas state public health official. He was at the cutting edge of applying progressive ideals to public health governance and took his ideas and talents to the national level. In the postwar era, Melvin Hatcher and Murray Stein advocated for water quality and, from the local to the national level respectively, struggled to protect it with weak regulatory tools. In the final chapter, we come to more recent times and follow Vicki Richmond, an organizer of river cleanup events. After leaving behind her corporate job, she has devoted the last twenty years to getting Kansas Citians, especially the young ones, down along (and sometimes in!) the Missouri River. Each of these people is part of a tradition that believes in a democratic public health, connects human and environmental health, links the local and regional, and, lastly, each saw the river as a common force that accomplishes all this.

The city cannot be seen in isolation. We must connect local neighborhoods to the bistate metropolitan region, and then to the multistate watershed in order to understand the complex social and ecological system of which the Kansas Cities have been a part. Everyone was an upstreamer; everyone was a downstreamer. The river has connected people and placed them in wider environmental and social relationships—particularly through public health. The relationship between the river and the city is a simplification of a complex system, one with cycles and feedback loops, which allows us to connect micro and macro. No single resident of the basin was isolated; every person within the city, and every city along the river, was located downstream or upstream from everyone else. In using the river as that which ties everyone together, we gain a social and ecological microcosm. The river becomes a metaphor, a method of understanding complex processes and relationships, to unite the social and environmental.

PART I

CITY

URBAN INNARDS

A resident of Kansas City [Kansas] is under the control and supervision of four governments. . . . But the one which touches his daily life most closely is the city government. When he washes his face at the bathroom faucet, when he turns on the electric light in his home or office, when he takes his family for a picnic in one of the parks, when he turns in a fire alarm or calls the police, he is in contact with the city government. The paving of the streets, the sidewalks, the sewers, the drives, the public swimming pools. . . .
    Kate Cowick, The Story of Kansas City

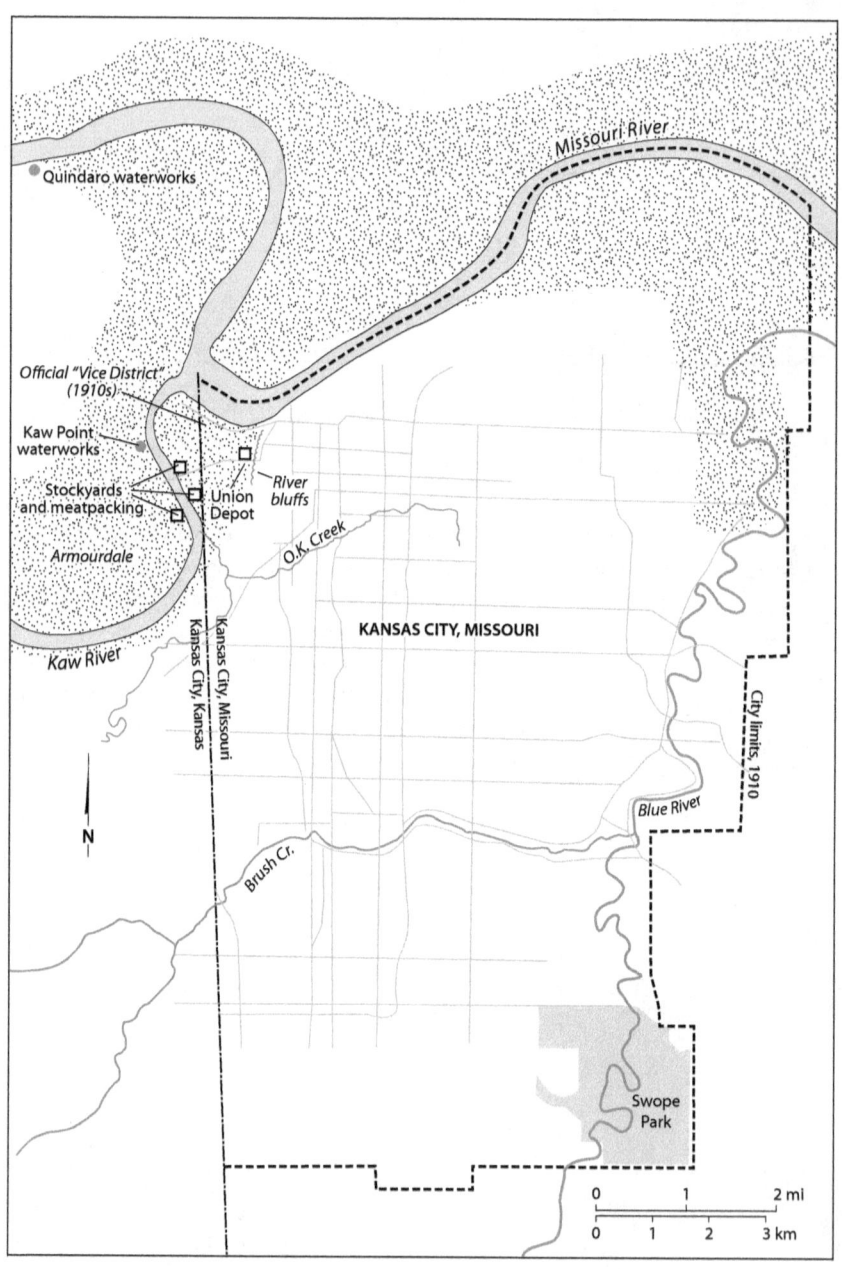

The Kansas Cities in the early twentieth century. Cartography by Bill Nelson

A VIEW FROM THE BLUFFS, 1903

2

Kansas Citians expected flooding in the spring of 1903, but they were not prepared for an epic disaster. The drama unfolded over a span of several days in late May and early June. After as many as a dozen inches of "incessant spring rains" fell in the Midwest, the lower Missouri basin was brimming.[1] The enormous volume of water rushed downstream, headed for the Kansas Cities, which sat at the confluence of the Missouri and Kaw Rivers.

At first the flooding seemed distant to Kansas Citians. Residents heard news about flooding upstream in Kansas, but soon reports circulated that the Kaw nibbled at its banks closer to home. Despite the fact that the Kaw River was in a higher flood stage, it was unable to find release in an already swollen and more powerful Missouri River and its roiling waters backed up into the urban bottoms.

Subscribers to the *Kansas City Star* read about an African American man who drowned. As the paper put it, his "Weed City" shanty "drift[ed] away like a handful of baby's blocks thrown into a bucket of water."[2] The shacks in the bottoms seemed far away to many middle-class Kansas Citians who infrequently ventured down from the bluffs and into these impoverished and industrialized neighborhoods on the front lines of flood risk.

Things were serious enough on Saturday, May 30, that many business owners in the bottoms anticipated flooded basements and moved merchandise up onto main floors. The Kansas Cities' bottoms were the heart of the meatpacking industry. Slaughterhouse workers moved dressed meats onto higher floors and into coolers; many workers were kept on the job so late that their safety was endangered. That same afternoon, calls went out that boats were needed to rescue people in Armourdale, a low-lying working-class and industrial community named for the Armour meatpacking plant along the Kaw River.[3]

By Sunday, May 31, the Chicago & Alton Railway reported that the water was rising as fast as one foot per hour. With as many railcars moved to safer ground as possible, the railway employees finally abandoned the yards and those from the freight house made their escape by raft. That same day, Union Depot, where Kansas City train passengers arrived and departed, had six feet of water in the building. Haunting photos show a waiting room and ticketing office devoid of travelers but filled with water. By Tuesday, at its peak, the flood was in the windows of the depot and up to the awnings of all the businesses lining Union Avenue. This part of town that Union Avenue ran through was, ironically, called Kansas City's "wettest block" because of the number of saloons it harbored.[4] The postcard of the flooded block with that very phrase scrawled on it is one of the most widely distributed and remembered images of the 1903 flood.

The river bottoms were home to the rapidly industrializing Kansas Cities and were a midwestern railroad nexus. Specializing in meatpacking, the Kansas Cities were a western "cow town" grown into a city, second only to Chicago in meat production. But the expansive stockyards and packing district were now "beyond recognition."[5] Stockcars were derailed and the acres of fencing from the stockyards looked like piles of matchsticks. Slaughterhouse workers relocated many animals by railcar, but thousands of cattle and hogs were drowned and just as many survived for days up in the stockyard chutes above the flood. The occasional hardy dog stood atop flotsam as rooftops and giant oil refinery tanks floated downstream.[6] Debris of all kinds hit the bridges and collected, the pressure compromising the structural integrity of these critical crossings. Chaos seeped into this industrial district with the floodwaters as smoke filled the air with explosions of combustibles. Even if firefighters could reach the flames, there

Flooded West Bottoms. The intersection of West Eighth and Mulberry was near Atlas Oats, Admiral Hay agricultural implements, and People's Ice—necessary for meatpacking. Armour is in the background. The 1903 flood dramatically exposed the vulnerability of the industrial district. Missouri Valley Special Collections, Kansas City Public Library, Kansas City, Missouri

was no city water to put them out, and on Sunday afternoon the Kansas City municipal water supply ceased to flow for the first time.

On Monday, June 1, the men of the Kansas City Commercial Club called a "mass meeting." It was held at ten in the morning at their headquarters atop the signature bluffs of downtown Kansas City, Missouri. The influential white men of the club were recognized as the economic and political leaders of the city. Many of the three hundred attendees had not slept due to their flood work. Their experiences informed the alarm already circulating in newspapers. Commercial Club members went into action, authorizing special committees for food, clothing, and transportation. The meeting proceeded quickly, collecting donations for relief, signing up more volunteers, and organizing wagons for supply distribution.

At the end of the meeting, Kansas City, Missouri, Mayor James A. Reed spoke "at length" about the needs of the city, saying that he wished to decline outside help and rely instead on Kansas Citians to address their own crisis, a tack with which the club agreed.[7] This self-reliance to weather the flood crisis was at the origin of what would come to be called the "Kansas City spirit." While citizens were asked to shoulder responsibility, the city promised to protect the economic sector. These men of means had less and more at risk—their homes were on higher ground, but their investments and businesses were closely tied to the flooding bottoms. Newspapers do not record the response of the working classes to this proclamation of self-reliance in the face of disaster.

Mayor Reed assured club members that commercial portions of the city were a priority for protection; every policeman was on duty and federal troops from Fort Leavenworth stood watch in the business district. The mayor ordered that looters be shot, perhaps fearing working-class disorder amid the storehouses of the bottoms. He had even requested that conductors and other employees from the Metropolitan Street Railway Company serve as a temporary security force. With trolley operators deputized to protect commerce, the mayor urged residential communities to organize neighborhood watches.[8] The city closed all saloons. Erroneous reports of hundreds dying in the flood circulated.[9]

Fueled by rumors that food supplies were dwindling and delivery was limited, there was a "rush" on Monday as people stocked up on fresh vegetables, meat, eggs, and flour. Reports circulated about unscrupulous entrepreneurs charging exorbitant prices for basic needs, suggesting a fear of greed. Before disbanding, the club members passed a resolution condemning price gouging of necessities—one of which had become drinking water.[10]

In addition to a near freezing of transportation, the services of gas, electricity, telephone, and water were all lost in the flood. One of the African American newspapers commented that the city was "extremely wet in one way and exceptionally dry in another, it being very near impossible to get even a glass of water to drink. Water was sold through the street for as much as 10 and 15c a gallon."[11] Only one of the seventeen bridges across the Kaw River withstood the flood of 1903 and it was *not* the one that carried the flow line for the Kansas City waterworks.

The complex regional system that provided drinking water for the cities was at risk in flood time. Drinking water was piped from an upstream location on the Missouri River in Quindaro, Kansas, through Kansas City, Kansas, across the Kaw River and to a station in the bottoms of Kansas City, Missouri, where it was then distributed through hundreds of miles of pipes into homes and businesses. In addition to being severed from the waterworks, the city's Turkey Creek pump station was inundated. Workers waded frantically through floodwaters, trying to keep the pumps running until they finally failed, the machines clogged with mud. The average resident did not know how long it would take to repair the infrastructure, but it was worrisome; electricity, streetcars, and gas were not essential to life the way water was.

The month of June began without city water and, early on, engineers and the city water department warned that it would take more than two weeks to restore service. The lack of water affected many aspects of daily life, from public sanitation to being able to get a shave. Lack of water meant no power because steam engines operated on city water and submerged railroads stalled coal delivery. Toilets could not be flushed—although in 1903 only the wealthy and some of the middle class had plumbing in their homes. The dangerous conditions of unflushed water closets led city health officials to ban their use, and to urge the use of lime as a disinfectant. Temporary privies were contrived by placing tents over sewer manholes.

The Turkey Creek pump station became the city's only water source for two weeks, not counting wells and cisterns. City engineers and the Metropolitan Railways Company fashioned an intake to pull water from the Kaw River and settle it in the Turkey Creek reservoir. When the pumps precariously started on June 4, they sent contaminated floodwaters into the system. The pump station was named for the urban waterway it sat near the base of, Turkey Creek, which was a major destination for Kansas City, Missouri's sewage and the host of a main sewer outfall. Unfortunately, as it pulled water from the river below its sewer outfall, Kansas City became its own downstreamer.

Kansas City's residential and industrial wastes emptied into the Kaw and Missouri Rivers. Urban streams like Brush, O.K., and Turkey Creeks and the Blue River carried the wastes of the city, returning drinking water to the river in an altered form and completing a cycle that meant the river ran through the city. In flood time, the normal cycle was altered as rising floodwater plugged sewer outfalls and the waste backed up. With a semisewered metropolitan population well over two hundred thousand, many millions of gallons of raw sewage and industrial waste mingled with the wreckage of the flooded bottoms.

On Thursday, limited water was available from the pump station but with the caveat that it was not for quenching the city's thirst. Health officials warned that the water was "not fit to drink." It was good for flushing toilets and fighting fires; if, out of desperation, the water had to be consumed, it should be boiled and allowed to settle. The public was daily reminded of this fact: "WARNING! The present supply is full of disease germs, sewage and other impurities," the newspaper blared.[12] Even with conservative consumption, the reservoir was scarcely filled and the water supply was "shaky."[13] At least one time the pumps broke down and the stored water was reserved for firefighting and protection of property. Each day the water department reported the difficulty of providing a supply and relayed information about work being done to reestablish the connection to the Quindaro water plant, including building a temporary bridge for the flow line.[14]

As many as twenty-three thousand people in the West Bottoms evacuated—most of them from the Kansas side—and thousands of workers were unemployed.[15] Even before the flood, the bottoms were the least desirable location in which to live. These polluted, impoverished, poorly serviced, older parts of the cities housed and employed the working classes—including immigrants, African Americans, and the underclass. It was congested and full of railroad tracks, manufacturing, saloons, the red-light district, and tenement enclaves where virtually all residents were renters. The most vulnerable populations lived in the most vulnerable places. As waters rose, Mayor Craddock of Kansas City, Kansas, threw open the doors of public institutions. Churches, schools, and the convention hall in Missouri welcomed evacuees and, through the generosity of others, supplies found their way to those in need. While the city leaders met at the Commercial Club, their wives participated in relief work for

Sheltering flood refugees at the newly rebuilt convention hall. Kansas City Convention Hall Records, State Historical Society of Missouri

the refugees. Women's most publicized role in relief was that of seamstresses: they sewed and donated clothing to replace all that had been lost in the flood.[16]

The 1903 flood exposed the social underbelly of the Kansas Cities. A women's club printed a poem about the flood that reveals social divisions of the disaster. Here is the first stanza of "On the House Roof—A Memory of the Kaw River," by Ellen G. Parkhurst:

> Three nights on the house-roof,
> The sky black an' drear,
> An ev'ry whar the water,
> The water creepin' near,
> An' all the night I'se callin',
> O, Lord cain't yo' hear?[17]

The poem is written the way a stereotypical uneducated person might speak, showing that for the middle-class white clubwoman, the typical flood victim was likely poor, uneducated, and maybe black. African Americans were significantly affected by the flood, and snippets and vignettes in the newspaper confirm the association of "refugees" with either nonwhites or poverty. At the end of the poem, the person on the rooftop is finally rescued by a capable oarsman. The newspapers depicted many similar scenes in which horse-drawn wagons and boats plucked families from trees. In public shelters, refugees were required to have passes to get in and out. One "negro refugee," the *Star* said, did not have his pass to return to the hall. When told all refugees needed passes, the man replied assuredly that he was indeed one of those "rivehgees," combining the words *river* and *refugee*.[18] Such articles reveal that black Kansas Citians were assumed to be ignorant and, like other poor people, they were at the mercy of both the floodwaters and public aid.

Within a week sympathy for flood sufferers was waning. Most evacuees were cared for in private homes, but five thousand homeless people were still staying in the convention hall and schools and the city wanted to free up these initial emergency facilities. An official tent camp was established on the Kansas side at Twelfth and Orville Streets, laid out and run with strict order, requiring passes for entry. The racially segregated camp was advertised as "quite an attractive place" and claimed to be the most sanitary spot in the city.[19] Employers looking to hire visited the camps, and the media portrayed slaughterhouses and other businesses as eager to put the people of Kansas City back to work at cleanup and rebuilding. One week after the river peaked, those considered "hobos" were evicted from shelters because they were thought to be idle, unwilling workers (rather than itinerant and seasonal laborers at the margins of society, as they most likely were). By mid-June financial help for refugees receded with the floodwaters. Yet, Kansas City, Kansas, still had two thousand homeless evacuees living in the tent camp.

While many factories and industrial workers were on the Kansas side of the border, the owners of wealth were overwhelmingly on the Missouri side. On a day-to-day basis the cities operated as a unit, but during the 1903 disaster the political boundary became a convenient excuse for ignoring the crisis. Because of the demographics of wealth and poverty, Missouri had the capacity to bounce back more quickly than Kansas. In the

"Armourdale cleans up." Hard-hit Armourdale, closely associated with meatpacking, was a largely immigrant community on the Kansas side. Harrison Photograph Collection, State Historical Society of Missouri

midst of the crisis, the usually magnanimous *Star* editorialized that the corporations and wealthy of Kansas City were "shirking great obligations" and not contributing enough money—especially pertinent to the urban unit because the Missouri side continued to decline outside help.[20] The general relief committee had collected $60,000 at that point, a paltry sum compared to need. A couple of days later, in response to outside aid being sent overwhelmingly to the better-known Kansas City, Missouri, the *Star* reported, "Kansas City, Kas., needs all the money it can get," and encouraged Missourians to refer potential donors to Kansas. The *Star* continued: "It is not understood in the East that, while only a small percentage of Kansas City, Mo., people are sufferers, one-third of the population on the Kansas side was drowned out."[21]

No donations were turned away on the Kansas side. By the middle of June, Kansas governor Willis Bailey's flood relief committee had taken in nearly $40,000.[22] Although money was raised for victims in the entire Kaw valley, the bulk of the need was in Wyandotte County because Kansas City had five thousand needy families. One old-timer living near the bottoms compared the 1844 and 1903 floods and summarized: "Then it was nature ruined, now it is man."[23]

Since the bridges were out of commission, the river became a gulf between two formerly connected banks. Stories of oarsmen who worked continuously, their hands blistered, and of boats swept out of control and needing rescue peppered the news. Horses and wagons were in high demand. With streetcars out of service, horses were called upon to work even harder for rescue and beyond. Fifty teams at a time were requested for hauling debris. Without water in troughs, exhausted horses could be seen around Kansas City with their tongues hanging out in thirst. Municipal services, transportation, and communication were severed but still workers, inspectors, and those doing cleanup had to find ways to traverse the river.

The flood highlighted the united environmental, social, and economic aspects of the Kansas Cities. Political boundaries were daily transgressed by the river, workers, and commerce, although the different states and municipalities struggled to recognize these connections. A combination of self-interest, pragmatism, and generosity were at play in these critical days. Recovery was not a well-oiled process, yet most public assessments were positive and agreed with the *Star*, which editorialized that "the state line exists only on the map, and Kansas City, from Argentine to the Blue, is one splendid, harmonious, helpful municipality."[24] Most commentary praised how the "one Kansas City" operated as a unit, particularly the fact that Missouri graciously redirected aid to the more-suffering Kansas side. The *Kansas City Gazette*, published on the Kansas side, lauded the "generosity" of the Commercial Club's $2,500 gift to Kansas City, Kansas. The *Gazette* also slipped in the fact that needs on the Kansas side "if not greater, were more urgent."[25] Flood-time unity aside, Kansas City, Missouri, liked to distinguish itself when convenient.

Kansas City, Missouri, widely proclaimed itself to be an "All-American" city because it had the lowest percent of foreign-born people of any major city in the nation. At the turn of the century, amid high rates of immigration, native-born Anglo Protestants of northern European origin felt threatened. Their view, in Kansas City and elsewhere, was that the United States risked losing its identity amid a wave of impoverished and non-Protestant immigrants from southern and eastern Europe who were flooding American cities. But the truth was that Kansas City, Missouri, was primarily native-born because most immigrants lived on the other side of the state line, down along the river. The variety of newspapers that were on newsstands relays the diversity: Jewish, Italian, German, Spanish, Catholic, Swedish, and African American. Therefore, the All-American city was a political concoction; immigrants were an important part of the "one Kansas City" and an important sector of the industrial labor force. The state line was a useful tool for the business class of Kansas City and the distinction they made is telling.

In the days following the flood, the community of Argentine, along the Kaw River, began to vocalize its concern about recovery. Argentinians, including Croatians, asserted that the forgotten poor, including immigrants, had no one to speak for them and that Kansas Citians were ignorant of the dreadful situation in their community. Because it was a separate city, Argentine had been left out of the Kansas City, Kansas, flood zone that was getting special financial help.[26] Argentine was stricken with poverty, labor strife, and industrial decline even before the flood and its citizens leveled the accusation that political boundaries were a guise to cast their needs aside.

Argentine was economically intertwined with the wider city; its residents worked in and patronized the Kansas Cities and considered their community a "suburb." Argentinians were returning to work in the businesses of Kansas Citians, so, asked one resident, why couldn't political divisions be "obliterated" in charitable donations as they had been in industry?[27] The city governments tended to give freedom to economic interests in order to promulgate trade and industry, and this raised the ire of Argentinians when they perceived that this borderlessness did not extend to people. These working-class complaints reveal that—as with Mayor Reed's pledge to protect commercial interests and with the city's saving scarce water for the protection of property in case of fire—the Kansas Cities prioritized economic wealth over public health.

While risk was unevenly distributed, those on higher ground and in suburban and southern parts of the city did not escape the impact. Importantly, even if their homes stood intact, not having water, gas, or transportation was significant. Concern for food supplies subsided by midweek as meat and other provisions were shipped in from the surrounding region. The Heim Brewery, one of the city's major employers, was hit hard and stopped producing beer for many days. Other businesses fared better and adapted their advertising to a postflood era. Grocers, millers, and other food producers advertised their abundant supply, normal prices, and the fact that their lights were on and their elevators worked. Economically stable Kansas Citians browsed newspaper advertisements for home water purifiers and shopped at department stores that were having flood sales. These same Kansas Citians supported the women's relief efforts through the recently organized Council of Clubs and attended performances of Shakespeare to raise funds for recovery. African Americans who turned to the back page of one of the city's black newspapers found a furniture advertisement appealing to those who lost furnishings in the flood. The company accepted cash or credit and reminded readers that it was "a friend of the poor class."[28]

Though social status was a strong indicator of flood experience, all people had to contend with extraordinary public health issues in flood time. With so much water but none potable, it was a situation comparable to Samuel Taylor Coleridge's "The Rime of the Ancient Mariner": thirst in the middle of an ocean. City officials warned of typhoid fever risks associated with the temporary water source, but there is no data because health organizations had little capacity to gather statistics. Dr. Samuel J. Crumbine, then secretary of the Kansas State Board of Health, recalled that the flood of 1903 left "in its wake terrific epidemics of typhoid fever, especially severe in Topeka and the two Kansas Cities, all of which was undoubtedly the inevitable result of the prevalent insanitary [sic] conditions in the flooded area." Lacking official records, he estimated that as many as 1,500 cases and 135 deaths from typhoid fever may have resulted along the flooded Kaw River.[29] The city gave smallpox vaccinations because, as Dr. E. H. Thomas said, "Many of those refugees come from districts which have been infected with smallpox." The health official suggested that the working-class community was unhealthy to begin with and that the wider public feared evacuees could spread illness. Therefore, evacuees were required to be vaccinated if they wished to leave the shelter, unless proof of a "good scar" from a former shot could be shown.[30] Kansas City, Missouri, which responded to the social and environmental crisis by passing strict sanitary laws and deputizing one hundred new health inspectors, saw firsthand how effective the enforcement of sanitary laws could be with a sufficient staff.[31]

After a few days, the "Big Muddy" subsided, trains resumed using tracks, and cleanup began.[32] The six to eighteen inches of rich mud left behind was great for agriculture but not for machinery, household goods, or neighborhoods. In The Patch, a poor neighborhood in the bottoms, the mud became a joke: houses had one and a half stories because homeowners did not dig out.[33] Stagnant pools of water bred mosquitos and the stench of decaying plants and animals hung in the air.

Despite all the havoc it wrought, the river was essential for cleanup: the Missouri was the city's free garbage can—and not just after floods. An "endless line of wagons" driven by men of all stripes dumped flood debris in the river at the foot of Broadway Street near the Hannibal Bridge.[34] The health board was upset that the stockyards were putting carcasses in the river as part of their cleanup. The dead animals were supposed to be burned, for sanitary reasons, but instead they ended up caught in the debris by the bridges, still rotting and causing problems.[35] Farther downstream, as waters receded, Missouri farmers reported carcasses from the stockyards on their land.

The bottoms were at the crux of health and wealth in the Kansas Cities, although the flood reveals ways these interrelated categories were sometimes at odds. In the first days of the flood, people were galvanized to remedy problems. But only so much could be done with a system that was already inequitable. New infrastructure could not be constructed overnight, and the homeless working classes could not quickly right their losses. Therefore, the flood reveals to us a system of power that already existed and would continue to exist. No muckraking journalist came forward to defend the residents and workers of the bottoms. Businessmen were portrayed sympathetically for what they had lost and heroically for what they gave. Many of them contributed to the "Kansas City spirit," and one gets the sense from newspapers that the general community identity of Kansas City grew positively through the flood experience.

In sketches and stories, the newspapers relayed information about the flood and the gradual process of cleaning up, the return of gas, electricity, and water, and the number of flood deaths—ten confirmed in the Kansas Cities after one week. Not all newspapers were able to continue publishing through the flood as did Kansas City's bulwark *Star*. For example, on June 6 the local German newspaper *Reform* caught up on the news of the last week, its headline reading "The Dreary Hours in the History of Both Kansas Cities."[36] The African American paper the *Liberator* had a two-week publishing gap, returning to print on June 12. The offices of the *Rising Son* newspaper, also African American, were in a flooded district on Sixth Street and the paper was out of print for three weeks. Its editorial thanked the flood committee for helping the blacks of Kansas City when there were also so many whites in distress. It was a respectful acknowledgment, as well as an indication of racial difference and tension in the city.

North Kansas City and Harlem, though not yet very urbanized, were submerged, with only the Winner Building surviving.[37] The East Bottoms had a large number of homes washed away and truck farms that were inundated. Sheffield, the community at the mouth of the Blue River, was similarly devastated. All told, the Missouri River rose to about forty-six feet.[38] Early assessment of damages in these "tragic scenes" was estimated to be in the millions.[39] Years later, property damage for 1903 was calculated to have been $450 million, the largest portion of which was in the Kansas Cities.[40]

When families in the bottoms fled the rising waters, their neighborhoods were surrounded by industry and already in decline; like many other boardinghouse tenants, immigrants, and packinghouse employees, they would not return to the bottoms in great numbers. Over the twentieth century, the floods served as catalysts for demographic shifts as fewer and fewer people made their homes in the bottoms.[41]

The river's power fascinated people. "The city is daily thronged with thousands of sightseers and they give the place a congested appearance," the *Star* reported.[42] Although the city issued fifteen thousand flood passes the first week, neither sightseers nor people with cameras were allowed

"Crowd gathers to view the flood from the bluffs." A fire breaks out as floodwaters spread across the bottoms, drowning the stockyards, rail yards, and the homes of the most vulnerable. Missouri Valley Special Collections, Kansas City Public Library, Kansas City, Missouri

access to the flood zone.[43] For most onlookers, it was from the safety of the bluffs that they saw the panorama of a river whose shapeliness spread for miles across the floodplain. The first time Nebraska author John Neihardt saw the Missouri River, he was standing on a Kansas City bluff. He was a boy and the sight of the swollen river—"my turbulent friend"—filled him with fear and wonder. "Shacks, stores, outhouses suddenly developed a frantic desire to go to St. Louis," Neihardt recalled of the overflowing "yellow" river that captivated him.[44] To get a better view of the river, some people paid one dollar each to ride through the valley atop coal cars on a still-open rail line. Within a week photos of the flood were being sold as souvenirs and those postcard images of bobbing furniture in awning-high water have stuck in the craw of our memory.

The flood affected other lower basin states as well, but it was the hard-hit Kansas Cities that epitomized the regional disaster. As water continued downstream it flooded fields and towns, swept away homes, redeposited river banks, lapped at railroad beds, and stopped train traffic—thus delaying deliveries of mail, milk, and passengers.[45] The disaster was capped by less-severe flooding in 1904 and 1908, and Kansas City emerged at the turn of the century as the leading lobbyist to engineer a river so rational that it would overcome the region's vagaries of flood and drought. The desire for predictable river levels initiated a culture of worry about the river, necessitating constant political vigilance in search of the millions of dollars that might protect and enhance the wealth of the cities. This economic rationalization of the river highlighted the Missouri's liabilities, not its assets.

DRINKING THE WATER

3

Montanan Robert E. McDonnell—or "R. E.," as he was often called—was born in the upper reaches of the Missouri River basin. He trained to be an engineer at Stanford University, where he was a classmate of Herbert Hoover. In 1898 McDonnell and his recently graduated friend Clinton Burns, an Iowan, met in Salt Lake City, where they pored over maps and demographic data to decide where to establish themselves as engineers. According to the lore of the firm they established, the young men determined that fast-growing Kansas City sat at the heart of a potential engineering boom and, within a night's train ride, they could access many small cities that would need water, sewer, and electric infrastructure.[1] Burns and McDonnell made a reputation for themselves as municipal engineers and began consulting for the Kansas Cities. By 1920 Burns & McDonnell had built over two hundred drinking water plants in the region and was helping plan a new waterworks for Kansas City, Missouri.[2]

In addition to influencing the physical infrastructure of the region, R. E. McDonnell would become a spokesperson for progressivism in Kansas City. The ideology and politics of progressivism deeply influenced the historical development of city services, including the drinking water supply systems of the Kansas Cities, which underwent such a significant transformation that the water system of 1870 and that of 1930 bore little resemblance to each other. The

Robert E. McDonnell. Burns & McDonnell Library

growing desire for water and then sewerage evolved from public health campaigns begun during the Civil War and new sanitary habits driven by the middle classes. These cultural shifts were infused by progressive ideology and Americans linked the "pursuit of cleanliness," as one historian calls it, to good public health.[3] By the turn of the century, new notions of health and cleanliness and advances in science and medicine demanded the expansion of safe, dependable water supplies. As the urban historian Nelson Manfred Blake wrote, "The new passion for civic cleanliness obviously demanded a liberal use of water."[4]

Kansas City's "exaggerated" growth, so attractive to Burns and McDonnell, meant that more people and more factories needed more water and they needed it quickly.⁵ Urbanization and industrialization coupled to transform the daily lives of a growing number of Americans. Water acquisition was an individual or local responsibility that evolved into a municipal responsibility; the family cistern was replaced by the city intake. With the introduction of domestic water, American water usage per person increased from two or three gallons per day to fifty or one hundred gallons.⁶ Because development threatened critical resources like drinking water, cities responded to the very sanitary challenges they created by inventing new ways of testing, treating, distributing, and regulating water. In these urban environmental shifts, Kansas City mirrored national trends.

Quantity and Quality

The water-supply story for the Kansas Cities is about a constant search for quantity and quality. Threats like fire and epidemics compelled it to actively protect health by expanding access to water. The first waterworks for Kansas City, Missouri, drew on the Kaw River—despite Captain William Clark's nose-wrinkled observation in 1804 that "the waters of the Kansas is verry disigreeably tasted to me."⁷ The city granted a private company, the National Water Works Company of New York, a twenty-year franchise in 1873 to develop and conduct all aspects of water supply. Private water companies were more common in the Gilded Age, when plumbing was a luxury afforded by few. By 1875 the firm had a maximum capacity of five million gallons per day, its water flowed through twelve miles of mains, and it served three hundred customers. Service was limited by what was profitable; therefore, the private company had little incentive to develop a comprehensive system for all people and locations in the city.⁸

This first waterworks was on the less-sediment-laden Kaw River and called Kaw Point Station. The intake was below the outlet of Turkey Creek (a relatively unpopulated watershed at the time) but upstream from preexisting pollution sources.⁹ The Kansas City Livestock Exchange and three important packing plants, including Armour, relied on the river. Daily, thousands of animals came in on the hoof and out on a hook. In just over a decade, the Kaw—its banks populated by the growth of industry, packing plants, and the sewage from Turkey Creek—became too polluted.¹⁰

Early waterworks of the Kansas Cities. This drawing was part of a National Water Works Company advertisement found in *Kansas City: Its Resources and Their Development*, 1890. Missouri Valley Special Collections, Kansas City Public Library, Kansas City, Missouri

The private waterworks company was forced to switch water sources and began to draw from the Missouri River in 1887.[11]

The new Quindaro Station on the Missouri drew cleaner water because it was a few miles above the confluence with the Kaw and upstream from the pollution of the cities. Named for the Underground Railroad town once located there, Quindaro was on property the National Water Works Company purchased, north of the city limits of Kansas City, Kansas.[12] In the 1890s the city council pressed the company for lower rates and improved services. The city claimed that the company was not performing as contracted: "To distribute in and throughout the said City of Kansas, pure, well settled, and wholesome water." Motivated by profit, the private water company did not have the incentive to supply the majority. Kansas City, Missouri, wanted to take over the waterworks before the franchise ended, but the private company's location in the state of Kansas complicated matters.[13] In the heat of legal and financial disagreement, Kansas City refused to pay its bills to National Water Works. In 1893 the company "served notice on the city" and threatened to shut the water off. When the franchise finally ended in 1895, Kansas City, Missouri, voters approved $3.1 million in bonds to purchase the waterworks.[14]

The city's purchase of the waterworks reflected the Progressive Era trend toward municipal ownership. Health statistics bore out that municipally owned waterworks provided better service, and nearly all cities had switched to public water. As historian Martin Melosi says, water supply was "the first important public utility in the United States and the first municipal service that demonstrated a city's commitment to growth."[15] Municipal ownership of garbage, electricity, railway transportation, and gas services were other debates that were soon to follow, but the waterworks was the only utility Kansas City, Missouri, acquired during the era.[16] Looking back on this decision, R. E. McDonnell commented that at the turn of the century municipal ownership had been criticized as "idealistic" and "socialistic" but within twenty-five years it had proven effective.[17]

McDonnell became a national advocate for public ownership of city utilities, especially because they promoted civic and public health. The firm Burns & McDonnell took about ten years to get established and the partners lived slim while they picked up jobs here and there—some in Kansas City, most in the region. Their first municipal water plant was in Iola, Kansas, in 1905. As their work in water plants increased, McDonnell, who was the front man for the firm, developed a pamphlet that revealed his zeal for public ownership—*100 Reasons Why 100 Cities Approve Municipal Ownership of Their Public Utilities*. McDonnell even instructed cities how to run a capital bonds campaign to finance public works.[18]

After purchasing the franchise, Kansas City owned the network of urban innards—waterworks, mains, and hydrants—which came under the domain of the Fire and Water Commissioners, a politically appointed board. Upon inspection, the city thought it overpaid, as the Quindaro settling basins were in poor condition and filled with several feet of caked mud, the Turkey Creek and Holly Street reservoirs needed maintenance, and mains needed repair. City officials congratulated themselves for having saved the city from certain disaster at the hands of a private franchise and declared success when they lowered water rates.[19] Editorials and letters to the editor suggest that the euphoria soon faded. By 1905 critics claimed that municipal management wasn't living up to expectations, water rates were comparatively high, and costs exceeded revenues.[20] Furthermore, no filtration plant had yet been built, and some Kansas Citians knew they were drinking a risky river brew.

To keep up, the city renovated Quindaro Station in 1905 and installed a new river intake in 1912.[21] The concrete-and-brick intake swallowed the river into wells covered by screens and gates, which kept out tree branches or the itinerant beaver. Connecting the intake to the bank were suction lines powered from an onshore pump. These lines could suck up to 124 million gallons daily.[22] Kansas Citians did not share Mark Twain's humorous proclivity for the slush, gruel, and pancake-batter-like Mississippi drinking water, which, he said, "comes out of the turbulent, bank-caving Missouri, and every tumblerful of it holds nearly an acre of land."[23] To keep Kansas Citians from drinking mud, near the pump station there were five sedimentation basins, four built in 1887 and one in 1910. The basins coagulated sediment, clarified the river, and held up to fifty million gallons.[24]

After sedimentation came transportation away from Quindaro. The settled river water was pumped southward in thirty-six-inch-diameter pipes (first one, then four) through Kansas City, Kansas.[25] The water had to cross the Kaw River to get to Kansas City, Missouri. The first pipe crossed on a bridge—that was the flow line that washed away in the flood of 1903 and left the city without drinking water. With a temporary flow line in place, Kansas City readied plans to tunnel the water main under the Kaw River and stave off future danger—an engineering feat.[26] The brick tunnel opened in 1905 and was buried sixty feet under the riverbed, continuing across the bottoms, under warehouses and railroads for over a thousand feet on its way to the Turkey Creek pump station. The vibration of trains above occasionally caused water main breaks in iron and steel pipes.[27] By the time river water from Quindaro Bend arrived at the pump station on the Missouri side, it had already traveled about six miles.

Turkey Creek pump station in the West Bottoms was the heart of water service, receiving the river and pumping it through Kansas City's major arteries; it was an integral part of the supply process and the destination for both Kaw Point and Quindaro water. From Turkey Creek, the water was distributed into mains that spread across dozens of square miles of the city. Buried four feet and safe from frost, these arterial pipes decreased in size with distance, going from thirty to sixteen inches, with twelve-inch veins connecting laterally. In addition to a massive extension of pipes in the first two decades of the century, most of the nineteenth-century pipes were replaced so that by the 1920s the majority of Kansas City's water infrastructure would be relatively new.[28]

The Kansas Cities share a drinking water history even though they were two political entities with different water systems. Most importantly, Kansas City, Kansas, depended on a hand-me-down waterworks infrastructure from Kansas City, Missouri—first Kaw Point, then Quindaro. Missouri never purchased the Kaw Point Station, which continued to unsatisfactorily serve parts of the Kansas side. In 1909 Kansas City, Kansas, voters—dogged by typhoid fever epidemics—agreed to spend one million dollars to build a municipal waterworks at Quindaro Bend, next to its sister city's intake on the Missouri.[29] The Kansas State Board of Health reported in 1912 that a "thoroughly modern" water plant and settling basins were being built that would use coagulation and rapid sand filters where no filtration had existed before.[30] As Kansas was building, Missouri was remodeling and upgrading. In sum, Kansas City, Kansas—working class and dominated by industry—was not on the technical cutting edge, nor did it show political leadership to advance water supply, because it operated in conjunction with its larger, wealthier sister.

In times of emergency, the Kansas Cities cooperated. Kansas City, Missouri, had installed a surface intake, while Kansas City, Kansas, drew from the riverbed, resulting in siltier water. When the Kansas intake clogged in 1913, Missouri offered to share its larger intake.[31] When animal carcasses dumped in the river introduced typhoid-fever-causing bacilli to the water supply, the cities dealt with the threat differently. Kansas City, Kansas, was not using chemical treatment, so residents resorted to boiling water to kill the bacilli. In contrast, chemist Dr. Walter Cross assured Missourians that their water was safe to drink because the city treated it with a purifying agent—hypochlorite and lime.[32] Side-by-side waterworks had advantages—like when an emergency backup line connecting the pump houses was installed in 1914.[33] But there were also risks—like when the Missouri River threatened to change its course and leave both waterworks high and dry in 1915. The river's habitual meandering, in this case occurring a mile upstream, meant the Quindaro intakes could end up poorly placed to draw from the river. Both cities cooperated to fund revetment work on the opposite bank, grading, piling, and constructing a six-hundred-foot dike to keep the river trained to flow past the waterworks.[34]

The fact that Missouri's waterworks were located in Kansas caused decades of contention, including a property tax row. From 1909 to 1911 Kansas City, Missouri, mayor Darius Brown contested and refused to pay property taxes for the Quindaro waterworks, arguing that because his city was outside of Wyandotte County and the state of Kansas, it did not owe property taxes. In December of 1913, the US Supreme Court let the Kansas Supreme Court's decision stand, which ruled that Kansas City, Missouri, must pay Wyandotte County $83,000 in accumulated taxes.[35] It did not end there. Kansas City, Kansas, later tried annexing the Quindaro property into its city limits. Despite Missouri's opposition, the Kansas Supreme Court ruled that Quindaro could be annexed. Because the waterworks would become the largest taxpayer, a newspaper article stated that Kansas City, Missouri, must "help run" Kansas City, Kansas, as if it were the more responsible big sister.[36]

Foreshadowing the relocation of the waterworks to the Missouri side as early as 1914, Kansas City, Missouri, mayor Henry Jost announced to civic organizations: "The waterworks system of the city now has its headquarters in Kansas City, Kas., and it appears that by taxing us they hope to raise most of the money for the expenses of their government.... The day will not be so far distant when a large plant on this side will have to be built."[37] If the waterworks had been publicly owned from the start, it might have been located in Missouri, but there were several reasons why the intakes were located in Kansas. Since the river was needed for carriage of municipal and industrial wastes, the waterworks had to be located upstream to ensure cleaner water. Spanning the Missouri River was a serious consideration; the flow line would be required to cross the river by bridge or tunnel from North Kansas City, which was not settled until the 1910s. A look at a map shows that, owing to the state boundaries and the bend in the river, the most efficient way to get drinking water was upstream on the Kansas side. Despite the border conflict, drinking water was largely a regional issue that resulted in more cooperation than contention between the Kansas Cities.

Water Pressure: The Reformers

Not long after the city purchased Quindaro, Missourians issued calls for expansion.[38] In 1908 a city newspaper forcefully criticized Quindaro, saying that everyone knew the Kansas City water supply was "INADEQUATE" and "SO FAR BEHIND THE TIMES."[39] This pressure for greater quantity and quality of water had multiple sources. Politicians, professionals, sanitarians with both engineering and medical leanings, social workers, women's organizations, industrialists, civic boosters, taxpayers, and reformers of all types—including Robert McDonnell and his wife, Georgia—were among the diverse advocates for abundant and safe drinking water. These pressure groups and individuals were active in the Kansas Cities particularly from the 1890s to the 1930s. Most reformers drew on the strains of progressivism, although many were not affiliated with the Progressive Party, which has lent the era its name. Because the push for water quantity and quality continued into the middle of the century and because there was such diversity among the interest groups, I use the term *reformer* to refer to people who optimistically worked to shape the urban landscape in the first decades of the twentieth century.

Once settled in Kansas City, R. E. McDonnell married Georgia Howlett in 1904. Georgia's family had relocated to Kansas City from Texas and seems to have established themselves well; Georgia's mother had been in charge of distributing donated clothing after the flood of 1903. Robert and Georgia started a family and moved to West Fifty-Third Street Terrace on the corner of Wyandotte, in the Countryside section of a J. C. Nichols development—not far from where the Country Club Plaza would soon be built. Their house, planned in 1914, was lauded for its healthful engineering.[40] It would be plumbed for city water and have flush toilets. Nichols's developments, renowned for influencing urban development, were the model of refined, middle-to-upper-class living. These well-planned neighborhoods included restrictive covenants and self-perpetuating deeds that prevented African Americans and Jews from buying homes—one of the ways that class and race segregation was built into the city.[41] Like J. C. Nichols, the McDonnells took advantage of the southward extension of city infrastructure.

White, middle-to-upper-class native-born Americans like the McDonnells were most visible in reform campaigns. These male and female reformers came to see drinking water as a democratizing force and as an avenue to better public health for working-class, immigrant, and minority populations—common targets of progressives' campaigns. Many of the housing and health improvements, for example, necessitated abundant access to clean water. Because public health was central to the goals of progressive reform, it is important to understand water, its advocates, and their ideologies.

As reformers exerted pressure for improved health, they had to grapple with the related outcome of economic development—which could be coterminous or in tension. A number of American aphorisms address the relationship between health and wealth. For example, Benjamin Franklin popularized pearls of wisdom like "public health is public wealth."[42] The order of Franklin's words suggests that he thought health led to wealth, and many sanitarians and reformers agreed, believing that investment in public health would create economic wealth. Others—like transcendentalist Ralph Waldo Emerson, who said, "The first wealth is health"—believed that the definition of wealth was, in itself, good health. In contrast, some economic boosters believed that a growing economy and financial wealth would enable good health.

Even though industrialization threatened urban bodies, public health reformers hitched their wagon to the philosophy of economic boosters. Historian John Duffy has noted that public health officials were fond of arguing that "health and prosperity were closely related."[43] Mimicking the powerful rhetoric of economic boosters, sanitary reformers portrayed public health as a noble goal, equating it to economic development. For example, as a consultant, McDonnell lectured that an expensive system delivering pure water was good for business.[44] While the role of the government in advancing commerce was well defined at the turn of the twentieth century, the role of the government in public health was not. One public health official repeated, "Statesmen are prone to waste oratory on national finance—is not the national physique and health of equal importance?"[45] Despite their seeming adoption of boosterism, when reformers lobbied to put public health on the agenda, they did so because waiting for wealth to create health was not a viable option.

McDonnell illustrates the reform era well—he believed that the application of knowledge by professionals was the best way to solve problems. McDonnell was a member of national organizations like the American Water Works Association and served on the editorial board of the journal *Municipal Sanitation*.[46] He also represented reform organizations in Kansas City, notably serving as a director and the president of the Citizens' League, the nonpartisan civic organization that opposed the Pendergast political machine. He was president of the Civic Research Institute, a private group committed to studying and solving problems in Kansas City. Other organizations in which he served in a leadership role included the Kansas City Club and the local Citizens' Historical Association.[47] McDonnell gave lectures, presented papers, and authored dozens of articles related to engineering, often using local issues for context. For example, making an on-air "Civics by Radio" appearance to discuss the importance of drinking water was typical of how McDonnell blended his professional career with community service.[48] McDonnell spoke more than most engineers about public health issues. He believed progressive ideals and engineers could build a better, healthier quality of life for Kansas Citians, and this led him to emphasize, among other things, the benefits of municipal ownership.

William Rockhill Nelson, the owner and editor of the *Kansas City Star*, made his political viewpoints known as well. His media outlet (one of several newspapers in the cities) became a voice for progressive reform and supported the Republican Party. In the 1890s Nelson supported the national "city beautiful" movement. This movement of professionals and reformers believed that artfully constructed parks and greenspaces could positively influence individuals and society to be physically and morally healthy. Using gardens and monumental architecture, the movement put an American twist on European public spaces and hoped to undergird civic virtue. By controlling urban design, the movement sought to mold people.[49]

Nelson engaged the German-American landscape architect George E.

Kessler, who had recently moved to the Kansas City region. Following in the footsteps of Frederick Law Olmsted, Kessler designed the park and boulevard system in 1893 for which Kansas City, Missouri, has become so well known.[50] Groups came from all over the country to survey the model of Kansas City's planned parks and boulevards. Kansas City established Kessler's reputation and career. Being wealthy, well connected, and something of a visionary gave Nelson a national reputation, and, because his ideas and wealth left behind a physical legacy (such as the Nelson-Atkins Museum of Art), Kansas City has not forgotten him.

When accused of being a busybody in city affairs, Nelson used his sarcastic editorial power to respond humorously:

> Under the malign direction of Nelson, the *Star* has kept things constantly stirred up. It has made tenants dissatisfied. They never used to complain about light and air. Now they won't look at a house unless every window opens on a flower garden with a humming-bird in it. The *Star* won't let anybody alone. It insists on regulating the minutest detail of people's lives. Its regulations are pernicious and extravagant. Its preaching about more parks and boulevards and breathing spaces and supervised playgrounds . . . and swat the fly, and housing reforms, and a new charter, and art galleries, and keep your lawn trimmed, and take a lot of baths, and throw out the bosses, and use the river, and cut the weeds on vacant lots . . . and for God's sake don't build such ugly houses, and make the landlord cut a window in the bathroom, and put goats in Swope park, and why mothers risk their babies' lives by bringing them up on bottles . . . and what's the use of lawyers . . . and build a civic center, and put out houses for the birds, and walk two miles before breakfast, and why are Pullman cars so hot in winter, and go to church, and cut out the children's adenoids, and build traffic-ways, and the square deal, and sleep with your windows open, and smash the saloons, and pooh-pooh on factories that employ women . . . and move out to the suburbs, and build hard-surfaced roads everywhere, and all the other things, [that] have increased the cost of living and given people inflated ideas, and pretty nearly ruined the town.[51]

Nelson's retort neatly summarizes the position of progressive reformers in Kansas City by mentioning many of the major concerns and campaigns as well as the attitudes. Notice how a number of his reform interests related to urban environment, health, drinking water, and the river.

On the Missouri side, the Pendergast machine—one of the longest-running political machines, spanning five decades—colored the urban Democrats. Based in a saloon in the river bottoms, Jim Pendergast began building his political powerhouse among the working classes in 1892. After Jim's death in 1910, his already-groomed brother Tom Pendergast inherited the political structure. "Boss" Pendergast became a significant informal power within the city, though he never held office. The machine success in the 1910–1920s came from focusing southward and winning votes among the suburban middle classes of Kansas City while continuing to count on the river wards for support. The machine raised the ire of many reformers over the years, including the McDonnells, but the reign of Tom Pendergast did not come to an end until 1939, when he was indicted and briefly imprisoned for tax evasion.[52] Within Jackson County, the Democrats had two factions, known locally as the urban machine "goats" and the county "rabbits," that represented nonurban interests. Newspaper articles were sprinkled with goat and rabbit references, as if Kansas City politics operated in Wonderland.

The interactions between political parties, the machine, and reformers were complex. All entities exhibited progressive goals at one time or another. Democrat Jim Pendergast and Republican William Rockhill Nelson, though in political opposition, supported similar improvements like the expansion of the city water system and engineering of the Missouri River. Race, class, and neighborhood also figured prominently in Kansas City politics and influenced the urban environment, including access to drinking water infrastructure.

The preeminent "booster" was the Kansas City Commercial Club, which held much political and economic power in the cities and tended to be Republican, or at least antimachine. The club was composed of white, well-off men dedicated to advancing their own economic interests—which they conflated with the good of the entire city. The Commercial Club espoused the philosophy that economic wealth would lead to good public health. The club members' personal wealth and interest in civic affairs resulted in expansive influence over urban reform. Club member Frank Faxon created the catchphrase of Kansas City at the turn of the century: "Make Kansas City a good place to live in."[53] For decades this civic ideal would guide the Commercial Club, public officials, and average residents.[54] The club, renamed the Chamber of Commerce in 1918, promoted river navigation and was the linchpin in securing federal funds to engineer the Missouri River for economic uses.[55] Its boosterism employed the rhetoric of growth and development and, through both municipal and national affairs, its clubmen shaped the physical infrastructure of the rivers and cities.

Women were responsible for key aspects of progressive reform, particularly for improving public health and adapting the urban environment to ameliorate the extremes of industrialization. The phrase "municipal housekeeping"—referring to the extension of domestic responsibilities into the community—encapsulated much of the reform work that women's organizations involved themselves in. Women were familiar with the daily life of the city and its people, and city needs seemed to match their cultural skills as homemakers, caretakers, and healthcare workers. Women's reform efforts shaped the city by stimulating citizen participation and a more active government—impressive because women usually operated at the fringes of formal political power.[56] Jane Addams and Hull House in Chicago are, perhaps, the most famous example of this.

Women supported city, state, and, eventually, federal programs and regulations to address various problems. Through community organizations like the Kansas City Woman's City Club, women advocated for playgrounds, better garbage service, public health facilities, a pure milk supply, and "social hygiene." The Susan B. Anthony League in Kansas City went so far as to "drop sewing to talk utilities" and concluded that whether it was water, light, or gas, there were benefits to the city and consumers in public ownership.[57] From their positions as charity volunteers, private welfare workers, and then professional social workers, women pushed government to take more responsibility for the social issues they worked on.[58]

Georgia—or Mrs. R. E. McDonnell, as the press and publications referred to her—was one such woman. She was a member of a few clubs, was president of the Kansas City Athenaeum, and would go on to become the president of a national organization committed to parliamentarianism. Her activism often dovetailed with her husband's interests and those of his engineering firm. She, for example, was a proponent of a new Kansas City waterworks and used her position to reach out to other women, as well as represent women's views in a wider forum. Like other women, she found a public voice by speaking on behalf of families and city health, but did not go so far as to question the political or economic status quo that contributed to the very challenges female reformers took on.

The Kansas City Board of Public Welfare—the first of its kind in the nation—employed women reformers. Formed under the 1908 city charter, by 1910 the board provided social workers a professional space to apply themselves to the health of the workplace and neighborhood.[59] For example, in 1912 and 1913 board employees made detailed investigations of neighborhood conditions and social affairs of the city.[60] The board compiled its research into two significant reports. With these statistics, descriptions, photographs, and maps, Kansas City reformers advocated for programs and legislation to improve urban health and environment. The board's work focused on the river wards. The North End—with its tenements and boardinghouses, and its working-class, immigrant, and minority populations—was a focal point for reform campaigns in the 1910s.

Using settlement houses and public health campaigns, the influential white middle class taught immigrants and the poor how to be clean, healthy, and "American" (in their own likeness). Mirroring reform efforts nationally, a combination of private and public interests worked to educate residents on sanitation, and to bring order and better conditions to the city.[61] Kansas City government was composed of several departments and boards that contributed to reform, in part because they responded to outside pressures. The water and street departments were responsible for building and maintaining the physical infrastructure that delivered municipal services like water. The Hospital and Health Board wished to see all segments of the population gain access to the cleansing and healthful

"Free Shower Bath in the Street." Kansas City, Missouri, Board of Public Welfare reformers used water as a tool for hygiene and Americanization. Due to poverty and weak housing regulations, neighborhoods in the North End often lacked running water. *Social Prospectus of Kansas City*, 1913

benefits of water, and supplemented the work of the welfare board and water department. The city's Pure Milk campaign was carried out by a combination of the health department and women's volunteer groups. Their joint work to eradicate illness transmitted through milk brought order to the dairy industry and sanitation to the farm, marketplace, and kitchen. Such campaigns involved and affected all parts of the city.[62]

Water was an important tool for reformers. Social workers, city departments, and sanitarians advocated constructing public bathhouses in the Progressive Era. The need for bathing facilities in public parks was demonstrated in a 1912 survey, when the Board of Public Welfare calculated that one bathtub existed for every 162 people in the North End.[63] Baths, showers, and wading pools, some combined with swimming pools, were services that parks provided. The clientele lacked access to bathing facilities or water at home. It was recommended in 1907 that a park in the North End's cramped McClure Flats forgo the recreation pool and have shower baths instead because it would be more sanitary for the immigrant residents.[64] The neighborhood kids were probably disappointed. The reformers saw the bathhouse as a way to improve public health but just as important, as a space to educate their target audience, to clean up the children of immigrants and make them American by sharing middle-class white ideals. City boards advocated that fountains, representing civilization and health, be in parks and on city streets for both aesthetics and slaking thirst—the beginning of Kansas City's legacy of fountains.

For all the talk of health, the threat to property was the most influential lobbying force for expanding the waterworks,[65] The public's fear of fire was justified, as many cities (built with wood and lacking building codes) experienced disastrous fires—famously Chicago and San Francisco. In 1900, when the Convention Hall, the pride of Kansas City, burned down soon after being built, a front-page headline blared: "Water Pressure Was Very Poor."[66] Another newspaper read: "Valuable time was lost before anything like adequate water pressure could be obtained."[67] Incidents like this were an impetus for the Commercial Club, having influence with city government, to lobby for increased water capacity. The club decried the lack of water pressure in city mains as more people tapped into them—not because they wished to keep people from accessing water, but to encourage the city to lay larger mains or expand pumping capacity.[68] Although the business elite's public language regarding the fire danger articulated a more civic duty, among themselves Commercial Club members talked of the need to protect their interests.

Commercial interests and industrialists wanted to protect property and bring insurance rates down, and, joined by newspaper editorials like those of Nelson, there was a unified call among the politically powerful for more and cheaper water on behalf of economic interests.[69] In 1905, when an insurance rate hike went into effect for warehouses but not residences, the club lobbied assiduously to improve the waterworks in order to bring rates down. In 1924, with the same waterworks still in use, the National Fire Underwriters Board, which influenced insurance policy, would conclude that Kansas City's water supply and infrastructure were barely making the grade.[70] Luckily, there were no major fires in the interim and, to the benefit of industry, the publicly funded improvements would bring insurance rates down.

Industry was the largest user of water in the city, it paid reduced rates, and it disproportionately benefited when residential taxpayers paid for waterworks expansions through municipal bonding. Meat processing, the city's signature industry, required a lot of clean water and about one-fourth of Kansas City's water was consumed in the West Bottoms, which had exclusive use of the Holly Street Reservoir.[71] The meatpacking plants reengineered their environments along the essential rivers.[72] Because packing plants generated a lot of waste, reformers and public health officials blamed these industries for negatively affecting water quality.

Interest groups supported increased water supplies at the ballot box, but water concerns were experienced individually too. Commercial Club member Colonel Fleming, who lived near the McDonnells, complained of water service in his section of the city. In places, residences had such low water pressure they could not get water in the bathrooms, he reported at a club meeting.[73] Even in (and especially in) this wealthy, new section of the city, residents lobbied for water. The connection between a safe, healthy, wealthy city and an abundance of water was crystal clear. Public health, a lessening of fire risk, economic growth, and convenience were a few reasons that groups and individuals pressured the city for more and better water.

McDonnell's many pamphlets touting the benefits of water utilities served his interests by garnering contracts for his firm from which his family profited. However, McDonnell really did take his philosophy to heart, as it became his life's work, not only at the firm but also in political life. In 1927 he penned for his alma mater, Stanford's, magazine: "What can be more romantic or thrilling than to be engaged in a profession that saves human lives and adds to the health, happiness, and prosperity of your own community?"[74] McDonnell is a typical quandary for historians who study progressive reformers whose ideology and actions advanced the betterment of society as well as their own interests—sometimes at the expense of others.

"What comes in must go out." This 1912 caricature of McDonnell in *Kansas City in Caricature* emphasizes how he saw drinking water and sewerage systems in tandem. Missouri Valley Special Collections, Kansas City Public Library, Kansas City, Missouri

New Paradigm

The maturation of a new medical paradigm revolutionized the way drinking water was conceptualized. The miasma theory of disease was replaced by the germ theory; no longer could consumers of the rivers judge water by silt content or smell. A bacteriologist with a microscope could discover elements in a glass of clear drinking water to see what the consumer could not. Professionals like the city chemist, sanitary engineer, and board of health physician protected the public from invisible dangers. Bacteriology enabled a profound change in the understanding of human health and caused a change in the perception of responsibility for it.[75] Armed with germ theory and the "new" public health, as it was called, sanitarians could prove the effectiveness of expensive monitoring and preventative measures, but the political system did not yet prioritize the new medical paradigm. Consequently, public health officials found themselves woefully underfunded and unable to carry out their mission.[76]

At the turn of the century, the Missouri State Board of Health monitored drinking water supplies and compiled statistics for typhoid fever but lacked enough funds, humanpower, or equipment to do so effectively, and this greatly frustrated staff. With only one laboratory, officials traveled long distances around the state responding to problems and testing water. In 1902 Missouri's state bacteriologist and chemist reported his concerns to the board: "Only a small amount of money was available" and the University of Missouri in Columbia "very kindly gave us space, and some apparatus and supplies with which to begin our work . . . until sufficient funds are available." Of the samples sent to the makeshift laboratory, one in three was contaminated with bacilli of tuberculosis. "It would seem that it is necessary," the chemist continued, "to call the attention of the public to the importance of protecting our water supplies." Without testing, he warned, it is impossible "to know that a certain water is polluted until the family or town is stricken with typhoid fever." In arguing for the funding to undertake systematic statewide water protection, the chemist's report concluded: "The people have a right to know that they are not drinking diseased germs."[77] Because the state board was weak, Kansas

City, Missouri, assumed more responsibility and in 1908 empowered its city health board to do the work that the state did not.

Neither of the Kansas Cities was a pioneer adopting the technologies that reduced typhoid fever rates. Chlorine was first used in the United States in 1908 and some cities in Kansas began to experiment with it—but not Kansas City, Kansas, which was remarkably unreformed. The city was so ineffective at providing pure water that it had a decade-long average of over sixty-six typhoid fever deaths per one hundred thousand—a terribly high statistic, especially for its consistency.[78] City water was not always the source of illness but clearly it contributed, because rates dropped once new treatment processes were adopted.[79]

Kansas City, Missouri's waterworks did a better (but not great) job of purifying the river. Quindaro's settling basins and coagulation did not filter the "Big Muddy" enough to the taste of some, and its typhoid fever statistics were unenviable.[80] Newspaper editorials called for improvements and as early as 1905 R. E. McDonnell had a letter to the editor published advocating filtration to bring down illness rates.[81] Kansas City, Missouri, relied on settling and was slow to institute filtration and chemical treatment. As early as 1907 Kansas City, Missouri, had started treating its water with lime and alum to quicken sedimentation—a process that local engineer Wynkoop Kiersted invented. Sulfuric acid was used to control acidity. Shortly after the typhoid fever epidemic in the winter of 1910, the city began using hypochlorite seasonally to kill bacteria.[82]

The Kansas Cities focused on waterworks technology and drinking water treatment to secure safer water. City officials and waterworks engineers did not consider sewage treatment as a way to improve water quality; they assumed that contamination by fecal bacteria was uncontrollable. This question, whether US cities should treat only drinking water or should also treat sewage, was unsettled in the years before World War I. The debate was between two camps, of physicians and engineers respectively. The Kansas, Iowa, and Ohio state boards of health, which were dominated by physicians, were keen on sewage treatment. Physicians believed that sewage pollution should be avoided in order to protect water quality. Engineers believed dilution of sewage in waterways followed by downstream water treatment was sufficient. Engineers prevailed by the 1910s, and cities gladly accepted dilution because it was cheap and easy.[83] During the course of research on this book, a waterworks employee remarked to me that if all cities were required to put their intakes below their outfalls, the water quality of our rivers would be very different.

In the 1920s, cities in the lower Missouri basin began widely using chlorine for disinfection, which sold for an inexpensive seventy-five cents per pound.[84] Considering all that occurred in the process of bringing the river into homes, water prices were relatively inexpensive—a household might pay two dollars per month.[85] By volume, Kansas City residents were drinking 3,700 trainloads of water every day, and "no other commodity on earth," McDonnell proudly stated, "is furnished and delivered into the homes at such low cost as our drinking water."[86] This inexpensive and broad distribution network would underwrite public health advances. After the technological and bacteriological revolutions of the late nineteenth and early twentieth centuries, basic treatment did not change much; chlorine and ozone were still used, as were settling and filtration.

Typhoid Mary, Tom, Dick, and Harry

The measure of Kansas City's water quality in the early twentieth century was the incidence of typhoid fever. Cities bragged about having lots of good, clean water and flaunted the ever-improving health of their residents. Competitive, growing cities did not want to acquire reputations for being on the statistical bottom rung. As in other cities, poor typhoid fever statistics motivated Kansas City health officials, engineers, and the public to reform their laws and infrastructure to benefit public health. Typhoid fever was spotlighted because it was fairly well understood, communicable, easily transmitted—and preventable.[87]

Typhoid fever cases among the working poor—like "Typhoid Mary" in New York City—as well as sensational cases among the rich and famous—like the suspicious death of Thomas Swope in Kansas City—served as warnings that anyone could fall ill.[88] Like disease in general, typhoid fever was gendered and classed. The "gospel of germs," as historian Nancy Tomes terms it, was the cultural aspect of the new public health. Both nineteenth- and twentieth-century understandings of health placed a great deal of responsibility for healthfulness on the home and on women. The domestic scene, personal hygiene, and the cooking and cleaning practices of women were considered the most important influences on health. Middle-class homes were more able to afford the accoutrements that reduced disease, which contributed to the belief that working-class and minority parts of cities were unsanitary in environmental and social aspects. Middle-class white families came to identify disease with poverty, immigrants, and nonwhites—assumptions reinforced by scientific racism. Doctors reported that insane asylums seemed to have a large number of typhoid fever carriers, and that most carriers were women, many of whom worked in food service.[89] Like Typhoid Mary, these were working-class women. On the one hand, these middle-class medical beliefs could increase prejudice, but on the other hand, the new public health also proved that people were not isolated; the interconnections of the community and environment threatened to render the most fastidious housekeeper's efforts for naught.[90] This forced urban reformers to be concerned about the health of everyone.[91]

There were multiple pathways of exposure to the fecal bacteria that caused typhoid fever; flies or handshakes could spread it, but the most common way Kansas Citians contracted typhoid fever was through contaminated milk and water. Bacteriology offered new explanations for illnesses, tracing them back to environmental sources like water supply.[92] Spring and well water in the city was suspect by the first decade of the twentieth century; the city had grown too big and generated too much waste, corrupting both open and ground sources of water. When mapped, typhoid fever cases clustered around springs and wells. The city water system, statistically, was the safest source of drinking water and more people were gaining access to it. But even city water could be contaminated.

An obvious source for typhoid bacilli was the rivers. The Kansas Cities were only one link in a chain of river cities. Just as water from Quindaro had passed through cities upstream, so the Kansas Cities redeposited their share of the river as waste for those who drew from the Missouri downstream. The bacteria that caused typhoid fever were in the river and could enter the urban innards for distribution in the water supply. Typhoid fever became more significant at the turn of the century because the population of the lower Missouri basin saw immense growth in the Progressive Era. Similar to the rapid growth of the Kansas Cities, American watersheds like the Missouri, Ohio, and Passaic had reached a critical mass of urbanization and industrialization, and sewage and pollution reached intolerable levels.

The early 1900s saw an impressive number of health campaigns in which *Musca domestica*, the common fly, figured prominently. Recent studies had found that typhoid fever, as well as the dreaded contagion tuberculosis, could be spread by household flies. This fact was widely disseminated, and citizens of Kansas City and elsewhere were entreated to "swat the early fly." In its first report, in 1909, the newly formed Kansas City, Missouri, Hospital and Health Board mentioned the hullabaloo about flies among sanitarians and in the press. Responsible for municipal health, the board turned its gaze to the Missouri River: "Along our river front sewers empty. Deposits of filth are certain to collect on the banks which breed millions of flies, and disease germs may be carried from these places of deposit to the homes of inhabitants."[93] The health board drew attention to the riverside and connected it with disease, industry, dump sites, and poverty. Flies migrating from the urban bottoms, not preventable pollution, were the focus for the board.

Late summer and winter tended to be times of higher typhoid fever incidence and the winter of 1909 and 1910 was particularly bad in the Kansas Cities, with Kansas City, Missouri, reporting about fifty cases in February. That same year, Kansas City, Kansas, reported over fifty cases in the month of August.[94] Before 1910, Missouri's water department had already been searching for some solution to water contamination. The Hospital and Health Board said that some sanitary engineers had been wondering if "the city water supply should not be unreservedly condemned for drinking purposes unless it had first been boiled or sterilized," which was occasionally recommended. Because an unreliable city supply might tempt people to go to riskier sources, the board stressed that it was essential that people trust the city's treatment of the Missouri River. In 1908 and 1909 the chemists did experiments and looked into possible water sterilization technologies, but the department decided they were too expensive.

Available technologies were changing quickly and by 1910 a Kansas City representative (possibly Kiersted) was sent to see how other cities were employing hypochlorite. By 1911 the city had cheaply and easily begun sterilizing river water on a systematic basis, and chlorine was a staple by the end of WWI.[95] Typhoid fever rates were lower in 1911, and the health board boasted that the danger was removed from Kansas City, Missouri's water.[96]

Despite water treatment, the city's typhoid fever rates remained above average and the Hoxies, another progressive power couple, worried about these statistics. Dr. George Hoxie was the dean of the University of Kansas Medical School and then the health commissioner of Kansas City, Missouri.[97] Mrs. Hoxie was a director of the Citizens' League alongside McDonnell and chair of the milk commission for the Kansas City Consumers' League. She spoke nationally about her work to protect milk supplies from typhoid-fever-causing bacteria—an expression of the "gospel of germs," and a larger transition to a regulated food supply. In 1922 Dr. Hoxie rhetorically asked, "Why the typhoid mortality in Kansas City?" as he revealed investigative results. Milk, flies, and children swimming in sewage-tainted water were avenues, but the commissioner posited that a "universal agency" was at work. Obvious to the point of being humorous, Hoxie asserted that "city water should be improved so that it does not contain sewage products."[98]

Like the McDonnells, the Hoxies believed that city government should bear more responsibility for public health. In the shift to public ownership, water suppliers enhanced service, and typhoid fever statistics dropped from an average of 30 deaths per 100,000 in 1900 to 2.5 deaths per 100,000 in 1925.[99] Kansas City's health commissioner reported enthusiastically that none of the December 1925 tests contained B. coli, the bacteria responsible for transmitting typhoid fever. "All were O.K.!!!" he said.[100]

While the dangers of typhoid fever were on the wane, McDonnell remained concerned about politics in the water. In Kansas City, the water department was a nest for the Pendergast political machine. If water department positions were political appointments, then the department would not focus on efficiency, because "a Water Department never can attain its highest function so long as politics has anything to do with its operation." McDonnell mourned fickle partisan appointments, preferring professional control over city departments. "The public pays dearly for this high mortality among Water Works Superintendents!" he said.[101] Accordingly, he advocated nonpolitical water and sewer boards in order to maximize efficiency. He also recommended "advertising" the accomplishments of such city entities in reports, opening the doors of the facilities to the public for tours, and designing aesthetically pleasing facilities the public could be proud of. McDonnell wanted a sort of "sunshine policy" for municipal services, reflecting his progressive ideals and vision of American democracy and also his belief that transparency would improve service and therefore public health. He also felt that municipal waterworks were duty bound to supply high-quality water to citizens and believed that courts should aggressively enforce responsibility to public safety.[102]

The Third Waterworks Is a Charm

As Kansas Citians filled their homes with tubs and toilets, advertisements for soaps and toilet paper proliferated and water consumption increased such that the waterworks was hardly able to meet the growing city's needs.[103] In 1924, when the National Board of Fire Underwriters made its report on the status of Kansas City, Missouri, in the face of fire danger, it pointed out a number of weaknesses with the waterworks system, from intake to distribution, calling it "barely adequate for present needs" and with "many unreliable features."[104] For example, the nine thousand hydrants that dotted the city were useless without good water pressure.[105] Facing a juggernaut of threats to safety, expanding need, and jurisdictional complications, the political discussion regarding a new waterworks boiled over. Additionally, the waterworks would be caught in the crossfire between political factions in Kansas City.

Reformers like McDonnell sought better water service as a way to improve health and "good government." Because the Pendergast political machine exerted such an enormous influence over city government and was consolidating power, reformers—particularly of the Republican persuasion—organized more assiduously to cap the power of the machine.[106] In the early 1920s the waterworks would be at the center of a clash between reformers and the political machine that would eventuate in both a new city charter and a new waterworks. McDonnell and his firm were an important part of this process.

At least five different proposals for the new waterworks had been presented by 1920. Some involved retrofitting and expanding Quindaro, while others involved a new location and a new facility altogether. Surprisingly, a filtration plant was not a feature of every proposal. Three of the proposals offered filtered water and a fourth would have filtration except during high-consumption parts of the day or year, when settled but not filtered water would enter the system. A fifth proposal, the one that was eventually adopted, did not initially include a filtration plant but would significantly expand settling and purification capabilities in a new facility in North Kansas City, and to that end, the city arranged to purchase a large piece of property on the Missouri side of the river in 1918. The range of plans focused more on expanding quantity than on improving quality.[107]

In November of 1921 voters were presented with a number of reform initiatives that included a new city charter and an $11 million bond to finance a new water plant. These initiatives were an attempt to wrest control from the machine. The Board of Fire and Water Commissioners was a decidedly partisan entity, so one of the reforms on the ballot would split the old board into two nonpartisan commissions. This was significant because the mayor—at that time, James Cowgill, a machine Democrat—directly controlled the water board and its projects; reformers loathed the idea of putting millions of dollars into the hands of the machine. Voters already knew who the members of the nonpartisan commission would be: former Republican mayor George H. Edwards and Democrats Alexander Maitland, Louis Rothschild, and Hughes Bryant—all prominent men. The measure to make the board nonpartisan succeeded, but voters failed to approve the bond issue needed to build a new waterworks. Arguably, voters desired a responsible body before they approved the funds.[108] Maitland, an engineer and chair of the proposed water commission, praised the decision to "filter politics from the water."[109]

The next spring, voters revisited the question of waterworks bonds. To make the case, engineer Maitland and reformers of the Citizens' League painted a bleak picture of the water supply, especially harping on the fact that Kansas City did not measure up to other cities. Maitland berated the "antiquated" Quindaro, calling it "patched and added to."[110] It was the largest city in America not filtering its river water, argued Maitland. With modern filtration the incidence of waterborne illness declined in other cities, leaving Kansas City an embarrassing statistic atop the typhoid fever risk graph. The American Medical Association (AMA) ranked Kansas City's water supply at the "bottom of the list" due to high typhoid fever rates and occasional boiling orders. "There are fortunately now few cities in the United States that periodically have to make this humiliating announcement," the AMA journal remarked of the boiling procedures.[111] Maitland asked rhetorically, "Do we in Kansas City have to wait for an epidemic before we realize the seriousness of the situation?" McDonnell added the assurance that "good health can be purchased." It was cheaper to *prevent* than treat illness by building a new water plant. "The science of water purification," McDonnell said, "has developed to such an extent that it is possible to secure a pure, safe, sparkling water out of the Missouri river."[112] The reformers roused support for waterworks bonds by appealing to women, cost- and health-conscious individuals, and "every citizen [who] is a stockholder in this public service corporation."[113]

Women, whose roles as voters were seen as family oriented, were also concerned with water supply, and bond proponents targeted them by touting the comfort, convenience, and economy of softening treatment. "Every woman in Kansas City is interested in filtered water, not only for the health of her family, but for bathing and laundry purposes," said one advocate.[114] "Hard" water had more minerals and left rings in the tub; it shortened the lifespan of fabrics, encrusted pipes, and wore out machinery. Mary Miller of the Citizens' League put it this way: "Soft water requires less soap, lathers more freely and saves both time and labor for every housewife." A consulted chemist reported that the Missouri River was a good candidate for softening. But because it required additional facilities and the use of chemicals like lime and soda ash, softening increased the cost of the new water treatment plant.[115] Georgia McDonnell leaned in to help justify the cost, stumping for the new waterworks on behalf of women and reformers at the league—and because it benefited her family. "All the women I have talked to here are wild for the soft water plan," Mrs. McDonnell told a reporter at a rally.[116]

With the security of knowing the $11 million would be in nonpartisan hands, Kansas Citians in all wards, particularly in the newer southern districts, responded affirmatively, voting nearly four to one in support of the waterworks bonds.[117] But victory was short-lived. Reformers had gnawing new concerns because in June 1922 a legal challenge to the nonpartisan board was mounted and while it languished in the courts, machine Democrats managed to challenge the new water commission in another way. Two choices would appear on the November ballot: option one would reaffirm the new nonpolitical board and option two would revert back to the political board. Citizen reformers redoubled their efforts to keep municipal water apolitical and the voters agreed.[118] Shortly after, rebuffing reformers yet again, the Missouri Supreme Court ruled the nonpartisan commission unconstitutional because it was not in alignment with the city's charter. This December court decision "ousted" the voter-approved commission, putting the bonds in the hands of the water board chosen by the new machine-influenced Democratic mayor Frank Cromwell.[119]

Just before Christmas, the *Star*'s editorial cartoon commented on the turn of events in this power struggle over the waterworks. Titled "The Mayor's Christmas Tree," it pictures an evergreen, its top weighted down by a gift "For the Boys," as the tag reads. The tree's hefty ornamentation is a treasure chest of $11 million, indicating that control over the Fire and Water Board, and the money that would build a new waterworks, was the most prized gift of the year.[120] This struggle over the waterworks was the galvanizing event in Kansas City's reform movement. The day following the decision, J. C. Nichols announced that, after a flurry of telephone calls, he and a committee of citizens had resolved that "something must and can be done to see that the will of the people, in respect to the nonpolitical expenditure of the water bond money, is carried out."[121] After twenty years, disparate reformers effectively joined to combat the machine by advocating a new charter with a city manager form of governance.

The Citizens' League, an umbrella group established in 1921 with connections to good government leagues and the Chamber of Commerce, led the charge. The league leaders were familiar names: McDonnell, Edwards, Rothschild, Volker, and so on. In 1924 reformers succeeded at electing Albert Beach, a reform-minded Republican mayor, and leading voter approval of the city-manager form of governance and a new city charter. The victory felt hollow for reformers because the Pendergast machine unexpectedly supported the reforms and subverted them.[122] In the meantime, the Cromwell administration moved forward with the new waterworks and reformers tried their best to contain the project within their parameters.[123]

The first report for the North Kansas City waterworks was made by the firm Fuller & McClintock in 1920, and the city gave project oversight to Fuller and Maitland—George W. Fuller was a prominent sanitary engineer based in New York and Alexander Maitland Jr. was a Kansas City engineer. These consulting engineers would then work closely with the city and let contracts for pieces of the project. The first paragraph of the 1923 waterworks contract gives an overview of the urban innards:

> The ENGINEERS shall plan, design, direct and supervise the construction of a new waterworks supply system for Kansas City, Missouri, comprising a new intake with adequate pumping stations, settling basins, filter plant, river protection work, with all necessary appurtenances and appliances therefore, to be located on the Missouri River in the vicinity of Kansas City, Missouri, and also the necessary tunnels, conduits, appurtenances and appliances for conveying necessary supply of filtered water to the present Turkey Creek pumping station and to a high-lift pumping station to be built in the East Bottoms; also all necessary reservoirs, feeder mains and other appurtenances and appliances for obtaining controlling and delivering a sufficient supply of filtered water into the distributing system of the Kansas City waterworks.[124]

As construction proceeded, the cities continued to rely on Quindaro. Joint operation of the plant by the two Kansas Cities had enabled an agreement between the states of Kansas and Missouri that exempted Kansas City, Missouri, from taxes in Kansas—an unusual interstate agreement.[125]

In the spring 1924 mayoral race, Republican Albert I. Beach defeated Cromwell; the future of the waterworks had been central to Beach's

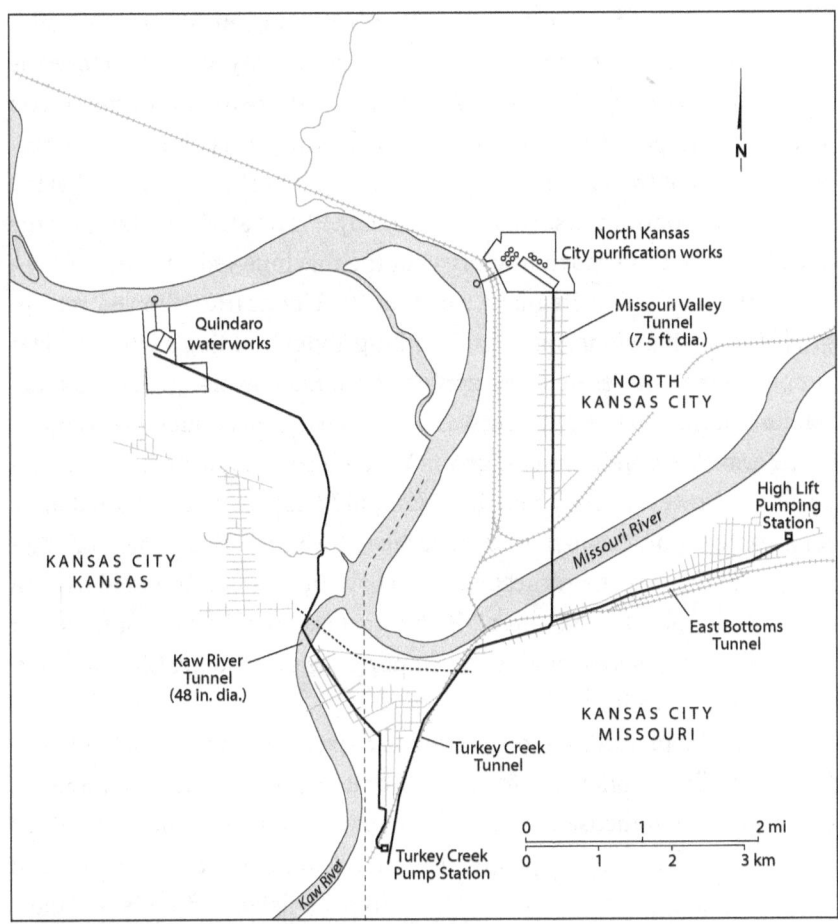

Plans for the North Kansas City Waterworks system, circa 1928. Cartography by Bill Nelson

campaign and the hopes of reformers skyrocketed as they envisioned controlling municipal affairs.[126] Instead, the defeated incumbent administration accelerated the process of letting contracts and decisively tying up bond money. The Republican-leaning *Star* accused the machine-influenced Fire and Water Board of being in a "mad rush" to get "under the wire" before the new administration came in.[127] Beach would take office on April 21 and in the meantime reformers tried to prevent last-minute decisions like the multimillion-dollar contracts for the river tunnels made earlier that month.[128]

Trying to slow the political board on behalf of reformers, five plaintiffs moved the controversy into the courtroom. Appearing before the federal district judge Arba Van Valkenburgh, they sought an injunction against the Kansas City Common Council, which was about to confirm waterworks contracts.[129] The judge granted the injunction, finding that the waterworks contract deserved further study.[130] The *Star* reported that the decision halted the "spending orgy" of the machine-led city council.[131] A key contention plaintiffs made was that the city charter allowed for upgrades and expansions, but it said nothing about building a *different* water supply system, therefore overstepping chartered powers. In addition, detractors warned that the project was on course to cost about twice as much as the available bonds, and Judge Van Valkenburgh concurred.[132] "I am not prepared, however, to say that the city should not spend 20 million dollars for a water supply," said the judge, because "I share with the mass of other citizens of this city the earnest desire to have our homes provided with an adequate supply of good water."[133] The legal issues, whether legitimate or not, had to be weighed against the importance of good health and drinking water for the city.

Joining the plaintiffs in opposition, a citizen committee, including reformers William Volker and former mayor Edwards, argued that the North Kansas City plan focused too narrowly on water quantity and maintained the status quo for water quality. If the plan were not reoriented toward health, drinking water would continue to pose risks, and Kansas Citians would continue to boil muddy tap water. The committee predicted a water shortage and interrupted service. Hoping to influence the outcome, they measured the water purity by clarity and not bacterial counts. Neither the existing Quindaro nor the planned North Kansas City waterworks had a filtration plant, but relied instead on settling and chemically treating the water. Edwards and Volker said that Quindaro could provide everything the new waterworks would and stay within the bonded $11 million.[134]

When Beach took office in April, his first tasks were to perform an audit and appoint a new Fire and Water Board.[135] Its bipartisan members—William Buchholz and George Edwards, Republicans, and Hughes Bryant, Democrat—were well-known and respected men in the business and political communities. Edwards and Bryant represented the shadow of the now-unconstitutional nonpartisan commission.[136] The engineers Fuller and Maitland, appointed by the Cromwell administration, remained at the helm of the stalled project.[137] Almost a month after the federal court first enjoined Kansas City from moving forward, the judge ruled the contract legitimate and within the city's powers.[138] Political warfare characterized the waterworks project, now four years along. Believe it or not, the war did not end in the courtroom.

In June, the Beach administration assembled an all-star committee of consulting engineers from local firms like Burns & McDonnell and Black & Veatch. A purification facility was now planned. The advisory board included McDonnell, N. T. Veatch, Wynkoop Kiersted of the Kansas City Water Department, and city engineer Louis R. Ash—all local engineers with good reputations who were invested in Kansas City.[139] Burns was also tapped for service. McDonnell, Veatch, and Ash were all on the board of directors of the Citizens' League—the organization's letterhead tellingly read: "Those who wash their hands of public concerns are as responsible for municipal misrule as are the men who are in politics for revenue only."[140] Beach obviously intended to infuse progressive politics into the waterworks planning process, because this advisory committee was chock-full of reformers concerned with public health and clean politics.

Ground had been broken in North Kansas City and the site was being graded when Kiersted came forward to push for expanding the Quindaro location instead. Kiersted's experience commanded attention from the mayor; he was a long-time engineer in Kansas City, the superintendent of the water department, and a member of the new waterworks advisory committee. When Beach was elected there had been speculation about whether the location might change—the place itself may have been chosen for partisan reasons.[141] Kiersted, assuming he had a chance to alter the plan set in motion, said that Quindaro had a "superior natural advantage," including river flow, likelihood of ice, and bank stability.[142] As another engineer warned, with urban growth, the planned waterworks would soon be "within the heart of Greater Kansas City" and therefore too polluted.[143] Despite Kiersted and others' rally to salvage and upgrade Quindaro by replacing the settling basins and building a filtration plant—all within budget—Mayor Beach swayed but did not bend.[144]

The mayor's arguments were not those of an analytical engineer; the politics mattered more than the physical function of the waterworks. Beach replied to the insurgents in his camp that the project was already under way and that honoring the city's commitment was important, adding that the engineers had agreed to take less money, dropping the total cost of the project. Finally, jurisdiction was important. The "complications and inconveniences" of having the waterworks in Kansas and the need to buy land in Kansas to expand Quindaro were distasteful requisites. In contrast, the undeveloped land of the north waterworks was in the state of Missouri and was easily purchased and annexed. The rights of way would be easier and the proximity closer.[145] Despite its burgeoning cost, the city must "control [the] situation" of its waterworks, Ash wrote to the Board of Fire and Water Commissioners.[146] Fuller and Maitland responded to Kiersted point by point, putting primacy on engineering aspects, not on political jurisdiction as Beach had, nor on water quality as Edwards and Volker had.[147] So, Kansas City continued to move forward with the new waterworks in Missouri, despite its technically inferior location and greater cost.

In the Beach administration, reformers intended to wrest power from the machine and capture city affairs. Stymied, reformers like those in the Citizens' League pushed forward with a new city charter that divided city governance between a mayor and a nonpartisan, businesslike city manager who would prioritize efficiency. The mayor would control little, whereas the city manager would appoint city boards—including the one overseeing the water department. The influence that reformers hoped this would have on cleaning up city politics was illustrated in the *Citizens' League Bulletin*. In this depiction, a bolt of lightning from the clouds, labeled "City Manager Plan," drives away a rabid, emaciated dog scavenging on bones.[148] Many American cities, including Kansas City, Kansas, had already made this political transition.

Kansas Citians approved their new charter in 1925. But Tom Pendergast, realizing the charter would be approved, got behind the reform effort and succeeded in having his men elected to five of the nine seats in the new city council. Reformers' hopes were dashed when the council's first hire for city manager was aligned with Pendergast. Unable to alter the waterworks plan much upon coming into office, Mayor Beach would further lose supervision when the new city manager, Henry F. McElroy, took office in 1926.[149] The window of opportunity for reformers was narrow.

The new waterworks required water tunnels, which were the most challenging aspect. On the ground, the complication was securing the rights of way, and under the ground, geology and construction proved an impediment. Tunnels three and a half miles long would come from the North Kansas City site, cross under the Missouri River, and then send the water to the two pump stations in the East Bottoms and at Turkey Creek. The Missouri Valley and Turkey Creek Tunnels, as they were called, were first bored in 1925. City officials took a tour of the shafts that year, but plans changed over the course of the next two years. The "TC3" tunnel shaft to Turkey Creek was unstable, abandoned, and relocated owing to geology, causing expenses to go up.[150] Today, Burns & McDonnell is careful to note that their firm was *not* involved in that aspect of construction.

The waterworks was not the financial fiasco reformers feared.[151] But it was no picnic by the river either. Developing drinking water infrastructure was a long-term process; it took about twenty-five years before a modicum of consensus was reached, engineering plans were laid out, finances were secured through voter-approved bonds, land was purchased, and construction was carried out. The new waterworks opened in 1928.[152] The new plant, fully completed after two more years, was the city's largest municipal project and it brought an end to decades of piecemeal waterworks improvements. The new plant could settle, treat, and pump up to one hundred million gallons of water daily and had the capacity to expand.[153] The plant is still in operation for the metropolitan area today, one step in the process of the river running through.

After the new waterworks opened, McDonnell reflected on the technological improvements and the contributions of public facilities. After thirty years of work, he was still repeating his mantra about clean water being the single most influential element of good health. By the 1930s Kansas City's waterworks provided tangible health benefits for the average person in Kansas City and McDonnell considered this democratization of an essential resource to be a "monumental achievement" of the early twentieth century. Urbanites began to take water services for granted; what they "once considered luxuries are now necessities," commented

McDonnell.[154] The abundance of reliable water had the unintended outcome of making the water utility invisible to the public—and thus veiling the presence of the river.

McDonnell realized how infrastructure was connected to larger social and political contexts. He never doubted his profession, but he exhibited a complicated understanding of the role of the engineer in society. Whereas engineering journals tended to speak of measurements and processes and put technology and data in the driver's seat, McDonnell emphasized the role of the consumer and the state in defining the waterworks. It was the public that raised the bar, he said: expecting high drinking water standards, the public demanded that waterworks adopt better technologies and offer safer service. At the same time, he credited city and state health boards with improving overall sanitation.[155] McDonnell continued to use the phrase "municipal housekeeping" into the 1930s to refer to the civic duty of professionals like him to improve the city and its services.[156]

Kansas City, Kansas, took over the Quindaro waterworks—once again, a hand-me-down facility. The two cities had already had their water plants next to each other, so the move was a short one for Kansas.[157] During the Progressive Era, Kansas City, Kansas, had been more enthusiastic about adopting municipal ownership and a city commission form of government than its Missouri sister. Consequently, Quindaro was immediately put under the control of an elected nonpartisan Board of Public Utilities, which lowered water rates. With the expertise of Burns & McDonnell, Kansas City, Kansas, expanded Quindaro in the 1930s, increasing its intake capacity of the "difficult" river.[158] From the 1890s to the 1930s the Kansas Cities underwent a significant transformation of their urban innards; drinking water, formerly available to a minority, was now accessible to a majority, and the quality of the water had improved too.[159]

In the years that followed, the Kansas Cities showed off their mastery of the Missouri with pride.[160] When the 1951 flood hit the Kansas Cities, both waterworks stayed in service. McDonnell, who had a penchant for photography, had pictures of Quindaro that showed floodwaters approaching the levees temporarily dozed around the plant. These photos are preserved in the firm's library. In one photograph, the photographer's shadow stretches across the ground as a duck is captured darting into the brush and toward the waterworks and the rising river. Perhaps it is the shadow of a man whose firm has been key to generations of waterworks in the Kansas Cities.

The Kansas City water supply, and thus the rivers, were the backbone of the city. By midcentury the public drinking water system was dependable, safe, and inexpensive and reached most urbanites regardless of social status. This security caused the issue of drinking water to fade from public view, as did the politicized nature of resource control and allocation. But for all these years, the Missouri River has continued to provide health and wealth for the city. When author William Least Heat-Moon made his way up river at the end of the twentieth century, he commented with sadness: "We pushed up along the Quindaro Bottoms, past the Kansas City water intake sucking in nine thousand gallons a minute to fill their fountains and make it easy for citizens to forget the river."[161] The importance of water remains, and the place "where these rocky bluffs meet" has been one of commingling between the people and their river, the river and its people.

SEWERS

THE WASTE STREAM

4

> *I maintain you can tell a lot about a city from its sewers. For instance, you can divine the intelligence, wisdom and foresight of the founding fathers of the city by the configuration of the sewer plan. You can judge the political, economic and social climate of the decade in which the sewers were laid. You can also tell where the "swells" of the time lived and where they were expected to move.*
>
> Paul Hohl, Kansas City Magazine (1979)

Author Richard Rhodes recalls exploring the storm sewers of Kansas City in the summer of 1948. In dry weather a boy could get around in the large underground pipes. His brother Stanley "took over the sewers" carrying a Boy Scout flashlight and a BB gun for rats. Stanley "wandered for miles, emerging at the corners of major intersections to peer out at the world through the drain gratings." Sometimes Stanley "popped up in the middle of streets, cautiously tipping up a manhole cover" and then "clanging it back down" when cars approached.[1] The Rhodes brothers remind us that a world lies beneath our feet.

After water was sucked from the Missouri River at the waterworks intake, it flowed through the city and its citizens, was transformed into waste, and then entered the sewer system and returned to the Missouri River, completing the cycle. Sewers contain rivers of household sewage, gray water, industrial waste, and storm water that connect humans to urban and natural surroundings. More than just bricks, clay, and concrete, sewers are technological and social artifacts that contain as much Kansas Citian as they do river. A sewer pipe (or the lack of one) is a sign of economic, social, and political power, and a measure of health.

Prior to the building of sewers, Kansas City relied on vaults, privies, and cesspools, which were private and decentralized. A family might have an outhouse in the backyard or an apartment building might have a shared vault that a night soil collector emptied for a fee. As the city grew in population at the turn of the century, the informal system was inefficient. Sewer construction sped up and an increasing percentage of people gained access to the waste disposal system. Sewers emptied to urban streams like Turkey and Brush Creeks and connected households to the watershed. The process of sewerage—engineering and building a sewer system—was capital- and labor-intensive. In other words, improving health required the commitment of public wealth.

Sewerage Dilemmas

Kansas City's long history of dependence on the Missouri River for sewage disposal was first set in stone (and brick) with the Main Street sewer.[2] Begun in 1860 but delayed by the Civil War, the Main Street sewer was completed in 1871 and measured 4,400 feet. By 1884 Kansas City had constructed nearly thirty miles of sewers—a direct correlation to the new waterworks and increased consumption.[3] Prevailing methods and technology offered two choices of sewer systems: combined or separate. Whereas separate sewers kept sewage distinct from storm water, combined sewers accomplished the goal of removing sewage and runoff in the same system, speeding flow and diluting waste.[4] Doctors, sanitarians, and engineers debated which sewer system was best.

Prominent engineers stepped forward to advocate for different systems in Kansas City. In 1883, the civil engineer Octave Chanute took up the topic of Kansas City sewers. Though best known nationally for his contributions to aeronautics, Chanute made a mark as a railroad engineer, designing the stockyards of Chicago and Kansas City. He also designed the Hannibal Bridge, the first to span the Missouri River. Chanute explained why Kansas Citians suffered from more "zymotic" diseases—what today are called "contagious diseases," like typhoid and scarlet fever—saying that they were caused by "imperfect ventilation and drainage, and especially by the exhalations of large sewers." Applying the miasma theory of disease prevalent in the nineteenth century, Chanute placed responsibility for illness on "noxious gases, which physicians tell us are dangerous to life."[5] Doctors warned of the dangers of "sewer gas," which could build up and pollute the air, exposing those who breathed it to disease.[6] For Chanute, sewerage was critical and he argued that a separate system would be healthier and cost-effective for Kansas City.[7] Chanute estimated that a separate system could save one million dollars, assuming that a complete set of storm sewers would never be needed. He made this argument in 1883, when Kansas City was smaller and there were few impervious surfaces.

Like Chanute, the medical field tended to favor a separate system, even as it adopted germ theory. Today, medical knowledge holds that while smell might be indicative of something risky, breathing sewage odors alone does not make one sick. While the older miasma theory of disease explained some things, it did not stand up to the rigors of the microscope and scientific medical knowledge. This new understanding of disease changed conceptions of pollution, community health, and the urban network.[8]

Supporters of combined sewer systems argued that public health and cost were in their favor. Wynkoop Kiersted, a Kansas City waterworks and sewerage engineer for many years and highly respected nationally, forwarded the argument in his 1894 book, *Prevailing Theories and Practices Relating to Sewage Disposal*, that the best available technology was found by harnessing nature.[9] "On the whole, it may be said of sewage disposal by dilution that it is the simplest, most economical, most expeditious method of sewage disposal that has yet been tried; and it is as fully efficacious as any when the self-purifying powers of water are duly regarded," Kiersted wrote.[10] Storm water, creeks, and rivers provided a natural way to dilute and carry waste away and, like most large cities, Kansas City committed to a system composed primarily of combined sewers.[11]

Engineer Robert E. McDonnell spoke in favor of sewage treatment, regardless of the system adopted. By the Progressive Era, water pollution had become a serious issue and the sheer volume of waste brought the theory of dilution into question. Harnessing the restorative and cleansing powers of rivers was effective—unless there was not enough time and distance downriver, there was too little water, or there was too much sewage. Sewage dilution, espoused by Kiersted and other engineers, versus the need to treat waste before releasing it to the river, became the most important sanitation decision at the turn of the century. Kansas City chose not to treat its sewage, but not without challenges from those concerned about public health, especially sanitarians with medical backgrounds.[12] City budgets were stretched, unable to meet changing infrastructure demands and explosive growth, and spending millions of dollars on sewage treatment could not compete with dilution, which was seemingly free.

Both gravity and the law favored urban streams becoming sewers. Because of gradient, waste ran naturally into streams, and the city's streams were the property of the city, making right-of-way acquisition easier. In time, the construction and extension of sewer systems entailed enclosing streams. Today, most trunk sewers are located in former creek beds, or alongside streams. Because most streams are no longer visible, this underground system disguises the watershed characteristics of the Kansas Cities. The sewers of the Kansas Cities eventually found their way to the Kaw and Missouri Rivers.

Kansas City, Missouri, had two main watersheds and therefore, in the early twentieth century, two main sewersheds were developed. The first was O.K. Creek and Turkey Creek, which gathered primarily from the older, western part of the city and flowed to the Kaw River, crossing into Kansas to do so. This stream picked up half of Kansas City's sewer discharge at the turn of the century, a significant waste load.[13] The largest tributary, O.K. Creek, became a closed sewer by 1912 and Turkey Creek followed in the 1920s. The Brush Creek and Blue River watershed drained southern and eastern Kansas City and developed later. By midcentury, the Blue, with Brush Creek as its largest tributary, carried over half of the city's waste load. It emptied to the Missouri River and never became a completely enclosed sewer. Every sewer connection meant a fundamental change in the way Kansas Citians interacted with their watercourses. Each manhole cover is evidence of this network of invisible waterways.

Sewers were a critical layer of Kansas City's urban innards. Infrastructure included what could be seen aboveground—like roads, street lamps, fire hydrants, and sidewalks—as well as the invisible city below—like mains, lines, pipes, and wires. Sewer construction usually accompanied street improvements; it was a city policy not to lay water mains if streets were not graded, and water service generally preceded sewer service.[14] Sometimes, owing to lack of funding, streets were graded and paved before sewer pipes were laid, causing the expense of ripping out the streets later.[15] Retrofitting was difficult and expensive. New neighborhoods platted in the twentieth century, when plumbing was common, tended to install infrastructure all at once and had the best access to urban innards.

Sewers were made of various materials, depending on when they were built. The Main Street sewer was walled with lime, roofed with logs, and had a large bricked gallery.[16] In the 1880s most sewers were stone or brick. They could clog because the walls were not smooth and greasy globules built up—like a blocked artery. As nineteenth-century sewers began to break down, they were replaced and often enlarged. In the twentieth century, clay and then concrete were the most common sewer materials. Kansas City used many clay sewer pipes because Walter S. Dickey, Commercial Club member, Missouri River navigation booster, and owner of Dickey Clay Manufacturing Company, had a large sewer pipe business. Six or seven percent of Dickey clay pipes were used in Kansas City, and the rest shipped by rail to the West, Mexico, and Canada.[17] Newspapers in the 1910s accused Dickey of having a monopoly in the city, making it impossible to lay concrete pipes.[18] By the 1930s Tom Pendergast, the second brother to lead the Pendergast political machine, had a concrete company that received most of the contracts in Kansas City, including the paving of Brush Creek with concrete to aid in the carriage of sewage.[19]

City building in the Kansas Cities was arduous because of the combination of bluffs and bottoms. Jokes about city topography told of people who fell out of their yards and landed on their own roofs. One visitor, speaking of her trip up the stairs from Union Station to the elevated trolley, asked humorously, "Why elevate anything in this Kansas city, when nature has done the job already?" The inclines were such that "no house floor is parallel with the ground it rests on," she said, and "women are always stepping on the front of their dresses or somebody is stepping on the back of them."[20] For all the aesthetic attributes of the landscape, carving an infrastructure into the limestone bluffs was difficult.

A Kansas City health official said that the "Lord knew what he was about when he built Kansas City." Cities without hills were "at the mercy of the artificial means of cleaning provided by more or less inefficient municipal cleaning departments."[21] But many references to the topography of Kansas City remarked on the difficulty of infrastructure, or "hewing a city from the hills," as historian Roy Ellis put it.[22] As an example of complications of molding an infrastructure to the topography, consider repairs on the Main Street sewer. Because it drained across the bluffs and down to the Missouri River, it was buried forty feet underground in places.[23] One sewer line coming down from the bluff to Beardsley Road above the West Bottoms was a trouble spot for decades. The sewer line leaked. In the winter months, sewage would freeze on the road, endangering automobile and pedestrian traffic. During the summer it smelled. Businesses and associations complained about it for years.[24]

While limestone bluffs and hills vexed some neighborhoods, the flat bottoms were a different challenge. Because sewers relied on gradient, the low-lying West Bottoms had sluggish streams and poor drainage, and proved dangerous to public health. Before good sewerage, Ellis recounted, the bottoms were stagnant sloughs that bred disease; the warm summer weather caused pestilence and the weeds overran everything.[25] A repeated story from the nineteenth century told of a part of the Main Street sewer that often blocked up and the subsequent pond that would form. Hogs thought it a glorious pool and rooted in it. A businessman at Fifth and Main Streets made a joke of the smelly wallow by sticking two boards into it covered with a pair of pants, making it appear that someone was diving into the putrid pool.[26]

The West Bottoms combined sewer system was first begun in the 1880s and both Kansas Cities managed the sewers. The two systems were connected; at least two sewer mains crossed the state border. Overall, the Kansas sewer system of the West Bottoms flowed in a northerly direction and emptied to the Missouri River, very near the state line.[27] Sprawling enterprises like the stockyards and packinghouses sometimes built private sewers, better able to service their needs than the city. As Kansas City's

"Main Street sewer." The city's first sewer was made of brick, stone, and wood.
Missouri Valley Special Collections, Kansas City Public Library, Kansas City, Missouri

oldest infrastructure broke down and collapsed, streets sank, sewage ceased to flow, and risk to residents increased. The West Bottoms sewerage system suffered from problems of design, age, gradient, and growth.[28]

Calls for Nasal Relief

In search of nasal justice, citizens wrote letters to the editor, formed committees, and griped to city officials. In one 1882 case, a property owner sued the city for damage, claiming that due to the discharge and smell of the Main Street sewer, he suffered soil erosion, lost rents, and devaluation. The courts ruled that the city's responsibility depended on where the natural or "ancient" flow of water was supposed to be. If the city had changed the flow, then the city was liable for property damage.[29] Court decisions like this one reinforced the city's adoption of "natural" watercourses for waste disposal. But this search for nasal justice in the courtroom was uncommon; unlike other cities, Kansas Citians left no long legal record of nuisance abatement.

Protests and complaints about sewers in residential neighborhoods reveal that citizens expected the city to protect their health in both public and private spaces. Although the "new" public health downplayed sensual experience, most people continued to believe that smells were dangerous. "Good Health" wrote a letter to the editor of the *Post* and complained about the "abominable" sewers: "One cannot pass a sewer in any part of the city without holding his nose, and undoubtedly disease is booming with the sickening odor from all of them. Isn't it about time our mayor and health department got to work and used some of our money in flushing these sewers?"[30] Citizens used the rhetoric of the city to demand better services. One group, led by Commercial Club president L. M. Miller, demanded that the Board of Park Commissioners abate a sewer nuisance in their neighborhood, saying: "Gentlemen—the accompanying petition calls your attention to the existence of an open sewer along 'The Grove.' As it is customary to show the improvements under your management to our visitors I would ask that this unsanitary condition receive your early attention because it detracts very considerably from your efforts 'to make Kansas City a good place to live in.'"[31] Citizens were fully aware that

Kansas City proudly showed off progressive achievements like the park system to delegations from other cities. Citizens reflected the language of boosters, politicians, and reformers back on the city in order to demand grandiloquence fulfilled.

Flushing was a popular remedy for malodorous sewers. Speaking of "notorious" sewers, George Neff of the Commercial Club confessed, "I hate to admit it, but I have to hold my nose every time I pass a catchbasin." To alleviate the problem, Neff recommended to a club committee: "We ought to have automatic flushing tanks, at least at the head of every street. With our own waterworks and abundance of water, with two rivers bordering the city, there is no excuse for this state of affairs."[32] Using fire hydrants to send bursts of water through sewers was a popular but temporary solution that required extra city workers. An alternative was the installation of automatic flushers that sent regulated amounts of water into the sewer at intervals, moving stagnant sewage and clearing blocked passages, especially in times of low rainfall.[33]

In 1904 city health officer R. P. Waring responded to complaints of foul odors, which stemmed from old catchbasins or trapdoors that did not close properly or were clogged with sticks, leaves, and other debris. It did not help that saloons and restaurants improperly threw organic garbage into manholes, where it decayed and clogged sewers. Given these challenges, the health officer explained that it was impossible to prevent the "sewer gas" of fecal matter from escaping, much to the annoyance of the person on the street and to the detriment of community health.[34] The street department's "gang" of sewer flushers only had enough staff to respond to complaints. Flushing could be comprehensive, urged Waring, if each fire station flushed near their firehouse. There was little unity among city departments: the public works department built and maintained sewers, the police department was in charge of preventing the dumping of garbage into sewers, the Fire and Water Board took care of the flushing, and the Hospital and Health Board's responsibility was public health.[35]

Kansas City, Missouri, addressed local agitation by laying out a comprehensive sewerage plan in 1905 that understood sewers as a public good and recognized the realities of topography. President of the Board of Public Works Henry M. Beardsley, an active municipal reformer who would become mayor in 1906, assembled data and a plan for "construction of a sewer system for *Greater* Kansas City" at "public expense." The estimated bonds were to be over a million dollars.[36] The plan included converting Westport's separate sewer system to Kansas City's combined system, condemning creek beds to create open sewers and prevent encroachment in less populated areas, making O.K. Creek into a closed sewer, proposing possible routes for a Turkey Creek sewer diversion, installing septic tanks and filter beds if sewerage was too difficult or expensive, and repairing and extending existing sewers. This was a thorough plan with foresight and became the blueprint for sewerage well into the 1930s.

Unfortunately, just having a plan was not enough. Political leaders needed money. In 1908 Mayor Thomas T. Crittenden considered creating sewerage districts and then taxing property owners according to the need of their district. Benefit districts would become a common method of building infrastructure. A few years later, Mayor Henry Jost told residents they needed to open their wallets and approve bonds to expand the sewer system.[37] City engineer Curtis Hill reported that only 20 percent of the 747 flush tanks worked. Improper installation caused expensive equipment to sit idle. On Hill's recommendation, Jost sent flusher gangs out to manually flush the sewers with water from hydrants.[38] The mayor gently suggested that if residents wished to deal with the problems of sewers, they must invest in infrastructure:

> It seems to me that the situation is one that should call to the attention of the public more vividly than anything else the need for sewer bonds. . . . There have been many complaints about these odors, and while the condition can probably be somewhat relieved by flushing in the newer sewers, there are certain sewers that must be rehabilitated in order to relieve the situation. We must have sewer bonds to do that.[39]

More plumbed households and an increase in water consumption sent more water to sewers. Self-flushing might have alleviated the problem of slow-moving waste, but the sewers could not always handle the volume.[40]

A functional sewer went unnoticed and maybe unappreciated, but a sewer that clogged, sank, was built at the wrong gradient, or improperly filled with garbage, manure, gravel, or industrial wastes made itself plainly evident and citizens had no problem making a similar stink.[41] In a letter addressing the city's "Trouble Department," one resident suffering from a sewer overflow complained, "If it were clean water it would not be so bad but believe me when you have to wade into Human Dung with a shovel and a broom, it makes you want to say things." The city sent out inspectors to catch plumbers who connected downspouts to sanitary sewers, people who illegally put trash or ashes into sewers, and businesses that leaked unwanted materials into drains.[42] When it came to sewerage and other infrastructure, squeaky wheels tended to get the grease. Politically and economically powerful men like R. J. Delano, J. C. Nichols, and W. S. Dickey already had the ear of the city and they received prompt responses to their complaints and requests.

Kansas City tackled the challenges of O.K. Creek, Turkey Creek, Brush Creek, the Blue River, and the Kaw River in different ways as they were added to the sewerage system. The Kaw River's watershed straddled both Kansas and Missouri. As will be discussed in the next chapter, the stream caused years of embroilment between the states. Brush Creek and the Blue River, discussed in this chapter, took even longer to resolve. In the meantime, all of the watercourses of the Kansas Cities came to function as either open or closed sewers. Importantly, all of these watercourses drained to the Missouri River and the city was cognizant of the fact that regulations might one day restrict sewage disposal in the river, something McDonnell and other sanitary engineers recognized.

The specter of sewage treatment threatened and loomed over Kansas City. Among the concerns of the city engineer in the 1905 comprehensive sewer plan was that Kansas City might be enjoined from dumping sewage to the Kaw River. The engineer correctly feared that Missouri's sewage emptying to a Kansas watercourse would become a problem, though it would be a few more years before the controversy developed. Also, the city feared that the federal government (via the US Army Corps of Engineers) might disallow dumping raw sewage into the Missouri River. Kansas City was already in disfavor because it used the river as its municipal garbage dump. Perhaps mayors tired of memoranda crossing their desks from chiefs and generals who ordered the city to stop this or that. Mayor

Crittenden contemplated sewage treatment in 1908. Amazingly, Kansas City briefly reconsidered building a separate sewer system and devoting existing sewers to storm drains. This "reduction plan" would make a treatment more feasible. The plan even considered the cost of building intercepting sewers to transport all the sewage from the main outlets to a central location near the Blue River's confluence with the Missouri, thus building only one treatment plant.[43]

In 1914, facing the same issues, Mayor Jost gravely said, "We do not know what day or hour the United States government may order us to cease emptying sewage into the Missouri river. When that day does come it will entail immediate action and a great expenditure of money."[44] In the 1920s the *Citizens' League Bulletin* ticked off the names of cities on the Missouri River, telling the story of urban relationships: there were upstream cities and downstream "victims." The Citizens' League editorialized that it was Kansas City's "moral obligation" to keep from polluting the water. The obligation was twofold: the safety of other people and the health of wildlife.[45] Despite the support of reformers, engineers, and sanitarians, addressing sewage in the Missouri River would have required Kansas City to consider its responsibility to the river community, and the city simply did not want to deal with the "knotty problem" of caring for sewage until forced to do so.[46]

It would be another six decades before Kansas City carried through, required to do so by the federal government. The sewage treatment plan instituted in the 1970s resembled the original proposed plans. In light of that, the fact that Kansas City discussed treatment plants as early as 1908 is impressive—and sadly comic.

Brush Creek and the Blue River—A Sewer Park?

The Blue River (Brush Creek is its largest tributary) drained thousands of acres of southern and eastern Kansas City on its way to the Missouri River. This region underwent rapid development in the first three decades of the century. The debate over the Blue River lasted many years, and it was the last watershed to be integrated into the sewer system. The plans for the Blue were vastly different from the other main urban drainage for five reasons: location in time, application of progressive ideals, fewer political boundaries, availability of funding, and terrain.

Sewerage and parks were often interrelated and planned together. Kansas City planned a comprehensive park system as early as 1893 under the direction of George E. Kessler, an engineer on the park board and one of the icons of American landscape architecture. Through the progressive years Kansas City established one of the most comprehensive park systems in the nation and Kansas City rode Kessler's coattails of foresight and planning for decades.[47] The Blue River was a facet of this park and boulevard plan.

Assuming the city would grow eastward, Kessler and the park board envisioned the Blue River as a parkway, its banks protected from urban encroachment. Brush Creek and the Blue River Valley were an essential part of the planned park system, argued park commissioners, because "the control of its internal waterways by a city is of such vital importance from a health and sanitary standpoint."[48] From the 1890s, then, the park board planned a sixteen-mile parkway along the Blue River. Supporters of the plan called for extension of the city limits to the Blue River in order to annex and condemn the creek beds quickly and cheaply.[49]

The Blue River Valley bore increasing amounts of sewage but the park board and other progressive park enthusiasts believed they could still preserve it by building intercepting sewers that would parallel the waterway and leave the Blue River a free-flowing, open, and clean stream. It was too late to do anything about Brush Creek, which already was a nuisance.[50] Because of development, the city could not dillydally. The work of creating the park "cannot be undertaken too soon," insisted Superintendent Dunn of the park board in 1913.[51] Plans were easy to make but funding was difficult to secure.

Should the people who would be served by the sewer pay, or should the entire city contribute to the cost of sewerage? Mayor Crittenden found himself "between these two fires." A "strong sentiment" for issuance of citywide bonds surfaced in 1908.[52] In the end, the city had more plans than money. "We are parked to death now," accused one political opponent who had opposed a vote to purchase properties along Brush Creek, claiming it was a "real estate scheme" pushed by developers and the "silver slippered crowd" who stood to profit from the park debt incurred by the city.[53] Some citizens opposed park bonds, fought condemnations of land, and entered injunctions against sewerage and parks.[54] These detractors saw parks as money makers for the real estate industry and as accoutrements for the wealthy that the city cunningly persuaded the general public to pay for. Social classes disagreed over what constituted a public good, how it should be paid for, and who would enjoy it.

In 1912, after three years of surveying and studying, George Kessler and the park board elaborated a plan for the Blue Valley Parkway that would incorporate the Blue River into a park while retaining its functionality as an important drainage outlet for Kansas City.[55] The main elements of the multiple-purpose plan were to be conservation of natural beauty, flood control, recreation, and the development of a boulevard alongside the river.

Upstream, the Blue River ran through Swope Park, the crown jewel of the park and boulevard system, donated to the city by real estate mogul Thomas H. Swope in 1896. It was a huge park in a then-isolated southeastern part of the city. This is not to say that nobody lived near the Blue River. The park board's report included photographs showing small homes and privies perched on the riverbank alongside impromptu garbage piles. There were even houseboats. That the park board would engage in playing proprietor to the poor is not surprising. As with New York City's Central Park, the Kansas City park board believed that condemnations and evictions would "improve" the land.[56] To reformers, a park could give returns in the form of social uplift.

The philosophy of the park board, first articulated in its 1893 *Report of the Board of Park and Boulevard Commissioners of Kansas City, Missouri*, is a noteworthy example of this progressive thinking. The report stated that not all citizens had the economic power "to select their place of residence" and that many people did not have the "opportunity to temper the daily recurring struggle for existence with a reasonable modicum of rational enjoyment and recreation." The proper improvement of a city, the board noted, depended "upon the wisdom, not less than upon the humanity, of those who influence and direct the policy of the government of a city, and of those that govern it."⁵⁷ Steeped in this philosophy, the board felt it was the duty of planners and the city to provide improvements for all classes. But the board hinted at its definition of the proper use of land when it referred to the need of the city to purchase lands with "temporary structures" and "eyesores." More directly, the board stated that locations with the potential for wonderful recreation areas were "now themselves disfigured by shanties and worthless structures, and in turn exercise a depressing effect upon the value of adjoining lands, better suited than they for private uses."⁵⁸ Clearly, people who lived on lands that the park board deemed worthwhile for "public concession to esthetic considerations," in the words of historian Roy Ellis in 1930, were worth evicting. Ellis described the creation of Penn Valley Park, which required destroying hundreds of "shanties," as a necessity to "restore its natural beauty."⁵⁹ The end justification was that the betterment of the city through public parks would increase property values and improve the lives of everyone.

Using the Schuylkill River in Philadelphia and the Charles River in Boston as examples of wise urban planning, the park board's 1912 report encouraged the city to push ahead with plans to purchase land alongside the banks of the Blue River because Kansas City had no other watercourse like it. The Kaw River "flows through an unsightly manufacturing, stock yards, and packing house district" and the Missouri "is muddy, unsightly, treacherous, difficult of access, and too swift of a current for pleasure boating."⁶⁰ The juxtaposition of the Kaw and Missouri Rivers proved how important it was to secure the Blue River and institute a plan of conservation. "Just now," the park commissioners urged, "before the ruthless and avaricious hand of man pollutes and disfigures the whole stream, before industrial development encroaches upon it, while yet the property . . . is comparatively inexpensive, this stream should be conserved, protected and improved."⁶¹ Superintendent Dunn warned that "in a very little while the sewage will be distinctly in evidence in the park."⁶² The park board considered the Blue yet unspoiled by encroachment and pollution; it was redeemable. The plan for the parkway called for straightening, channeling, and damming the river. A twenty-foot dam would make boating possible for nearly fifteen miles, all the way up to Swope Park—where the board did not intend to alter the river. Though the board was seeking land it perceived as lightly touched by humans, it still intended to shape the river to create an idealized version of nature.[63]

By 1915 Kansas City had spent somewhere around $15 million on parks and fifty-five miles of boulevards.[64] Public sentiment turned against liberal spending for the cause of "the city beautiful" and the Blue River was the marker. The bonds to improve Brush Creek and the Blue River did not pass the city council in 1915.[65] Voters failed to pass the same bonds in 1916. Years later, a commentator would note that the initial growth spurt in parks and boulevards had ended by World War I as Kansas Citians enjoyed their parks and stopped investing in them.[66]

The number of typhoid fever cases originating within Swope Park proved how urban growth was affecting the upper Blue River. The city health board reported that wells were contaminated, likely from sewage in the Blue River.[67] The park board told the mayor and Commercial Club that the Blue was polluted enough to endanger public health and "early action should be taken to correct the evil." Appealing to the club's prodevelopment progressivism, the board campaigned on the idea that preserving the Blue as a park would be good for economic health.[68] Despite the fact that wells and springs were condemned, people continued to use them. In 1925 health commissioner Herman Pearse was exasperated that condemnation signs were always disappearing and unwitting people were getting sick. Pearse requested that Mayor Beach assign a park police officer to the springs to prevent people from drinking the water.[69] Health officials called the lack of sewerage in the Blue watershed the "greatest obstacle to modern city sanitation."[70]

Downstream from Swope Park, the lower reaches of the Blue were industrial; Sheffield, the East Bottoms, and the Blue Valley industrial area used the river for its flat terrain and waste carriage. Fish kills proved that industrial pollution ravaged the Blue. The city health board often knew the source of the pollution but was either too weak or unmotivated to police industrial waste. Responsibility for action was passed on to the state and nothing happened.[71] As the second-most industrial part of the city, this eastern region contained a large working-class population. The area's largest employers in 1910 give a sense of the industrial layout of the lower valley: American Radiator, American Creosoting, Prier Brass

Manufacturing, Sheffield Car & Equipment, and Kansas City Nut and Bolt, which was an iron and then steel manufacturer.[72] These industries shaped the population living in the bottoms and the environment of the lower Blue River. It was mostly a native-born white population, though a quarter of the industrial workforce immigrated from Ireland, Germany, Sweden, and Canada. There were also immigrants with small farms in the area.[73]

In a 1912 survey of Sheffield and the East Bottoms, the Board of Public Welfare reported poor water service and discovered contaminated wells. The board blamed the general lack of city services among residents of the lower Blue Valley on both the city and landlords.[74] Representing a white working-class perspective from the East Bottoms, Mrs. M. J. Loftus wrote Mayor Beach to complain about unsanitary conditions in her neighborhood. Loftus described the lack of municipal services and her lack of faith in the city. The improvements her neighborhood sought—toilets, sewers, and gas lines—were blocked by propertied men with power who did not even live in the bottoms.[75] As an East Bottoms have-not, Loftus observed class warfare in the way city services were distributed. The city failed to provide residents with resources of social value, like parks. The industries may have been an "important asset," as one Blue Valley businessman put it, but working-class residents felt that industry and railroads had too much power to shape their Blue Valley neighborhoods.[76] Meanwhile, businessmen advocated building a protective dike around the East Bottoms to nurture even more industry.[77] Whereas white working-class residents valued municipal services like parks and sewers, city leaders, park commissioners, and businesses defined economic wealth as the most valuable contribution the East Bottoms had to make to the city.

A business-minded publication that supported the creation of a parkway all along the Blue snidely commented, "Eventually the Blue river to the Missouri river must pass into the possession of the city and be redeemed from the thralldom of commercialism and squalor which has it now in its grasp." Wanting the Blue to either imitate the genteel cultural aura of Swope Park or be given over to industry, the businessmen saw Blue River residents and recreationists as uncultivated and undeserving.[78]

The Blue River was already an active recreation spot serving the working-class area around it. A few bridges crossed it, including a footbridge, and

"Leisure on the Blue." This postcard shows recreation on the lower Blue River in 1910, when a parkway was still being discussed and before the Blue became an open sewer for urban and industrial waste. Missouri Valley Special Collections, Kansas City Public Library, Kansas City, Missouri

there were tents, houseboats, boats, and canoes crowding the banks. "Boats for rent" signs were painted across dockside buildings and people fished and hunted the banks. There were also theaters, pool halls, and places to dance, indicating that the river was part of a social entertainment scene.[79] The plans for the sewer and parkway could either hurt or help continued enjoyment of this area. Judge James E. Guinotte, formerly head of the Yacht Club located on the Blue, became a vocal opponent of the plans to lay a sewer. Judge Guinotte opposed the Centropolis sewer, which would empty raw sewage directly into the Blue River near Fifteenth Street, the center of recreation. Preferring an interceptor sewer to carry all the waste directly to the Missouri River, he began legal proceedings against the plan.

Guinotte imagined the Blue River as a navigable recreation area from Swope Park to the Missouri River and described its 1910s state as a "poor man's playground" that served thousands of Kansas Citians who could not afford to go to the mountains or lakes for vacation. The Blue was the only remaining location to boat safely, he said; the Kaw River had become too polluted and the Missouri required more experience of boaters. "The city is making a big cesspool out of the Blue," Guinotte charged, and if the Blue were preserved, people would soon appreciate it and want to restore it. Elliot Smith, another Yacht Club member in opposition to the sewer plan, had reminded the public of the situation downstream. The Blue had already become useless where a sewer emptied near Independence Avenue. These advocates wanted the Blue to be preserved for recreation, despite its functionality as a waste drainage for a growing city.[80]

In addition to recreation, the area along the Blue River housed a number of poor people, including African Americans who had chosen to live outside the city. They squatted on undeveloped pieces of land, on bluffs, and along waterways like the Blue. In these more rural locations, lacking plumbing was less of a hazard. Some social reformers applauded these housing arrangements because the squatters thereby escaped the "evils" of the congested city. Welfare board investigators reported that even if the squatters lived in shacks, their quality of life was better than in the rental properties of the river wards.[81] These squatters contributed to a scene of shacks and boathouses along the Blue River that more powerful white citizens did not like—white businessmen and the park board preferred to have a parkway controlled by the city.

By the end of the 1920s, the Blue represented a compromise. Despite the intentions and dreams at the turn of the century, the project would languish uncompleted for nearly three decades and would never be realized to the extent it had been proposed. More sewer than park, the valley did not match the grand expectations of planners. The interceptor sewer, accepting all the minor lines from creeks running into the Blue, was not intended to keep the Blue sewage free all the way to its mouth. With Swope Park upstream, industries downstream, and fifteen miles of sewer in between, the frog prince of a river park was suffocating under the waste load. The amended idea was to continue to allow the Blue to function for drainage but also to preserve its natural beauty for recreational enjoyment while engineering the waterway for flood control.

Overall, the 1920s were contradictory to the progressive planning identity that Kansas City earned from the 1890s to World War I. The 1920s had little vision for the future and funding was hard to come by as voters did not approve many bonds.[82] The building of public infrastructure had slowed during World War I and did not catch up. In 1915, 1924, and 1928 bonds for the waterworks, "improvement" of waterways, and sewerage either failed at the ballot box or passed piecemeal. An exceptional issue that met with wide public approval was the financing for hospitals, which commentators accredited to publicity about Kansas City's poor health statistics. The most expensive propositions were for the Blue River although there were dozens of smaller projects within the valley and city, too.[83] The city engineer warned that old districts desperately needed repairs and new districts needed sewers, and city health was threatened without both.[84] Areas like the Country Club district continued to build their own sewer systems that emptied to streams, while others lacked sewers. Topping the list of concerns for Health Commissioner Pearse was the lack of a main sewer in the Brush and Blue watershed and, until the city could propose, fund, and follow through with satisfactory plans, the entire east side would remain poorly sewered, he admonished.[85] Mayor Beach put a positive spin on all the sewerage delays, saying that the protracted debate was an assurance that the decision-making was nonpolitical and the outcome would be a better, more conservative proposal that would be amenable to the public.[86]

The Sewered and the Unsewered

Kansas City, Missouri, began the century without housing regulations, and what building codes were in place by the 1920s were limited. Interested in improving health and regulating urban space, reformers sought legislation that would standardize housing construction and upkeep. Expanding the sewer system's reach and connecting all homes, including rentals, was an essential part of reform. Because of a World War I shortage of materials and labor, home construction slowed and adherence to sanitary codes was lax. Due to the postwar housing shortage, "anyone able to build a house was urged to build it, and restrictions as to plumbing and sewage disposal were largely disregarded," a health conference reported. In the aftermath of this weakly regulated period, city inspectors had to check plumbing and require buildings to make sewer connections.[87]

Echoing debates over the Blue, some residents perceived codes and regulation as class warfare. In the 1920s, to encourage sewer connections, the city required permits for privies, vaults, and septic tanks, and anything that was not "modern plumbing."[88] Poor citizens resented such requirements, especially when the city did not force landlords to improve their rental properties. Amelia Stoeltzing, one such vocal resident of a neglected neighborhood, felt that city "czars" forced homeowners to install toilets, then failed to build adequate sewer infrastructure. Street improvements did not materialize and, due to prolonged sewer construction, her street was a "hogwaller."[89] Another woman felt that the city unfairly targeted her, a widow of little means, when the city required her to seal her privy vault.[90] Entire neighborhoods and districts had poor access to the urban innards that improved public health because these elements of class power, property ownership, and regulation combined to discriminate.

Newly built, wealthier districts found ways around limited city services. In advance of city sewers, the owners of "palatial modern residences" built vaults or installed septic tanks. When asked to abate leaky septic systems, they complied and the city promised to extend sewers to them within a year. The area along Brush Creek—for example, the Country Club district of J. C. Nichols—was rich and, despite sewage nuisances, the health board reported that it was an exceptionally healthy district of the city.[91]

The disintegrating infrastructure in the older sections of town devalued the housing, much of which was rental. Property owners did not want to be taxed for the cost of repairing or installing sewers. The North End, for example, was primarily a working-class district and its affordable housing came at the cost of aged or lacking infrastructure and services. The Main Street sewer came to serve an impoverished district and city engineer Hill proclaimed: "All the old section of Kansas City—which roughly speaking is the business section north of Twelfth Street—is in a deplorable condition."[92] Sewers collapsed and sewage flowed into basements. After thirty years, these sewers were dilapidated to the point of being rotten, and too shallow and small for the increased load of waste.

That social power influenced access to sewerage is significant because, similar to city water, sewer connections reduced risks and improved overall urban health. Despite the tremendous growth—seven hundred miles of sewers laid by 1924—a significant portion of Kansas City did not have access to these municipal services. The gap between the sewered and unsewered that existed before World War II was defined by class, race, and neighborhood; sewers followed money and power.

When the Chamber of Commerce carried out a health survey, it cited persistent gaps in services in the "north end of town" where "the colored and Italian districts have unsewered dwellings." The "Mexican section" as well was listed as a "problem area," among other areas that were white working-class districts. Licensed private scavengers were still responsible for the removal of wastes in these areas. The survey concluded: "The commissioner of inspection and sanitation maintains a spot map which shows many of these problems and complaints received. But the problems from a sanitary viewpoint are extensive, not easy of solution, and require trained personnel."[93] It is important to remind ourselves that the base infrastructure of Kansas City was built during a period of segregation and before the civil rights movement. These inequities were set in concrete and would remain influential in the coming decades.

Between 1900 and 1930 a majority of Kansas Citians gained access to both water and then sewers.[94] It was during the 1930s that the sewerage gap narrowed. In 1931, the Chamber of Commerce calculated that 98 percent of Kansas Citians had water and 75–80 percent had sewer connections.[95] By 1934, the number of Kansas Citians with sewer connections grew to 85 percent, showing the influence of Depression-era spending programs. In correlation, the number of privy vaults declined from fifteen thousand to ten thousand.[96] However, the health report considered the persistence of cesspools, privy vaults, and septic tanks "far from satisfactory for a modern city."[97] Nationally, an estimated 89 percent of urban residents were connected to sewers in 1930, so Kansas City was not far below average.[98]

The Ten-Year Plan and the New Deal

The fragmented and sluggish nature of building infrastructure had benefits and drawbacks. A long time usually elapsed between when needs were recognized, to planning and financing, and finally to construction. Protracted public debates and political turnover may have helped build the infrastructure that was most necessary and assured that the best engineering decisions were made. The disadvantage was that not everyone gained access to sewers quickly—especially low-income neighborhoods, which were often located in expensive places to sewer. With the Pendergast political machine in power, there was snide (and perhaps true) commentary about who the preferred sewer pipe supplier for the boss was, hinting that there was more than a little graft in the public works department. Powerful businessmen W. S. Dickey and Tom Pendergast both controlled companies that contributed to sewer construction—clay pipes and concrete.

Just as the Kansas City metropolitan area was thinking regionally for water supply in the 1920s, so it did for sewerage. McDonnell advocated metropolitan sewer districts for more sanitary control and better health outcomes. He cited other large cities that had done this and encouraged Kansas City to begin planning for these districts now, and not wait until need was dire. A regional sewer district would extend beyond the city boundaries and include about thirty thousand people who already contributed to the waste load of watersheds like the Blue River. All the "scattering but growing communities are in great need of being taken into a sanitary district if we intend to protect the health of the community as a whole," McDonnell wrote to a political ally in 1925.[99]

In 1927 Missouri passed the Ralph Sewer Law, allowing the creation of sewer districts in Kansas City and Saint Louis. Health problems arising from lack of sewerage were the impetus for the law, which put voters in charge of approving financing for the districts. The Missouri State Board of Health was not empowered to oversee this process, which weakened the law. In Jackson County there was a need for and discussion about a regional sewer district. Anticipating use of the Ralph Sewer Law, the Kansas City Public Service Institute investigated the county proposal for a district in 1930. It found that many outlying suburban areas were unsewered and threatening health in the Kansas City–Independence area. The city and county did not see eye to eye in planning the sewer district and the institute recommended hammering out amenable details, but it was too late: under protest from Saint Louis residents who decried use of eminent domain, the state legislature repealed the Ralph Sewer Law in 1931 and the Jackson County sewer district failed to be created.[100]

The city of Kansas City, Missouri, was having a difficult time getting voters to approve funds for municipal improvements. The antitax cultural sentiment of Missourians resulted in a state law requiring a supermajority vote of two-thirds to pass a bond. This meant that measures needed broad support and required a significant campaign to sell the plan and secure the votes. Heretofore, Kansas City's infrastructure had been funded with benefit districts—excising taxes on the property owners benefited by the infrastructure. Such taxation reinforced urban inequity because some property owners were willing and able to pay for infrastructure, whereas the landlords of the working-class districts resisted such taxation. General taxation through citywide bonds was needed and in the 1920s such bonds had strong support among reformers.[101]

The Kansas City, Missouri, Chamber of Commerce (formerly the Commercial Club) believed public improvements had become problematic. The city had reached its indebtedness limit and voters were not approving bonds. Consequently, the city resorted to special taxes that were assessed on property, a taxation that the real estate and business community disliked. Communities unwilling or unable to "bear the burden of special assessments" did not have adequate infrastructure, including sewers. The chamber stated that when the health of the community was in question, then everyone should contribute to taxes rather than leave some without, leading to the club's ideological commitment to a major citywide bonding campaign.[102] The club ran a national advertising campaign for spending on infrastructure that swelled Kansas City's civic pride and morphed into "Get It Done."[103] Running from 1921 to 1924, the "Get It Done" campaign did not achieve full success, but it was a national exemplar for encouraging civic improvements.

The reform campaigns paid off in 1925 when Kansas City adopted its city manager form of governance. Even though Pendergast got his man—Henry F. McElroy—installed as city manager, McElroy, along with some unlikely political allies such as Chamber of Commerce president Conrad H. Mann, Mayor Albert Beach, and others, helped assemble the Committee of One Thousand, whose directive it was to produce a well-researched, well-planned, multifaceted proposal for public improvements.[104] It was advertised that the members of the committee represented all geographic parts of the city and therefore the interests of different social groups. The "Ten-Year Plan" was born.[105]

A broad coalition created and pushed the Ten-Year Plan, an extensive and expensive plan for civic spending through general bonds. Unlikely allies—including the Pendergast machine, elected officials, reformers, and boosters—formed a well-organized, bipartisan campaign of hundreds of individuals and organizations that pushed for voter approval of a "unified plan" for city building.[106] Though initiated in the mid-1920s, the plan took years to come to fruition. The proposals received only a simple majority in the 1928 election, not the needed two-thirds. But in a May 1931 special election, the $40 million Ten-Year Plan passed four to one at the ballot box, in spite of the economic downturn.[107] The plan would become Kansas City's response to the Great Depression. Under the Ten-Year Plan, benefit districts declined in number while per person public debt rose.[108] Already committed to public spending and armed with a blueprint, Kansas City, Missouri, was poised to take advantage of the New Deal.[109]

William Allen White, the Emporia, Kansas, journalist, wrote admiringly of the Depression "monument" that the Ten-Year Plan would create. "Kansas City has the system," he said. Rather than wait for the federal government to act, "she voted bonds so that they [the unemployed] may be given jobs at useful and beautiful public improvements." Favoring Kansas City's "get it done" spirit, White glowed: "How much better it all is than the expensive soup kitchens maintained for idle men by private charity, or an equally demoralizing government dole!"[110] White respected the Ten-Year Plan for its self-reliance, its action, and its spirit. There is no denying that Kansas City leaders and boosters had done years of work to plan and rally support, but the ultimate success of the Ten-Year Plan was connecting to opportune federal funding. Kansas City succeeded at winning an above-average share of New Deal dollars.

In the 1930s the Pendergast political machine reached beyond the city and county and became influential in Missouri state politics. In the 1932 election for Missouri governor, the Democratic candidate that Pendergast chose to support died but the replacement candidate, Guy Park, turned out to be even more loyal, if not a yes-man.[111] This gave Pendergast unprecedented access to the state political system and Missouri was

SEWERS: THE WASTE STREAM | 95

Cartoonist S. J. Ray depicts the Ten-Year Plan in 1931. Courtesy of the *Kansas City Star*

rewarded for its powerful Democratic voting bloc for Franklin D. Roosevelt. The machine secured Matthew Murray a position as director of the Works Progress Administration in Missouri.[112] Murray was no stranger to Kansas City's infrastructural needs, as he had been the head of the city's public works department. From the state level, Murray was able to help Kansas City make the most of its already-approved Ten-Year Plan bonds by channeling matching funds from the Civil Works Administration, Public Works Administration, and Works Progress Administration. Kansas City "would reap a lion's share of the state's benefits" from relief programs, as two historians explained, because it had polished plans and political connections.[113] A Thomas Hart Benton mural in the state capitol in Jefferson City portrays the power Pendergast had without being an elected official. The 1936 painting, titled *A Social History of Missouri*, depicts Kansas City with scenes of meatpacking and a business meeting. A smoking man in a suit sits behind the lecturer at the dais. He represents Tom Pendergast, the power behind the curtain of business and politics.

Waterways were avenues for New Deal economic and social reform. In the 1930s the Missouri River, Blue River, and Brush Creek were Kansas City's most important pick and shovel projects. Overall employment on work projects peaked at 16,500 in 1939 and about one-fifth of city residents received some form of relief.[114] The Brush and the Blue got modified makeovers that had been in the works since the 1890s. Both makeovers stressed sewerage and flood control, but delayed the creation of parkways.

A prominent bond in the Ten-Year Plan was the Blue Valley Flood Protection and Parkway, which was based on the 1912 proposal designed by Kessler and the park board. Before the election, a newspaper commented, "This project, which would not only prevent a recurrence of the disastrous floods which periodically inundate the Blue Valley but would provide a great aquatic playground as well, has been under discussion for so many years and is so universally approved that it scarcely needs discussion."[115] The Ten-Year Plan proposed buying land between Fifteenth Street and Swope Park for a more modest parkway, the only proposed addition to the park system. Sewerage was one of the elements of the proposal. The Blue Valley Sewer would carry a large portion of the city's waste, including from the Gooseneck Sewer that drained much of eastern Kansas City.

With enormous help from the Works Progress Administration (WPA), an interceptor sewer was finally built and along with sewerage came the straightening, shortening, and widening of the river. Recent flooding of homes and industries in the Blue Valley made channel and bank changes for flood control amenable; these would serve as the "foundation" for a parkway.[116] After $100,000 in bonds, the WPA provided as much as $2 million. The Corps of Engineers worked on New Deal projects to remodel the industrial lower Blue River for flood control and by 1940 the "Big Blue

Bend" of the Missouri River was moved, requiring the Blue River to be "stretched" to meet the Missouri's new channel location.[117] The long-envisioned park never materialized.[118] By the late 1930s the only part of the river still suitable for a parkway was upstream from Swope Park. The final one and a half miles of the Blue remained an open sewer.

The Ten-Year Plan included a $1 million bond issue for "channel improvements, roadways and beautification" of Brush Creek and a future parkway along the creek, ending at its junction with the Blue.[119] With a mostly urban sixteen-square-mile drainage area, the Brush moved far too slowly for the waste load it carried. To discourage stagnation, a wider, straighter concrete channel was planned under the Ten-Year Plan.[120] Brush Creek was paved to prevent pools of fouled water from collecting and to speed runoff of mixed sewage and storm water. Between 1931 and 1938, with a combination of local bonds and federal funds, the Brush was transformed. The Chamber of Commerce marveled: "Its valley is protected against floods. The channel is paved, the banks walled with native limestone, landscaped and kept in orderly and sanitary condition.... A beautiful setting has been provided for an eventual extension of park drives along its banks across the city."[121] The Brush Creek engineering was significantly underwritten by the WPA and it put a lot of people to work by maximizing hand labor. "The channel of Brush Creek was alive with men," the Chamber of Commerce enthused.[122]

Brush Creek never disappeared underground as a giant storm sewer, nor did it become a clean stream. It continued to accept the waste of occasional sanitary sewers and other runoff but it did not warrant the complaints of decades past. The paved creek bed has been a source of Kansas City lore because it was one of the highly visible projects for which Boss Pendergast's concrete company won the contract.[123] Murray, the Pendergast appointee in charge of the WPA in Missouri, defended the Brush Creek paving project in 1936. "They may call it a roller skating rink," Murray said of critics, but "it is a good, clean place for children to play. In its former condition, it was a menace to the public health."[124]

But the improvement of Brush Creek exacerbated flooding, a problem identified even before the Ten-Year Plan. In the 1920s the park board warned the City Ice Company, located on Brush Creek, that "run off is

Brush Creek and the Country Club Plaza. Paved in the 1930s in order to speed the movement of sewage, Brush Creek became a parkway where it passed through the moneyed section of the city. Missouri Valley Special Collections, Kansas City Public Library, Kansas City, Missouri

increasing in all the Brush Creek area as the territory becomes more intensively developed." Impervious concrete surfaces sped the flow of water and the park board had hoped this channel improvement would remedy flooding. Instead, the engineer's report "conclusively" showed that the company's "menace" by flood was "great and constantly increasing."[125] As with much of the engineering along waterways like the Blue, Kaw, and Missouri, Kansas Citians assumed they were staving off flooding, yet their efforts to confine streams increased flood risk.

Brush Creek became the "expressway of the sewer system."[126] In 1977 multiple dangers visited themselves on Kansas City: impervious surfaces, aged and too-small combined sewers, and the paved creek bed. Brush Creek flooded, killing people and doing significant property damage. Memorable photographs show a traffic jam of seventies sedans in the creek. The Plaza, the iconic commercial district created by Nichols, was submerged. The only way to prevent such flooding would be to build enormous, expensive sewers that could hold unusual amounts of rainfall—an application of the "one-hundred-year flood" calculation to storm sewers. Such flood control measures were expensive but possible, the Corps of Engineers assured, but one civil engineer with the corps stated simply, "If people wouldn't build on a flood plain there wouldn't be any damage."[127] After decades of investment in infrastructure along Brush Creek, the cost and feasibility of completely avoiding such flooding was beyond the reasonable reach of the city and its residents.

Today, part of Brush Creek is still paved and another part has become a parkway. The carefully constructed parkway section has limestone walls and pools of water calculated to flow just enough to keep them from becoming stagnant. Near the Plaza is a popular recreation area with fountains in the creek that help aerate the water. Despite all the engineering, raw sewage still finds its way into Brush Creek, and signs posted along the creek warn residents not to come in contact with the water. Although this area was once the suburban domain of Nichols, the revamping of Brush Creek is now part of an urban revitalization project because today's new suburbs are located many miles away.

The Ten-Year Plan and the New Deal shaped Kansas City's skyline. The art deco of city hall, the county courthouse, and the municipal auditorium all attest to the power of the Pendergast machine to control Missouri's cut of federal programs and to bring those funds home to Kansas City. Thousands, most of them men, labored on city and county buildings, the airport, parks, expanded water supply, pools, fire stations, schools, sewers, roads, and other public infrastructure. The New Deal medium was the environment, including the urban environment, and its programs brought about "changes that had been wrought in the land itself," as *Fortune* magazine put it.[128]

Like the nation, Kansas City became better sewered in the 1930s. Kansas City was able to take advantage of federal financing because it had a ready-to-go civic improvement blueprint in the Ten-Year Plan.[129] The sewers built in the 1930s helped Kansas City fill the gap between supply and demand. When public infrastructure was funded through benefit districts, wealthy parts of the city could afford the property-tax-based improvements. Infrastructure like sewers was equally needed in poor places that could not pay higher taxes—or where landlords had no incentive to do so. Under the Ten-Year Plan, city spending took a different philosophy and used citywide bonds for which all residents would be taxed. Therefore, the Ten-Year Plan and the New Deal projects often paid for sewerage that would have been more difficult to fund under benefit districts because it also serviced difficult-to-sewer working-class neighborhoods.[130] These projects, it was argued, would be good for everyone because they would improve public health as well as provide multiple-use functions like parks and flood control.

Despite a deluge of new sewerage during the 1930s, the Kansas Cities built no waste treatment plants. Laying sewers might have made the city healthier, but it polluted the waterways more, which endangered those downstream. Therefore, the increased consumption of water from year to year, with capacity growing in the 1930s, correlated to increased sewage pollution.

Looking Ahead, Into the Blue

By the latter half of the twentieth century, the waterways of the Kansas Cities appeared, at first sight, to be largely industrial. Maps of the metropolitan area showed industry radiating like strands in a spider web along the Missouri and up the Kaw and Blue Rivers. There were not nearly as many parkways as had been envisioned at the turn of the century. In the 1960s a metropolitan planning commission imagined waterways, once again, as future parkways and open spaces.[131] By the 1970s the Blue River was described as a "miserable" stream hidden by the city and lacking the grandeur of the long-dreamt urban parkway. Jackson County began considering a parkway upstream from Swope Park, whereas previously the park was planned downstream.[132]

The Blue River interceptor sewer, built to keep the Blue from being an open sewer and to transfer waste to a central location, drew from one hundred square miles of land in two states and several cities by the 1970s.[133] At the base of the stream would sit Kansas City's Blue River primary sewage treatment plant. The "grit" removed by screening was placed in a landfill and the sludge was burned, its ashes put into lagoons. In 1971 the newly created federal Environmental Protection Agency described the Blue River near its confluence with the Missouri as:

> severely polluted by wastes discharged from the . . . sewage treatment plant, and industrial wastes including that from steel processing. Bottom samples collected from the Blue River were devoid of bottom animals which indicated gross organic pollution and possibly, the presence of toxic materials. Bottom sediments were composed of a gray-black organic sludge that had a strong hydrogen sulfide odor. The water was colored gray and pock-marked by bubbles of decomposition gases.[134]

Because of the "intensity of urbanization" the Blue River had complex and severe water quality issues, the report said. In times of heavy rain, the system could not handle the millions of gallons of flow, including sewage and toxic waste overflowed from sewers. Overflow bypassed the treatment plant, going directly into the Blue and then into the Missouri. One proposal had been to discharge the sewage directly into the Missouri River, skipping the Blue altogether. Other urban watersheds also contained industries that emitted dastardly stuff—ammunitions, industrial chemicals, agricultural chemicals, heavy metals, and concrete. Citizen groups and government studies continued to sound the alarm because the water quality problem was unsolved. In 2006 the US Geological Survey and Kansas City released a comprehensive multiyear study—one of the most detailed yet done on urban waterways—that showed the Blue River and its tributaries like Brush Creek were intensely polluted. The storm and sewer system was partially to blame, as the combined sewers occasionally overwhelmed treatment plants. The cost of making the Blue healthier by fixing the sewer system was estimated at $3 billion.[135]

Waterways began with mixed uses like drinking water supply, sewerage, and recreation—a testament to how interconnected the city and rivers were. At the beginning of the century, two different city departments dealt with water supply and sewerage; today, Kansas City's Water Department is in charge of both drinking water and sewage because it is part of the same process of the river running through. Owing to government regulation, the water quality of the Missouri River improved toward the end of the twentieth century.[136] But Kansas Citians continue to see their waterways as dirty, industrial sinks, in part because residents have lacked a riverside infrastructure that enables them to physically go to the river and learn otherwise. But Kansas Citians also perceived their waterways as dirty because being *downstream* indicated both social and environmental hazards.

PART II

REGION

BORDERS, BONDS, AND BODIES

When we try to pick out anything by itself, we find it hitched to everything else in the universe.
   John Muir

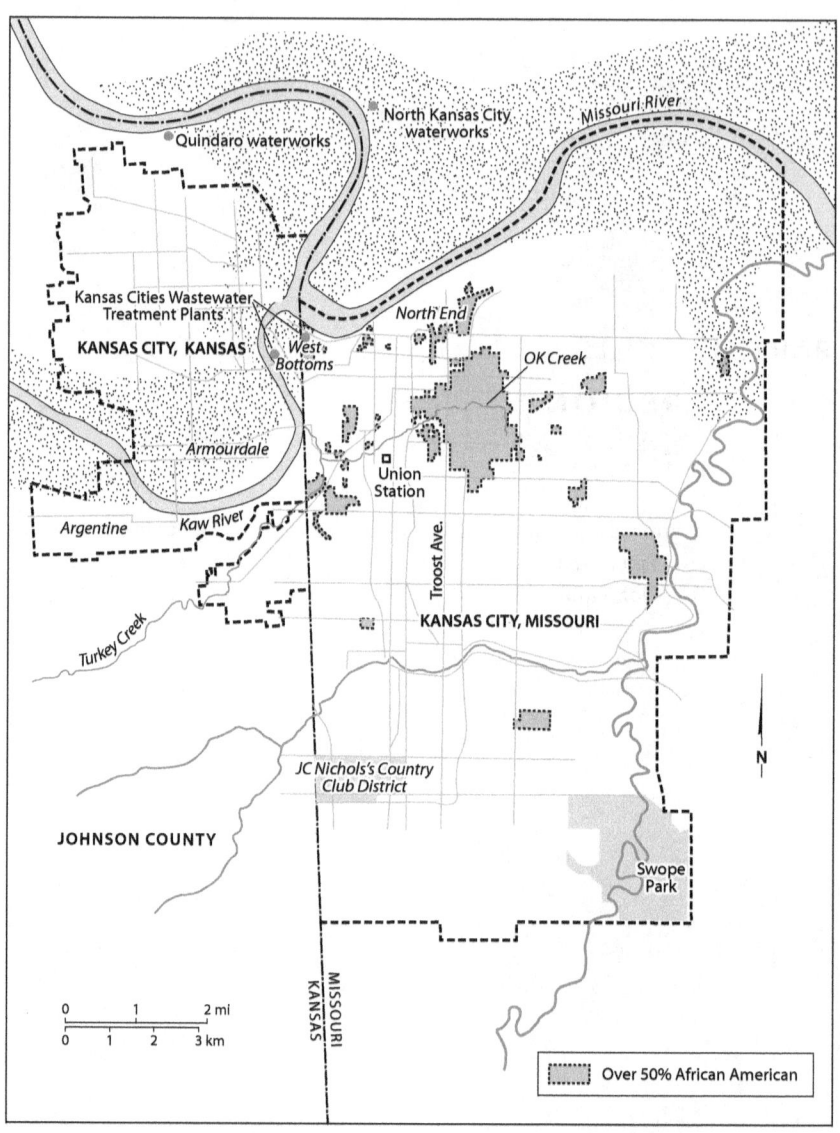

The metropolitan region at midcentury. Cartography by Bill Nelson

SISTER CITIES

5

> Kansas is neither northern nor southern, neither eastern nor western. It is the great central State, the hub of the Union. . . . As I have elsewhere expressed it, it is the rich, juicy meat in the national sandwich. But let me repeat that the great fact which stands out in this Conference like a promontory, is the fact of the mutuality of interests between the states.
>
> Kansas governor Edward Hoch addressing the 1908 Conference of Governors

This chapter broadens to the border-straddling Kansas City metropolitan region and to its two respective states, Kansas and Missouri. From the turn of the century the sister cities were an economic, social, and environmental unit. Despite their bonds, there was a cultural fault line between the two states. The mind-set of Kansas Citians, however, reflected aspects of both states, fusing the cities in a hybridized political culture.

Over the century, the cities increasingly cooperated and advanced regionalism. Environmental forces encouraged regional thinking, and infrastructure, like drinking water and sewerage, was significant in uniting the cities. One way to examine the region is through watersheds, which did not conform to political boundaries. At the beginning of the century, cities and states had few mechanisms for dealing with the cross-border complications caused by waterways and urban pollution.

Both watersheds and public health required looking beyond boundaries. Sanitarians and public officials such as

The Kansas City stockyards, 1920s. Missouri Valley Special Collections, Kansas City Public Library, Kansas City, Missouri

Dr. Samuel J. Crumbine encouraged cross-border thinking and worked to break down rigid political boundaries. The O.K.-Turkey Creek Sewer dispute shows how politics made managing sewage between the cities difficult in a shared urban waterway.

Touring the Wealth

In the 1920s, the younger Kansas City, Kansas, Chamber of Commerce put together a tour that showcased the attributes of its city. The tour began with city buildings but quickly moved to the economic heart of the Kansas Cities: the stockyards down by the river, which mainly lay in Kansas. These stockyards received around seven million animals a year.[1] Most of the cattle, hogs, and sheep that arrived by rail were slaughtered and processed in Kansas City, Kansas, near the Kaw River, making this packing district the second-largest meat producer in the United States. Next, the tour passed a huge creamery and the largest flour mill in the "Wheat State," then made

Industrial riverscape. This photo from the 1950s shows some of the heavy industry that accumulated in the bottoms, from Fairfax to Argentine. Missouri Valley Special Collections, Kansas City Public Library, Kansas City, Missouri

its way to the west side of the Kaw. All of these industries were located along the river because water was a necessity for industrial production and waste removal.

Passing through the rail yards, which were a national hub, the tour guide pointed out the largest structural steel mill west of the Mississippi River and a branch library in recently annexed Argentine, a working-class district with an industrial workforce. Next on the afternoon tour was the largest grain elevator in the world, then the famous potatoes of the Kaw Valley district, grown in the fine soil of the flood-prone bottoms. Finally, must-sees in the river's industrial district included one of Sinclair Oil's largest refineries and the plants of Procter & Gamble. Kansas City was the second-largest soap producer in the United States. The soap industry and meatpacking evolved together because soap manufacture used the fatty waste ingredients from animal processing.[2] The tour continued north, toward the Fairfax industrial district and airport, near the Quindaro waterworks. Along the way, there were churches, parks, and "luxurious homes."[3]

The tour showcased health and wealth, both the refined aspects of Kansas City, Kansas, and its smoky industrial complexes along the Kaw and Missouri Rivers. As the Kansas Cities pulled in the gas and coal energy and the grains and animals of the Midwest, they churned out products like steel, beer, flour, and meat. The Chamber of Commerce drove home the point that this industrial and manufacturing prowess of the Kansas Cities existed mostly on the Kansas side. Known for its huge manufacturing output, the Kansas side had more industry per person than all but one other city in the United States.[4] In contrast, Kansas City, Missouri, had a reputation for boulevards, fountains, exclusive housing developments, and other cultural features.[5] Kansas City, Kansas, produced more of the wealth whereas Kansas City, Missouri, held that wealth.

Kansans had long felt slighted because, in spite of their contributions, they were thought of as the little sister. Kansas businessman Carl Dehoney summed it up back in 1908: "We, on the Kansas side, are proud of the Kansas City on the Missouri side. We are proud of its fine civic spirit and progress, of its splendid parks and boulevards, hotels and business houses. We feel that we have a part in all these things. Meanwhile we want the world to know how much of Kansas City's greatness lies on the Kansas side."[6] Dehoney also shed some light on the origins of the "Kansas City spirit," locating its source a little more on the Kansas side. Many of the greatest contributions of the city have come from leaders with roots in Kansas, he said, singling out two men: Commercial Club members Frank Faxon, creator of the motto "Make Kansas City a great place to live in," and lawyer Herbert Hadley, who would become the Progressive governor of Missouri. Dehoney's friendly reprimand about "what Kansas City owes to Kansas" was not just about sharing credit, but about recognizing that the border was arbitrary and permeable.[7]

Early boosters suggested that it was "inevitable" a great city would develop at this confluence of rivers, in the heart of such agricultural and mineral riches.[8] Geography and providence provided for a great city, boosters proudly repeated, not intending division or divisiveness. Kansas banker Charles Brokaw said that cities like Leavenworth and Quindaro failed to develop, but "when men saw the finger of God pointing to the conjunction of the two most fertile river valleys on the earth . . . only then was the mighty metropolis of the west a sure prophecy."[9] Brokaw, believing the locale was naturally destined for a profitable city, created the "One–Kansas City Idea." Like Dehoney, Brokaw advocated for more unity between the cities because "what helps one helps both."[10]

Businessmen better recognized that the cities were an economic unit than a social unit. Thousands upon thousands of people lived in the bottoms and even more worked there.[11] People bridged the Kaw River daily as they traversed the state line for work, consumption, and leisure. Businessmen believed that one of the assets of Kansas City, Missouri, was a large native-born white population, whereas they didn't brag about demographics on the Kansas side.[12] A greater proportion of immigrants, transient workers, and working poor lived and worked in the bottoms. Kansas City, Missouri, statistically segregated itself with the state line, congratulating itself for its larger native-born population, accumulated wealth, and cultural institutions. This was an identity to which many middle-to-upper-class whites would continue to cling. In 1930 a Missouri industrialist referred to the industrial workforce in the Blue Valley as the "best of labor" because they were "native white American."[13] In 1940 the Kansas City Woman's City Club reaffirmed the city's "proud boast that it is the most American city in the United States." The club's Naturalization Committee made efforts to "Americanize" the "foreign" Italians, Mexicans, Jews, and Slavs.[14]

For African Americans, the community extended beyond the boundaries of any one city. The black press thought in terms of a "greater Kansas City," and individual black citizens saw themselves allied with other blacks more than with white neighbors. The black healthcare system reached across the state line. Before public facilities were created in Missouri, black Kansas Citians went to the Kansas side for care and surgery because the private healthcare system did not recognize the border. Similarly, entertainment was not bound by the border. When an African American fair was organized it appealed to both rural and urban, Kansans and Missourians. There were prizes for the best agricultural products for the truck farmers, as well as automobile and embroidery contests, highlighting the economic diversity of the community as well as the existence of a sense of regional identity among African Americans.[15]

The One Hundred Yards War: Turkey Creek

Over the years, Kansas City, Kansas, remained the lesser-known sister. Some came to resent Kansas City, Missouri's claim to being the bastion of the American white middle class while looking down its nose at Kansas City, Kansas. Kansans also chafed because much of the greatness Missouri claimed was produced on the backs of Kansans. Despite the border, the sister cities were economically, socially, and environmentally inseparable.

Nothing better illustrates the intertwined fates of the sister cities than the course of Turkey Creek and the decades-long border conflict over the sewage it carried. Turkey Creek—a "craggy" and "winding" watershed, in the words of a local historian—originated in Johnson County, Kansas, passed through the city of Rosedale, then crossed into Missouri.[16] In the short time that it ran through Kansas City, Missouri, it picked up about half of the city's sewage, including that from its largest tributary, O.K. Creek. Turkey Creek left Kansas City, Missouri, however, crossed the state line into Wyandotte County, Kansas, and ran for a paltry one hundred yards across the Kansas City, Kansas, bottoms to empty into the Kaw River. In summary, Turkey Creek started and ended in Kansas but, as it briefly crossed the border, it served as a major drainage for Kansas City, Missouri. Between cities, counties, and states, there were eight different political jurisdictions involved and, without any overarching policies to guide them, these entities found it difficult to look beyond political borders when it came to sewage. No laws existed to oversee water quality in interstate situations, making it nearly impossible to prevent pollution from upstream.[17]

Sewerage in this watershed was of such importance that in 1887 organized labor refused to support Kansas City, Missouri's charter based in part on the absence of an O.K. Creek sewer provision.[18] Under the successful 1889 city charter, the first debt included sewers for O.K. Creek.[19] Engineer George Waring made his government report on Kansas City at this time, foreseeing the O.K.-Turkey Creek drainage problem: "A very large part of the city, and one most difficult to drain, lies on the southerly

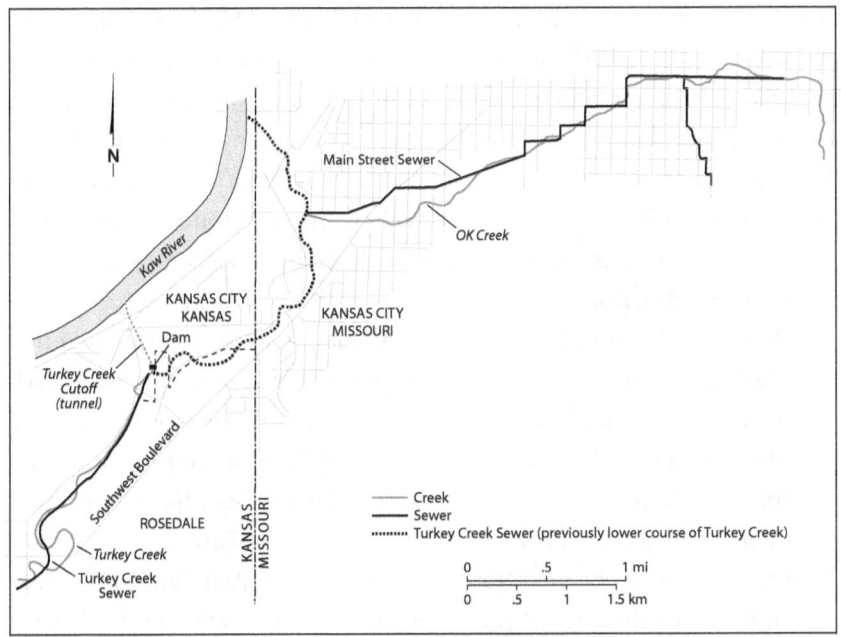

Cross-border sewerage and the Turkey Creek diversion project, circa 1920. Cartography by Bill Nelson

slope beyond the dividing ridge along Ninth and Eleventh Streets. . . . This small stream is quite insufficient to carry away the sewage from the large area naturally draining to it."[20] The quirky creek was a menace by the turn of the century and the city was making damage payments to streamside property owners for flood and sewage nuisances.

Industries and the railroads in the West Bottoms were not as concerned with pollution as they were desirous of flood control, because they had a lot invested in the bottoms. Following the flood of 1903, the Missouri River Commission and the US Army Corps of Engineers discussed drainage in the lower Kaw basin. Turkey Creek presented difficulties because its last leg was across the flood-prone bottoms. As a major sewage outfall, it was essential that its drainage not be blocked. In flood time, sewage would have to be pumped over the levee. Therefore, a levee would not be very effective unless it were built on both sides of the stream, all the way to the bluff. In addition, the elevation of levees would complicate the many railroad tracks crossing the bottoms. The commission explained it this way: levees are expensive and often are breached in flood, and "most important of all is the fact that in time of flood all sewers or drains of every kind, whether natural or artificial," invite flood waters to back up and overflow behind the levees. "If there be sewers they must in time of flood be closed and the sewage and storm water be pumped." It was an expensive endeavor with no guarantees of safety. In the absence of the perfect solution, the commission recommended lowering the flood plain around Turkey Creek as a first step.[21]

The political boundaries and environmental realities complicated this little bistate area in a number of ways. By the 1910s both O.K. and Turkey Creeks were committed open sewers. Rosedale, which was upstream, had the distinction of being the largest unsewered city in Kansas.[22] With the watershed as their connector, Rosedale and Kansas City, Missouri, worked together on what would come to be called the O.K.-Turkey Creek Sewer. Longtime Kansas City, Missouri, engineer Louis R. Ash had two proposals for the watershed by 1905: diverting O.K. Creek north to the Missouri River, or enclosing O.K. Creek and diverting Turkey Creek upstream in Kansas to the Kaw River.

The latter proposal, to divert Turkey Creek in Kansas, garnered the most favor from the outset. This accepted plan would divert Turkey Creek through a tunnel under Greystone Heights, up on the bluff of the Kaw and within the city limits of Kansas City, Kansas.[23] Because the only distinction between Kansas City, Missouri, and Rosedale was the state line, the smaller city would be connected to Kansas City, Missouri's infrastructure, even though it would eventually be annexed by Kansas City, Kansas.[24]

Kansas City, Missouri, pressed forward to address sewage in O.K. Creek because that stream existed wholly within its boundaries. By swapping land with the adjoining Belt Line Railroad, the city would acquire enough land to build a new channel and bury the creek underground as a sewer, which would be paid for in "special taxes" levied on the railroad.[25] Burying the creek would reduce erratic flows—a benefit to the railroad—and quell the smell, but it would also send O.K. Creek rushing in a beeline for Turkey Creek. Commenting on this possible speeding of drainage upstream from the Kansas bottoms, a Missouri newspaper reported: "It is thought that complications may arise from this."[26] The O.K. Creek Sewer was completed in 1912, and once the new nuisance of a sewer outfall was set in concrete, downstreamers were unhappy indeed.

People living along Turkey Creek complained that the conditions were "unbearable" and a "great deal of sickness resulted" from the sewage. In the 1910s Kansas City, Missouri, began channeling, dredging, and clearing the banks of the creek in pursuance of the larger plan for the watershed.[27] Safeguards against flooding had deleterious effects on those downstream and there were calls not only to abate the sewage but to remove structures like streamside retaining walls.[28] In the bottoms, commercial interests wanted flood protection and lobbied in 1912 on behalf of the Turkey Creek diversion project in Kansas.[29] With the exception of the streamside nuisance complaints, the engineering plans were publicly labeled flood control and not sewerage as a public health improvement.[30] Focused on wealth and not health, prominent economic interests talked up the flood control aspects; however, public health was the trump card and it was Kansas City, Kansas, holding the card.

It was in 1912 that the row between the sister cities first became heated. With the O.K. Creek Sewer built, the plan called for heading off Turkey Creek in Rosedale, Kansas. Before Turkey Creek entered the state of Missouri, it would be given a new outlet—a tunnel through Greystone Heights, leading directly to the Kaw River. With far less volume in the creek, the remaining bed could be made into a closed sewer and more easily carry the increasing waste load through the city. This would also mean that instead of a live stream in the West Bottoms there would only be a large sewer outfall, which in flood time would be easier to control by pumping. But Kansas City, Kansas, had no intention of approving this plan, and sewerage and flood control were stalled.

Kansas City, Kansas, mayor Charles W. Green, a businessman and civic activist from Argentine, prevented advancement of the Turkey Creek project for three years.[31] Since the project would be carried out partially on the Kansas side, within his city he had the prerogative to refuse. Kansans did not like the project because making Turkey Creek a permanent outlet for the sewage had distasteful public health implications. In the final one hundred yards of the stream, Kansas would have to deal with half the sewage of one of the fastest-growing cities in the United States. The creek was "already a nuisance," Green argued, and making it a permanent sewer would take away any agency on the part of his city, the ultimate downstreamer. Green preferred the other idea of diverting the creek around the bluffs and through Missouri's portion of the bottoms—and his public agreed with him. Missourians snidely remarked that the true opponents were wealthy people in Kansas City, Kansas, who did not even live near the creek.[32]

The crux of Mayor Green's argument was that the Turkey Creek Sewer was a public health risk and the Kaw River must be protected. Once he was willing to entertain the project, Green wanted the agreement to read that, should the sewer ever become a nuisance, then Kansas City, Missouri, would be responsible for abating it or ceasing disposal in the Kaw River altogether.[33] A city councilor from the Kansas side asked Missouri's mayor, Henry Jost, in 1914: "Do you want us to go ahead and pass an ordinance the people will not stand for?" Kansas leaders must "protect the future interests of this city," the councilor said, otherwise "the people will petition for a referendum on the matter."[34] In sum, Kansas City, Kansas, politicians fought to protect public health by refusing to allow the Turkey Creek Sewer to be built.

Kansas City, Kansas, and Mayor Green brought the Kansas State Board of Health into the debate. Led by Dr. Samuel J. Crumbine, Kansas had a strong board of health, whereas Missouri's board was weak and unlikely to weigh in on the issue. The Kansas board of health was legally empowered to approve or disapprove projects like these, which meant the Missouri city would answer to the Kansas entity. Naturally, this did not sit well with Mayor Jost, and the Missouri side declared the Kansas stipulations a deal breaker.

The Kansas State Board of Health was willing to approve the plans after an internal investigation by W. C. Hoad of the Division of Water and Sewage. Perhaps to the chagrin of Mayor Green, the health board thought the plan was an important sanitary improvement.[35] Therefore, the state board was willing to support the Turkey Creek Sewer plan because it benefited regional public health, whereas Mayor Green was opposing it because it threatened local public health. After the Kansas board of health declared itself in favor of the project, it did not relinquish oversight. Crumbine and the board ruled that if a health threat should develop, then sewage treatment must either "cease" or sewage treatment would be required.[36] The Kansas board of health retained control over final plans and specifications of the Turkey Creek project, as well as control of water quality once the sewer was built.

With the Kansas State Board of Health and all the other entities—like Wyandotte County, Rosedale, and the Kaw Valley Drainage Board—in alignment with the project, Mayor Green represented the only obstacle. Since the sewage was *already* draining into Turkey Creek, the mayors of the sister cities paid a visit to the sewer's outlet at the Kaw to investigate the status quo. They concluded that it was not currently a menace, but Mayor Green was insistent. Since Turkey Creek was already an open sewer, what bothered Green was the permanence of the engineering project. Once the sewer and outfall were constructed, there was no backing out, no redoing the infrastructure of the city. Once the upstream infrastructure was connected, Turkey Creek would be a committed sewer and Kansas City, Kansas, would be the permanent downstreamer.

Frustrated with Green, a Missouri delegation visited the Kansas side in 1914, making a matter-of-fact appeal that would be funny if it were not so true:

> Our only object in changing the course of Turkey Creek is to protect the region in the southwest part of the city from floods, and to make your flood protection plan on this side effective. . . . Kansas has not the prior right to the Kaw River. We don't dump our sewage in the Kaw. We dump it into Turkey Creek, a natural right of way, and it is carried down into the Kaw for a mile and a half, and then goes into the Missouri River and flows through our state for 300 miles. The sewage you dump in the Kaw River doesn't stay long in your town but flows on down past our city. All we are asking you to do is discharge the same amount and the same kind of water that is now entering the Kaw through Turkey Creek, except that we want to empty it at another point.[37]

The argument was simple: water quality and quantity would not change, and Kansas had no legal leg to stand on; Missouri had the right to dump raw sewage to O.K. and Turkey Creeks. Furthermore, as the waterways twisted around the bluffs, both cities ended up being downstream from each other!

Discontent ruled. Kansas City, Kansas, threatened to sue the Missourians upstream for building walls and other enclosures that "choked" the stream and exacerbated flooding downstream.[38] Because Mayor Jost felt that Kansas was showing no "spirit of reciprocity" for the project, he threatened to remove his support for a shared bridge over the Kaw.[39] Taxpayers in the benefit district complained. Since tax money for improvements was to be expended across the border, detractors interpreted it as Missourians being unfairly taxed to build sewers in Kansas. Neither the Missouri State Board of Health nor the federal government entered the discussion. The Missouri health board was legislatively meek, unable to use the law to create better health outcomes, and the Corps of Engineers was not likely to weigh in because Turkey Creek was not recognized as a navigable stream like the Kaw.

In 1914 Green and the council approved a plan for the Turkey Creek project that addressed their concerns and relied upon active involvement of the board of health. In their plan, Missouri was obligated to adhere to Kansas water law and the Kansas State Board of Health had the power to regulate the project after its completion, including a sewage treatment plant if ordered.[40] This would give one of the strongest health boards in the nation jurisdiction over a city outside its state, all by virtue of having the final hundred yards of a stream empty across the state line. The Jost administration rejected this proposal on legal grounds and the negotiations stalled again.

For the second time in 1915, Kansas City, Missourians financed the bonds to construct pieces of the project even though tax dollars would be spent in Kansas.[41] As soon as Kansas City, Kansas, gave the word, the Missouri side was ready to complete the Turkey Creek Sewer. Rosedale was on board. The wrestling continued. Missouri moved on with the other parts of the plan: the O.K. Creek Sewer and the diversion tunnel through the bluff at Greystone Heights. This lessened the flow of storm water downstream considerably. Rosedale could use the diversion, but only to divert storm water—not sewage.[42]

By 1916, Kansas City, Missouri, undertook the sewering of Turkey Creek within its city limits but left the remaining portion in Kansas undone until a legal resolution was found. They had come to an agreement about all the other aspects of this watershed plan—but not the interceptor sewer and outlet. The Kansas hangup was still public health. The remaining bed of Turkey Creek still received the O.K. Creek outfall and remained an open sewer to the Kaw River. In summer 1920 the Missouri Supreme Court ruled that it was legal to expend tax dollars in Kansas.[43] The price tag had increased significantly since 1905, now with an estimated cost of $2.5 million.

Kansas City civil engineer Alfred Ludlow specialized in sewers and had been hired to complete the Turkey Creek project in 1916. He summed up the dispute over the waterway: "The political boundary situation was involved almost inextricably because of the winding of the stream across state and municipal boundaries, to say nothing of the adjudicating and assessing benefits so that adequate funds could be raised."[44] Kansas City, Missouri, renewed its campaign for completion of the full Turkey Creek project. It had been four decades since the problem was first realized, twenty years since the plans were developed, and ten years working out border and funding issues. Three project engineers had shepherded the project—Louis R. Ash, Curtis Hill, and now Ludlow. The campaign for the Turkey Creek Sewer was still being sold as flood control and not sewerage—surprising, considering how much the district grew.[45]

The full Turkey Creek Sewer project finally underwent construction in October 1920. Railroads used reclaimed lands at a profit to the city and flood protection along the Kaw River could now continue, knowing a sewer and outfall would soon be laid. Rosedale prepared to hook itself up to the sewer, and to be annexed by Kansas City, Kansas. There would be pumps in the bottoms to prevent waste from backing up through the sewer when it flooded. The engineering world lauded the Turkey Creek Sewer as a concrete wonder; it was huge.[46]

Coming into office in 1924, Kansas City, Missouri, mayor Albert Beach inherited the Turkey Creek Sewer and diversion project from the machine Democrat Frank Cromwell. An important reason the reformer Beach won the election was a corruption allegation: a $1,000 check from Pendergast to the builder of the Turkey Creek sewer.[47] One of Beach's first tasks was to inspect the sewer project because, as the Board of Public Works stated, it was "common rumor that a flagrant disregard of specifications" existed.[48] Following its completion, the public and the Beach administration questioned the fiscal honesty of the project.[49] The newspaper referred to it as the Turkey Creek "scandal." There was a lawsuit involving the construction company and its work across the state line. Kansas City, Missouri, was forced to accept fault in a court case for underpayment and was making additional payments to Davidson Construction.[50] Rumors circulated that the 1919 project contract was suspicious and the construction quality poor. The investigative committee of engineers, including Burns & McDonnell and Black & Veatch, found a specious portion of the contract and the city refunded money to taxpayers, even though the project was finished.[51] One political insider suggested that the project had been "put over" on the public.[52] The fact that money was spent on the Kansas side still stung for some Missouri taxpayers and some feared that such border-crossing infrastructure would dilute the economic strength of Kansas City, Missouri.[53]

The "war" over the last one hundred yards of Turkey Creek resulted in years of cross-border antagonism, but a dispute does not best depict the relationship between the sister cities. Most of the time they cooperated to meet needs, as they eventually did on the Turkey Creek Sewer. Although flood control was an essential part of the plan—and one that helped gain the support of railroads, businessmen, and policymakers—sewerage was more important. The relationship between the two cities and two states—their rights and obligations across the state line—illuminates the difficulty of managing and engineering watersheds across political borders and the ways that public health got caught in the middle.

Both O.K. and lower Turkey Creeks ceased to be free-flowing streams and disappeared underground as sewers and storm drains, a part of the vast urban innards that connected residents to the rivers. The O.K. Creek interceptor sewer is the "granddaddy of Kansas City sewers," as one sewer aficionado called it. The O.K. Creek outfall was twenty-four feet wide and the subsequent Turkey Creek outfall at the Kaw was thirty-eight feet wide—so large that inspectors entered by boat.[54] By the 1960s the federal government reported that the conditions of Turkey Creek were poor due to household and industrial waste from the Kansas (not Missouri) side.[55]

A needed second diversion sent the waste to a treatment plant.[56] Kayakers on the Kaw still report that the swift outfall sometimes smells like sewage because of overflows.

Turkey Creek is still visible to Kansas Citians in its altered state. Upper Turkey Creek is still an open stream that now is surrounded by suburbs and can be spotted from interstate bridges. The streambed of the lower creek provides the flat gradient for Southwest Boulevard. One local artist recently noticed that the streambed lacked a stream. "That's the creek that should be in the West Bottoms. Where the hell does it go?" he asked himself. He parked his car and started walking. He discovered a rushing Turkey Creek that disappeared into the bluff, headed into the diversion tunnel instead of the old streambed. The artist went on to film a music video in the creek and to commit to making Kansas Citians aware of Turkey Creek. He says, realistically, "My goal is just to get people to say its name."[57]

Crumbinian Kansas

The boastful *Star* was willing to admit that Kansas City "owed" something to Kansas, as businessman Carl Dehoney had encouraged. The *Star*, detailing contributions made to the cities and the nation, called Kansas "an inventor of health laws" and singled out the work of public health pioneer Dr. Samuel J. Crumbine.[58] Crumbine's work stressed the importance of regional thinking, and he looked beyond the boundaries that restricted health improvements. He advanced the "new" public health and contributed to the growth of government responsibility in health matters. Like R. E. McDonnell and Herbert Hadley in Missouri, Crumbine epitomized progressive reform and professionalization. In fact, Kansas clubwomen referred to proper food safety and sanitation as *Crumbinian*, turning the actions of Dr. Crumbine into an adjective that described Kansas progressivism.[59] He was the leading force in the aggressive nature of the state of Kansas in issues of health and environment.

Crumbine was appointed to the Kansas State Board of Health in 1900 and became secretary and executive officer in 1904, at a time when "we were still on the frontier of public health administration," he said retrospectively.[60] Crumbine's hallmark public education campaigns made

Samuel J. Crumbine. Kansas Historical Society, Kansas Memory, www.kansasmemory.org

Kansas a national health leader.[61] He recalled, "I determined to go straight to the people, to undertake a program of health education to ensure the people's support."[62] His most-remembered campaigns included catchy phrases like "swat the fly," inspiring the invention of the flyswatter; banning the common drinking cup, which popularized the paper cup; and "don't spit on the sidewalk," memorialized in bricks. Under his leadership Kansas also passed some of the nation's earliest food and drug laws. Crumbine served the Kansas State Board of Health for two decades and was dean of the University of Kansas medical school, leaving in 1923 to take a job with a national health organization; in the early 1930s he worked on healthcare issues in the Herbert Hoover administration.[63]

Cartoons depicted Crumbine as a hard-nosed professional whose health standards would inflict pain on private business. In one 1907 cartoon, Crumbine holds down a druggist and is literally "Putting the Screws to Him." His devotion to inspections, eradicating the unseen germ, and the rule of law caused enough eye rolling among his critics that cartoons illustrated him as negatively affecting the Kansas economy because he prioritized health over wealth. One cartoon depicts Kansas—an Uncle Sam–like figure sitting in an easy chair—reading a newspaper with the headline: "Dr. Crumbine Reports this is bad and that is short weight, the other is tainted, the next is colored. In fact, the doctor rather discourages us in the use of anything made in Kansas." An already discarded portion of the newspaper sitting at the feet of the Kansas figure reads "Doc Crumbine makes a VERY LOUD report!" indicating that Crumbine was an effective progressive who did his research and disseminated it to the media.[64]

Crumbine was an Ur-progressive. He would apply science and knowledge, carry out research, and use the law as a tool to reorder public health. He was optimistic that he could reform society and the economy to be healthier and therefore wealthier. The new public health could be used to shape the citizenry. But like many progressives, he was not without detracting aspects. Screen and label the food, bat the rat, swat the fly, standardize the medicine, test the water, and regulate the sewage—these Kansas crusades were a part of the successful turn toward the new public health. However, prejudice could also be embedded in these reforms. For example, addressing clubwomen, Crumbine attributed the strength and health of Kansas to its low number of foreign-born residents, making it "most typically American," he proudly said. Along with its agricultural economy, temperate climate, and progressive politics, Kansas could bank on health and wealth because it was white, Anglo-Saxon Protestant. Kansas was "clean" because of "whiteness" and Crumbinian sanitary reform.[65]

In 1905 Crumbine began a vigorous antifly campaign because research connected flies to the spread of typhoid fever. Flies fed on waste, transported the bacteria, then came into contact with humans. Applying this research to Kansas, he noted that rural areas and small towns had higher rates of typhoid fever than urban areas, which were gaining "sanitary consciousness," he said. Crumbine roundly criticized rural areas for treating sewage lackadaisically—locating outhouses near wells, for example. Pamphlets and speeches entreated people to adopt the principles of sanitation and Kansas cities slowly built the corresponding infrastructure.[66] Crumbine advocated better personal habits but, importantly, he also recognized that entities beyond the individual must act to protect water supplies and create sanitary facilities for sewage and garbage. Crumbine recognized the healthiest county in the state by awarding a trophy inscribed *There is No Wealth Without Health*.[67]

Another campaign for which Crumbine was well known was the ban of the common drinking cup. In 1907 he observed practices on railroad cars where a common tin cup was used for drinking water from a dispenser. The cup was shared alike by people who looked sick and healthy, and sometimes the cup was used for purposes other than drinking. Although Crumbine does not elaborate, he recorded watching one man take the cup into the toilet and use it to "hold the injection that he used to treat himself for gonorrhea."[68] It was on this rail trip from Kansas City that he committed himself to reforming this unsanitary practice; medical and sanitation knowledge had evolved enough to understand that many illnesses were communicable, and the common drinking cup could become a method of transmitting disease like typhoid fever. His efforts to ban the common cup taught him about the importance of borders. Kansas could probably ban the unsanitary practice within its borders, but it had no control over the train once it crossed the state line. A passenger train traveling from Union Station in the West Bottoms to Topeka was part of

There is No Wealth Without Health.

GOVERNOR'S TROPHY.

To be given to the Healthiest County in the State of Kansas.

Governor's Trophy, 1916. Crumbine designed this award for the healthiest county, emphasizing the interrelationship between health and wealth. Courtesy of the Clendening History of Medicine Library, University of Kansas Medical Center

interstate commerce and, under federal law, no state can generally impede interstate commerce.

In 1907, Crumbine and his colleagues secured two important laws from the Kansas legislature: the "tight" Food and Drug Act and the "excellent" Water and Sewerage Law.[69] When he entered office as the state's chief protector of public health, he had no legal mechanisms available to control water supplies or sewerage. The Water and Sewerage Law of 1907 gave the state board of health the power to abate nuisances on behalf of public health. Kansas pursued sewage treatment because 60 percent of the population drank surface water. This necessitated river protection and made dumping raw sewage "absurd," as McDonnell had put it.[70] Consequently, Kansas tackled water pollution early on and the health board's experiences with the Kaw River were a precursor to abatement attempts on the Missouri. The University of Kansas engineering school worked closely with the health board, uniting the state and university in common cause. Especially active were the dean of civil engineering, Frank O. Marvin, and professor W. C. Hoad, who served as the health board's sanitary engineer and as an "advisor" to Crumbine. Both engineers organized state sanitary conferences and helped write the water and sewage legislation.[71]

Crumbine's grassroots strength came from women who strongly supported his work. The members of women's clubs in the Progressive Era were active in all sorts of political issues, including municipal and sanitary reform. These organizations, popular among white middle- and upper-class women, formed civic and health committees by the turn of the century. "In the last year or two a new spirit has developed," said the *Star*, writing in 1909 that area women's organizations "are giving earnest thought to playgrounds, street lighting, housing conditions, night schools, medical inspection of school children and other live topics." It was a woman's responsibility to "help in building the city and in keeping it clean," one club member said.[72]

Women became even more publicly active in the Progressive Era and they often came to their reform work in gender-specific ways. The activists who carried out Crumbine's campaigns did so because it was in their ken and role as women to be concerned with home, family, and health. At the turn of the century they increasingly extended their sphere of influence outside the home and into the community. The "municipal housekeeping" that these women were engaged in was "within the scope of their activity," as one 1909 club publication put it. Before, women were meeting to discuss literature, but "women view civic matters differently now than a few years ago," said one club member, describing women awakening to their power and influence in the community.[73] Women worked—usually as volunteers—to remedy the complications of urban industrialization. They were pro bono welfare workers who met community needs—services that were not yet institutionalized within city, state, or federal government. Larger umbrella organizations, like the Kansas City Council of Clubs and Kansas Women's Federation of Clubs, extended the work of women beyond the city.[74]

In several ways Crumbine's work looked beyond the borders of Kansas and contributed to regional public health planning. As mentioned in the Turkey Creek discussion, Crumbine and the Kansas state health board exercised control over Missouri's sewer project as it flowed through Kansas. In another example, he cooperated with the US Public Health Service to set up the first sanitary district in an interstate region. Crumbine sought this regional solution to address endemic tuberculosis among miners in southeast Kansas.[75] Crumbine's experiences with tuberculosis influenced his thinking about combating disease; frustrated by the restrictions of state borders, he sought regional cooperation.

Crumbine's influence followed the flow of water. Finding his state water and sewer laws limiting, he concerned himself with issues upstream and downstream from Kansas. Crumbine set his sights on the Missouri River, which affected the drinking water of several major cities in Kansas. In 1910 he initiated the Missouri River Sanitary Conference (MRSC) to encourage the lower basin states to adopt common interstate pollution regulations and embrace a vision of the Healthy River. The premise of the conference was that by cooperatively adopting water quality laws, cities along the river could protect drinking water supplies.

Crumbine was operating within the context of a national sanitary and conservation movement that began demanding water quality control. He was influenced by activities of public health officials in other river basins, notably the Ohio River and the Great Lakes. The Ohio River basin, with the Ohio State Board of Health in the lead, became a model for how stakeholders in multiple states could work together to protect water quality, exemplified by joint agreements on the Mahoning and Ohio Rivers. Along with the MRSC, these river conferences were some of the earliest legal and cooperative attempts at pollution control. The *Engineering Record* called this ongoing conversation for an interstate agreement to control water quality "the most important discussion in sanitary engineering since the separate system of sewage."[76]

The Bense Act, passed by the Ohio legislature in April 1908, legally bolstered the Ohio State Board of Health, giving it the power to "require the purification of sewage and public water supplies and to protect streams against pollution."[77] Communities requested that the board aid them in abating pollution from upstreamers.[78] The health board was pleased, calling the law their "most important event of the year," and said it was much lauded by engineers around the world. The board anticipated a legal challenge to its power under the Bense Act and, when it came, they were not able to enforce the law until the case was decided. In April 1912 the board was pleased with the final resolution: the Ohio Supreme Court unanimously upheld the Bense Act as constitutional.[79] This was such a victory for public health officials and their sanitary power that even in Lawrence, Kansas, there was celebration.

To celebrate the court having upheld the Bense Act, in 1912 Crumbine and his family dined with sanitary engineers in the Water and Sewage Laboratory at the university in Lawrence. The attendees drank from beakers and pipettes, served their salad on petri dishes, and ate a goose cooked in an autoclave. Although the Ohio state law had no direct bearing on Kansas or the Missouri River, Crumbine and the sanitary engineers, drinking their coffee from beakers, felt "a natural pride in any advance made by Boards of Health, for we had been pioneers," referencing the MRSC Crumbine had organized two years prior. Although the 1910 proposal for pollution control and public health standards on the Missouri River failed to be instituted, Kansas sanitarians saw themselves as part of a broader regional and national public health movement. To the diners, the legal empowerment of sanitarians to control water quality was worth raising a pipette for a toast in the laboratory where by day they tested water samples for sewage contamination. As Crumbine recalled the laboratory dinner, he concluded with a positive assessment of their work: "Our campaign was beginning to be a crusade."[80]

The idea of public health officials securing control over water quality was so important and yet so difficult to attain that even when sanitarians nearly a thousand miles away succeeded, Crumbine and his colleagues celebrated the victory for the whole profession. Following the MRSC, Crumbine was essential to securing a 1912 report by the federal government on the Missouri River, bringing its water-quality issues into the federal realm for the first time. His commitment to water quality was recognized nationally and in 1914 Crumbine joined an esteemed committee that established the first federal drinking water standards.[81]

As head of the state board of health, Crumbine was responsible for Kansas City, Kansas—a challenge because it was so connected to Kansas City, Missouri. In one case, for example, during World War I, Crumbine found that soldiers stationed in Kansas had venereal diseases. Upon investigation he discovered that soldiers were passing through Kansas City, Missouri, by train and visiting the vice district down by the river. In his memoirs, Crumbine tells of how he went first to the Kansas City mayor, who claimed it was a matter for the police. But the police told him that the courts were most influential in regulating prostitution. Finally, a court judge told him that the political boss Tom Pendergast ultimately had the power to control vice, the police, and the courts. Word got around quickly and Pendergast arranged to meet with Crumbine that same day. Pendergast said he would immediately have the courts crack down on prostitutes and thus venereal disease.[82] This meeting is a rarely documented testament to the power of the Pendergast political machine in Kansas City. It also displays Crumbine's power and persistence, even when it meant crossing a border.[83]

City Services and Planned Regionalism

Municipal services connected the Kansas Cities. Regionalism advanced through infrastructure, the pipes and sewers often serving as the drivers for extending urban relationships across borders. Historian Sarah Elkind describes the growth of regionalism as the "longer pipes" theory.[84] For example, Kansas City took the initiative to expand its infrastructure when it planned a sewer along Shuttle Creek, a Blue River tributary. A city engineer recommended annexing the watershed and thus lowering the cost of sewerage because both demand and tax base would be increased.[85] In the Kansas Cities, the drinking water and sewerage system was a network of the river that flowed across the state line.

From the start of the century, the Kansas Cities had cooperated and even partnered to provide drinking water because it was essential. Even though each city had separate waterworks, the two systems were connected and provided mutual aid in times of crises. The metropolitan area forged closer and closer ties through the century and shared water supplies. Around 1910, the Burns & McDonnell engineering firm made a survey on behalf of William Rockhill Nelson on the "feasibility of a water system for Jackson County." The survey deemed a countywide system difficult because the Kansas City waterworks capacity was limited and the county population was too small, making supply and need unequal: Kansas Citians needed the water more than supplying so few county residents justified.[86]

However, the regional population increased over the years and in 1921 Kansas City, Missouri, made its first water supply contracts with communities outside its city limits such as Raytown.[87] Rosedale, Kansas, also got its water from Kansas City, Missouri, because it was geographically better suited to do so.[88] Noncity dwellers were willing to pay 50 percent more for their water, which made the water department profitable but also helped absorb the costs of extending the service.[89] These agreements evolved to charge for the length of main the water traveled. This "distance-demand" was unusual, resulting in water rates that more closely matched the cost of water supply, instead of forcing Kansas City, Missouri, to underwrite the cost of suburbanization.[90] Consolidation of water supplies made fiscal sense because of the expense of erecting new and distinct facilities. The Kansas City, Missouri, waterworks that opened in 1928 was part of a regional move to supply water because the plant had a greatly increased capacity to provide both water quantity and quality.[91]

The 1920s saw the first significant spurt of regional planning ideas for the Kansas Cities. An early example of the turn toward regional thinking was when the Missouri park board produced a map of the parks in "Greater Kansas City" in 1914, which became a loose conceptual frame for a future regional park and boulevard system. Kansas City, Missouri, established a planning commission in 1919, but by the 1920s, when other large cities were establishing *regional* planning commissions, Kansas City had none.[92] Comparatively, the Kansas Cities did not develop regional planning early because the state line complicated cooperation.

In 1926 the Joint Regional Survey Committee formed. It was led by longtime Kansas City engineer and planner S. Herbert Hare, who had also served as a landscape architect with George Kessler on the park board. The survey focused on city planning in Greater Kansas City, including parks and sewers, and included five counties (Jackson, Clay, and Platte in Missouri, and Johnson and Wyandotte in Kansas). Hare defined regional planning in 1929 as planning beyond geography and political boundaries, and for "an area which has common interests and is interdependent by reason of topographical, economic or social conditions." All of the same issues before a city—water supply, sewerage, parks, and transportation—received consideration on a larger scale. Such joint problem solving had the greatest benefits for rural and suburban areas, Hare suggested. Harry S. Truman, then a Jackson County judge from Independence—the city next door to Kansas City, Missouri—became leader of the regional planning association.[93]

Hare said of the survey, "While this area falls within two states, the interest of this district and of all the communities in the district are practically identical with those of the main center of population at the junction of the Kaw and Missouri Rivers." He thought that the river, bottoms, and bluffs still formed the heart of metropolitan Kansas City's identity. Other urban regions that were offered as planning examples used industrial rivers as their regional "backbone." Hare thought this translated well on Kansas City's landscape, too. The bluffs, hillsides, and overlooks along the rivers, Hare said, "are of unusual scenic value, and in many cases are practically waste land, and should be preserved for public enjoyment in the years to come."[94] Hare recommended better utilizing the urban core, bluffs, bottoms, and rivers in the "heart of America"—a phrase Kansas City boosters popularized in the 1920s.

The pushes and pulls for developing regional thinking were environmental, political, and economic. The influential Citizens' League avidly discussed regional issues but the organization's preoccupation with reforming machine politics limited its regional influence. Next, private industry impacted regional services. For example, both Kansas Cities managed a sewer system in the West Bottoms—often letting the industries like the stockyards and packing plants take the lead in design and maintenance.[95] The majority of sewers dumping to the Kaw and Missouri Rivers were from private businesses, and these businesses often resisted being integrated in the city systems, fearing regulations.⁹⁶ The Ten-Year Plan, the influential city-building program that Kansas City pursued in the 1930s, was not a regional economic development plan although it certainly had regional ramifications. Overall, the dramatic infrastructure investment of the Ten-Year Plan and the New Deal contributed to regional growth in the Kansas Cities, because it subsidized projects like road building and extension of sewer and water services. Furthermore, federal entities like the Corps of Engineers treated the river basin as a regional unit (in terms of environment and economy), from upstream dams to downstream levees.

In 1935 McDonnell investigated a regional water supply network for Jackson County—this on the heels of the similar sewer districts intended with the 1927 Ralph Sewer Law. The suburban population had grown and the millions-of-gallons-per-day capacity of the new waterworks exceeded the needs of Kansas City proper. Developers were running private water lines into outlying areas and private water companies were cropping up. A countywide supply would be a win-win situation: if Kansas City offered its water services, it could earn revenue while county residents could get high-quality water. "The time has arrived when we should have a metropolitan water and sanitary district for the protection of the city as well as the county," said McDonnell. He advocated that the county promptly undertake a thorough survey. If the findings were to the affirmative, the city and county could press forward and take advantage of federal Public Works Administration matching grants.⁹⁷ All told, the cost of extending the established network of water mains was the least expensive option. Longer pipes won and Kansas City, Missouri, continued to be the heavyweight in the bistate region.

By the 1950s suburban growth was booming and Kansas City, Missouri, was profitably provisioning about ninety thousand people outside its city limits with Missouri River water. In addition, the city kept the upper hand by reserving the right to supply Kansas City, Missourians first, and fulfill the demands of its contracted users second. The drinking water region included municipalities on the Kansas side of the border as well as cities to the south, east, and north.⁹⁸ These newer, already-sewered suburban areas could easily be linked to the existing sewer systems of the Kansas Cities and therefore were attractive for annexation.[99]

By midcentury the region had clearly become the way to manage water quality and sewerage. Kansas City had become one of the larger metropolises in the United States and the Mid-America Regional Council (MARC) was formed in 1972 to finally coordinate the region. It was a regional board of elected officials that planned and facilitated policy implementation, but it was not a government. MARC helped the Kansas Cities look beyond the city, county, and state boundaries that had hampered problem solving, especially concerning environmental issues. By the end of the century, the agency united an urban and suburban population of about 2 million who used 160 million gallons of water a day and had 160 distinct sewer systems run by 51 different agencies.[100] Everyone from these two dozen different communities was drinking the river, serviced through the water plants of the Kansas Cities. The numerous political boundaries hindered regionalism, as did different political cultures and social classes.

SOCIAL INNARDS

6

> In the civilized life of to-day the contact of men and their relations to each other fall in a few main lines of action and communication: there is, first, the physical proximity of home and dwelling-places, the way in which neighborhoods group themselves, and the contiguity of neighborhoods. Secondly, and in our age chiefest, there are the economic relations,—the methods by which individuals cooperate for earning a living, for the mutual satisfaction of wants, for the production of wealth.
>
> W. E. B. Du Bois, The Souls of Black Folk (1903)

When the rivers tussled for dominance at the confluence and engulfed the bottoms in 1903, the floodwaters demonstrated that social factors influenced the citizenry's relationship to the river. Recall that as the river spread out for miles, filling homes and fields, stalling trains and halting industry, it made refugees of the largely working-class, immigrant, and minority peoples who lived in the bottoms. A flood victim wrote: "We are at home again but haven't got our goods at home yet, the water was only thirty-two inches in our house and it is time for me to go to work."[1] The city was united by the flood crisis, but within a week it was again socially divided. The relationship to the river depended on where one lived, and the social geography of the city reflected the power relations of races and classes. In addition to limiting drinking water supplies and backing up sewers, floods showed ways that race and class influenced access to these vital municipal services. While some residents fled homes and businesses, others were drawn to the bluffs as spectators. Up and down the Kaw and Missouri Rivers, from bottoms to blufftops, people were awed by the power of the river to affect their lives.

From the layers of pipes and sewers, the urban innards, we move to the Kansas Citians who resided at the ends of those connections to the river. The river was pumped into the city and ran through Kansas City bodies before it drained back into itself as waste. This chapter explores how bodies—the social layer—interacted with the landscape.[2] Like the circulatory system of the social body, the river flowed through neighborhoods, across borders, and between social groups to examine relationships of power in the Kansas Cities. Tracing the connections between the river and the people leads us to a broader view of the city and river and a broader definition of environment—one that includes the river, the built environment, urbanites, and their health. Mediators in this environment included municipal services, infrastructure, and housing, which depended on social status. Sources readily tell how the categories of race and class influenced people's connections to the river. From river intake to sewer outlet, we can see concentric circles of upstreamer and downstreamer relationships among neighborhoods and in the metropolitan region.

The River Wards: An Amalgam

Like the money and power in Kansas City, engineer R. E. McDonnell and his partner Clinton Burns moved their offices southward, away from the river. The river wards of both Kansas Cities were the most socially and economically diverse parts of the cities, housing African Americans, immigrants, itinerant laborers, and others in the working classes. In addition to being socially amalgamated, they were mixed-use commercial, residential, industrial, and leisure. It was in this part of town that the saloons, vice district, and theaters were located—places considered seedy by middle-class standards, but which those same middle classes might visit for entertainment or see while shopping, catching a train, or doing business at city hall. The river wards were the most environmentally hazardous places because of decaying infrastructure and industrial districts. The most socially integrated place in the Kansas Cities was probably the North End, the original bastion of the Pendergast political machine, which gained its early strength in the river wards. The bottoms tended to accumulate those who needed an urban "refuge"; this was the "humble rock" upon which the urban machine built.[3]

Some neighborhoods were associated with specific groups. The revealingly named Hell's Half Acre and The Patch were African American neighborhoods. Also in the bottoms were Little Italy and a part of town known for its beer gardens. Germans, Irish, Swedes, English, Canadians, Russians, Austrians, and Italians were the largest immigrant groups in 1900. Irish and German Catholic churches were found in the river wards in the early 1900s, as was a Greek Orthodox church. Communities of Greeks and Russian Jews were also growing, as was a Mexican population that lived in the bottoms and worked primarily for the railroads. By the numbers, the largest immigrant groups in the early century were northern Europeans or Canadians for whom language was not a barrier, and they would shift status from immigrant "other" to white by the mid-twentieth century. Evidenced by the location of Catholic churches, the Irish and Germans started moving south to higher ground and improved neighborhoods by the 1910s. The river neighborhoods they had lived in changed in social composition, shifting from native whites to newer immigrants and then blacks.[4]

Burgeoning urban growth meant that almost everyone was a newcomer, but Kansas City's newcomers came mostly from other states to the east, with the most heavily represented being Missouri, Kansas, Iowa, Illinois, Ohio, Kentucky, and New York.[5] Kansas City elites boasted that 55 percent of its white population were second-generation Americans, whereas only 20 percent of white Chicagoans were. Because many foreign immigrants in the Progressive Era wave of migration were not initially considered "white," Kansas City could also claim to be one of the whitest American cities, which it equated with being American.[6] The local newspaper suggested that the lack of foreign blood made Kansas City "an admirable laboratory for working out the principles of democracy."[7] Kansas City saw a shift toward southern European immigration at the turn of the century but, again, that did not dramatically change the population demographics and Kansas City would continue ascribing its "heart of America" identity to being a native-born white city.[8]

The claim to being "most American" was misleading because remember that a large number of immigrants lived in the East Bottoms and on the Kansas side of the city in places like Armourdale, Argentine, and most of the West Bottoms. Many working-class whites and African Americans lived on the Kansas side too. Since the Kansas Cities were a unit, it was biased thinking on the part of Missouri press and elite leadership to separate itself from the social and economic reality of the sister cities.[9] The river made the same point about these urban interconnections when it asserted itself in flood time and negated city and state borders. Even in dry times, these connections were visible at the prominent Kansas City Livestock Exchange where a line of colored tile across the floor demarcated the state line.[10] During flood time, Kansas workers helped herd the livestock into chutes above the watery fray. Some workers risked their safety by staying on the upper floors and caring for these animals on hoof until after flood cleanup when killing and processing could resume.

Kansas Citians associated the river wards and bottoms with poor and minority residents. This was already the case early in the 1900s, but such distinctions only became starker in the coming decades. There was a lot of turnover in the worst neighborhoods, and most people chose to find better housing when they were able. To move up and out of the West Bottoms was to move up on Kansas City's socioeconomic ladder. Recounting the history of the Irish who lived in the bottoms, Clifford Naysmith wrote in his unpublished history: "An Irish woman who had risen to social prominence and began to put on airs might be dismissed with the comment: 'She was just what you call: "Come up from the Bottom."'"[11] Naysmith's comment indicates that living in the West Bottoms marked a person's class and social status. The bottoms were not just physical places, but social spaces, and as the decades advanced, the Kansas Cities increasingly abandoned the river wards in both social and environmental respects.

Wealthy neighborhoods like Quality Hill that were once made up of large single-family homes gradually became part of the older city as mobile whites moved southward. Those big homes became boardinghouses or were broken up into multiple-family dwellings, reflected in the nickname change from Quality Hill to Quantity Hill. Most homes in these poor districts were rentals and, in proportion to income, housing costs were highest.

Although people in the river wards often lived among the innards that supplied the river's water and then carried it away, they did not have equal access to infrastructure or municipal services. The southern districts, on the other hand, were farthest from the river, the most insulated from its spring rises, and farthest from the polluting industries and vice district that crowded the banks. Navigation boosters who envisioned the Economic River also tended to live away from the river. Despite the geographical separation, these southern neighborhoods benefited the most from the river. They were upstreamers who had the best access to the river's healthy qualities.

A quick look at the 1909 Sanborn Fire Insurance maps that display city infrastructure reveals a network of water mains laid beneath even the poor districts. Often the potential for city water existed, but the connection and plumbing had not been installed. The infrastructure of the river wards was in need of upkeep, but still, tenants could have had access to water in their homes. At first, no law required running water or sewer hookups in homes, tenements, or even boardinghouses, leading reformers to argue that such slack regulations perpetuated unsanitary and immoral conditions. With the 1908 city charter, a law was on the books that required buildings with over twenty people to have a bathroom, but the law was not enforced.

In 1910 Kansas City, Missouri, had a black population of about twenty-five thousand, roughly 10 percent of the total population.[12] In Kansas City, Kansas, the percentage of African Americans was higher—around 16 percent—making the total African American population of the cities closer to 12 percent. African Americans had three communities in older, northern Kansas City, Missouri. The three main districts were in the West Bottoms, the North End, and centered at Eighteenth and Vine Streets, just to the east of Troost Avenue. The neighborhoods around the cities gained nicknames like The Patch, The Bowery, Hell's Half Acre, Belvidere Hollow, Hick's Hollow, Negro Hyde Park, and Quantity Hill. The names alone reveal how class and race invested meaning in space. The oft-flooded Patch, for example, was on the Kansas side near the Armour packing plant. In 1910 the city threatened to raze this community of a few black families and mostly immigrants from Serbia and Croatia. The city issued an evacuation

Privy vaults in an African American neighborhood. Black reformer and sociologist Asa Martin wanted to bring awareness to issues confronting African Americans in Kansas City. Asa Martin, *Our Negro Population: A Sociological Study of the Negroes of Kansas City*, 1913

order citing that the squatters in these shacks supported criminal activity like bootlegging.[13]

Despite segregation and the benefits of community, not all black Kansas Citians lived in these identifiable neighborhoods. In the Progressive Era, about one-quarter of this population was dispersed around the city.[14] In the early 1900s some African Americans succeeded at moving into communities and housing developments in other parts of the city, or they moved into shanties along waterways and places they could garden. Still other communities existed on the Kansas bottoms, in the East Bottoms, and in Rosedale, Kansas. Leaders in the black community preached the importance of homeownership. Housing developers advertised new homes for middle-class black citizens in more southern locations. "Just the place for a colored physician, music teacher, professor or any high class colored man," read one advertisement.[15] Another, near Turkey Creek and Rosedale, advertised that it was "high, dry and sightly," suggesting that black buyers were accustomed to risk.[16] Developers noted proximity to black schools and churches. These demographic patterns of both concentration and dispersal resembled western cities.[17]

The river wards contained the majority of the city's unhealthy spaces, and historian Asa Martin's 1913 sociological survey, *Our Negro Population*, confirmed that African Americans had the worst housing in the city. "The industrious Negro finds himself much handicapped, since there is only a small portion of the city where he is permitted to live and a still smaller section where it is possible for him to purchase property; these districts are naturally the most undesirable locations in the city," he wrote. Nearly 90 percent of black Kansas Citians were renters, subjecting them to whatever the housing conditions were in the districts they lived in. Martin noticed that "little effort was evident on the part of the city to enforce the law in Negro districts." Of the fifteen thousand black homes, only half were hooked up to city water. Consequently, African Americans sought other (usually riskier) sources. They shared water, went to springs, used cisterns, bought from saloons, or resorted to using hydrants, which might freeze in the winter and leave even fewer options. Only one-fourth of black homes had toilets—privies were the norm, in spite of the congestion.[18]

It was in the river wards that the city decided to sanction saloons and theaters and to overlook prostitution. Largely places of entertainment where shows could be seen and alcohol and sex could be purchased, these areas were called the "vice district" or the "red-light zone" by reformers, who referenced "bawdy houses" and "well known sporting houses."[19] Vice existed before an official district was created in 1908 and continued after it was officially abolished in 1913, but for a few years it had a boundary and was maintained through zoning ordinances and policing. The main idea was to control the location of vice and keep it out of the residential areas—overlooking the fact that people *did* live in the North End and its vicinity.[20]

In 1913 the Italian community was upset to have the red-light district as a border. To enlist allies, women in the Italian community went to the Kansas City, Missouri, Board of Public Welfare, the first such city agency in the nation. As immigrants, they likely had contact with social workers and may not have felt confident about accessing city decision makers. A welfare worker then went to the chief of police on behalf of the Italian community. Protests of the Italian community adopted the tack of protection of morality and family, especially girls. Organizing through the church, Italian men met and drew up a resolution to present to the police commissioners: "We, the Italians of Kansas City, respectfully protest the segregation of vice where it will be flaunted in the faces of our youth and we do not believe the city should take part in the bartering of souls." The Italian neighborhood had become a social "sacrifice zone."[21] The vice district was a dual threat: the city both jeopardized the souls of those who engaged in vice, whether innocent or not, and threatened the members of the Italian community.[22] Their neighborhood had been socially abandoned by a city committed to confining vice to the river district, away from wealthier and whiter neighborhoods.

Poverty and a poor environment weakened morality, the Kansas City Board of Public Welfare believed, and the vice district was an extreme physical manifestation of immorality.[23] Owing to the protests of the Italians, Jews, and a church federation, the officially sanctioned district was abolished, and the first day's report heralded that the North End was "silent as [a] tomb."[24] But vice would continue—so, where would it go? "Women to the River," the newspaper announced! The location of vice would be "not 'in,' but near banks of 'The Missouri,'" as the police commissioner put it. It was as if the city and police hoped that if they pushed the "disorderly women" closer to the river, they might fall in and be swept away, or perhaps be flooded out, disappearing downstream forever. "We are going to get them closer to the river if possible," said the police. "We can't drive them into the river, but we are going to get them as near its bank as possible."[25] The river, the ultimate sink for the city's waste, appeared to be an invisible location for prostitutes and other social waste.

The river wards were not prime real estate—socially or environmentally. These older portions of Kansas City were overburdened with a dense, impoverished population that lacked connectivity to municipal services and were proximate to industry and railroads, and all of this deleterious outcome on health served to reinforce inequality. The body (and its health) was often used as a metaphor for community. Making an argument similar to the Italians, middle-class white residents fought to keep their neighborhood from similarly becoming a sacrifice zone for a railroad. Led by area churches, they protested the noise and smoke pollution that would invade their space and questioned the safety of the track. The track would have to be "done over our bodies," a priest testified.[26] Due to segregation and discrimination, the solution to better housing and health was not as simple as moving away.[27]

Mexican immigrants in the Kansas Cities had economic, social, and environmental conditions similar to those of black and Italian populations. Mexicans' bad housing both created and reinforced a stereotype that they were "dirty." The earliest Mexican men who immigrated before or during World War I lived in boxcars. The railroads viewed their offer of boxcar houses as paternalism, because the boxcars improved the Mexican workers' standard of living. These were the lowest-paid railroad workers, performing hard labor like laying track, which was a mobile job. Because most did not have families in the Kansas Cities, they did not settle in permanent homes; to outsiders, the lack of women made them appear domestically insecure. Seen as shiftless because they were low-wage migrant laborers without families, the Mexicans in the Kansas Cities were so "other" that the Santa Fe Railroad company and other Kansas Citians hardly questioned the boxcar hovels down by the tracks.[28] Obviously, the boxcars had no municipal services like gas, water, or sewer.[29] Mexican tuberculosis rates were six times higher than those for whites.[30] Segregation of Mexicans in housing and healthcare was only slightly less rigid, but, like African Americans, Mexicans found community strength in their isolation because discrimination resulted in the preservation of identity.

Healthcare was racially segregated in the Kansas Cities, and more so on the Missouri side. Black physicians on the Missouri side had no surgical facilities, and their patients needing surgery had to go to Kansas City, Kansas.[31] African Americans serviced themselves in voluntary hospitals like Wheatley-Provident, and the medical profession was patronized by those who could afford it, leaving many wanting.[32]

Kansas City, Missouri, began discussing a new city hospital at the turn of the century and proposed that the old hospital become a black facility. The *Rising Son* editorialized that "race distinctions in so far as public

Staff at the black hospital. The *Kansas City (Mo.) Hospital and Health Board Annual Report*, 1914–1915, introduced the doctors and medical staff of the old hospital, which the city turned over to African Americans. Courtesy of Missouri State Archives

rights are concerned is not what we are looking for. If we are sick we want the same chance to get well that the white man has. Legislation for separate institutions (public) for Negroes means that we shall forever remain in the rear."[33] In 1908 Kansas Citians passed bonds for a new hospital, in response to a campaign highlighting their woeful healthcare system and regrettable health statistics. Indeed, its construction institutionalized segregation because the new hospital admitted only white patients. Patients with contagious diseases remained on the grounds of the old—now black—hospital. The next year, reports showed that the city spent $1.86 to care for each white patient but only 86 cents to care for black ones. The newspaper editor's fear that segregation would produce unequal healthcare seems to have been realized.[34] Neither private nor public healthcare was adequate for black residents, who had less money to afford private healthcare but suffered from an underfunded, segregated public system. Voluntary organizations like the Urban League helped to fill the gap in services that fell short in the public realm.[35]

Turning over the hospital to African Americans produced gains as well.

By 1915 a black newspaper stated: "The Sun does not believe in segregation in the strict construction of the term, but it believes, as [do] all progressive and fair-minded Negroes, that that day was blessed when Thomas M. Finn was made president of the Hospital and Health Board and the Old City Hospital was turned over to Negroes."[36] Black healthcare was largely an internal community enterprise serviced by a growing, educated, middle class of health professionals who gained power from segregation. The staff and nearly half of the thirty physicians in their spotless, starched uniforms were black. This segregated independence continued in 1915, when a new county facility for the black infirm and aged was built. It had all the modern conveniences that urban white middle-class Americans were beginning to consider basic. At its opening, the black community celebrated the institution's attributes, including plumbing and gas heat. Their pride in these facilities suggests that they were accustomed to something different.[37]

Race Space and Reform

In 1992 Kansas City, Missouri's first black mayor, Emanuel Cleaver, called Troost Avenue the city's Mason-Dixon Line.[38] Troost is the straightest north-south road and is a racial dividing line. On old maps, Troost Avenue was the only street that went straight to the Missouri River, and it was the street the garbage wagons gathered on to make their pilgrimage to the river bank. It was a physical, social, and psychological boundary: the majority of Kansas City's black population lived to the east side of the road. Even as black communities expanded southward, they did not cross Troost Avenue. Space was "racialized" in the Kansas Cities—that is, urban space had cultural and social meanings related to race.[39]

Progressive reformers optimistically believed in fixing the problems of the urban environment through a combination of research, education, regulation, and social welfare programs. Alleviating the ills of poor housing was a priority. The Kansas City Board of Public Welfare believed the city could build good citizens. "The age is beginning to appreciate environment as determinative in character," its report stated.[40] The Board of Public Welfare collected extensive data on housing and welfare conditions in the northern and river districts; these reports provided the research and photographic evidence to support their efforts to get laws and enforce regulations. In conjunction with the Hospital and Health Board, social reformers believed that housing and health would be improved by regulating plumbing, fire hazards, and lot size, among other factors.[41] A Department of Buildings, created in 1884, was supposed to supervise construction and enforce ordinances but fire insurance interests regrettably reported that the police department did not cooperate to enforce codes. A new municipal building code passed in 1908, but discontented reformers still wanted stronger codes that would be enforced.

In 1913 the Board of Public Welfare made a presentation about "bad housing" before the Kansas City Commercial Club, the bastion of wealth and proponents of the Economic River. The welfare board, composed of white, middle-class progressive reformers of both sexes, sought stronger housing regulations, and the club was the city's most powerful political entity. The stories and photographs the board presented to club members were reportedly shocking, but the "truthful camera" could not be denied. The report was reminiscent of Jacob Riis's photographs of New York City tenements.[42] Kansas City did not have expansive tenement districts as did the largest cities. In fact, Kansas Citians joked that Chicago housing was so cramped, the wallpaper had to be removed to get the furniture inside. Doubtless some Kansas Citians were reading Upton Sinclair's 1906 exposé of labor practices in Chicago's meat industry and made connections to their urban environment. It might not have been *The Jungle*, but Kansas City was still bad enough to motivate reformers. Thousands of residents in tenements and boarding houses did not have running water or sewerage; they drew on contaminated wells, and entire families packed into one room. The welfare reformers prodded club members to support new housing laws in order to make Kansas City "a good place to live in," drawing on the Commercial Club motto and watchphrase of the city.[43]

When reformers educated the broader public about poor living conditions in the river wards, it is possible that this "othering" reinforced prejudice and fear and increased segregation.[44] If any of the businessmen were ignorant before the welfare board's presentation, they might have walked away with the idea that poor and minority peoples were inferior and should be confined to the filth they created, or at least prevented from encroaching upon their own communities. The welfare board reported of Mexicans that: "As their standard of living is so much lower than ours, the landlord keeps them feeling perfectly at home by lowering the standards of housing to that to which they are accustomed. . . . Needless to say everything is filthy."[45] One-third of the Mexican homes in Westside had no indoor baths or toilets, and access to water and sewer was generally lacking.[46] One provision Secretary Leroy A. Halbert of the welfare board sought was to require tenement buildings to be connected to city utilities. These new housing codes were proposed in 1913 but, despite lobbying before the Commercial Club, the local real estate board, and the city council, they did not pass.[47]

Housing segregation was both institutionalized and informal. Historians have found that Kansas City, Missouri's only segregation law on the books was the result of an organized white community preventing a black college from locating near them. This happened during the years 1913 to 1915, the height of Jim Crow segregation.[48] The Fourteenth Ward Civic and Improvement Club was determined to stop black Baptists from building Western College and asked the city to condemn the college's property for a park. Justifying the racist actions of the white residential district, one resident said: "Nearly all the people of this entire district are home owners. They came out here when the place was almost a wilderness and either bought homes outright or on time. They bought them to live in and had a right to expect they would be permitted to educate their children here without their having to come into daily contact with the black race."[49]

City aldermen agreed but Mayor Henry Jost vetoed the measure, only to have his veto overridden. The club was well organized and persistent, and race figured into its definition of good housing. A common tool of segregation was the real estate covenant by which the homeowner's deed specified who constituted an appropriate homebuyer.[50] Otherwise, segregation in Kansas City was not formally codified, but the segregation was understood, practiced, and informally enforced on a daily basis.[51]

Housing was important to reformers because "environment" shaped citizens and reformers could shape environments. While hereditary traits were influential, reformers did not believe that race, ethnicity, and socioeconomic status were definitive in shaping people. Environment and therefore people, reformer Scott Child argued, could be improved through housing. The "character of lighting, heating, ventilating and plumbing of the industrial and living quarters in Kansas City contribute largely to lowered resistance, the development of epidemics and subsequent deaths," the reformer said.[52] Instead of believing that impoverished or ill people were inherently weak, progressive reformers believed they could build a better society by improving housing and the urban environment. Black reformers agreed; the "environment" for urban African Americans was the neighborhood where they lived and "heredity" did not count for everything. A good home, or a good environment, would produce a good person.[53] "Uplift" in black communities included a housing survey and cleanup campaigns. The *Sun* newspaper editor urged businesses to sweep the Vine Street sidewalks and remarked that the superintendent of streets should "daily" have his attention called to the poor conditions of the sidewalks.[54]

The African American newspaper the *Sun* reprinted an article that eloquently linked citizens and their surroundings: "What a city government amounts to in establishing the character of a city is more reflected in the gutters and crossings than the proud parade on the public occasions. A foul or dirty street, a bad and dangerous sidewalk, a gutter of standing muddy water[,] destroy civic pride and undermine true citizenship."[55] Indeed, the city "amounts to" its gutters, streets, and sidewalks. The commitment Kansas City made to its infrastructure was a sign of the social standing of its people; the municipality was daily with its residents in the form of its pipes and sewers. A good street was "more powerful than prisons or parks," the article continued, because the streets were "where the population lives." In sum, the article argued that a city could structure its citizens and create equity through infrastructure; respectable sidewalks made respectable citizens.[56]

This philosophy of social outcome through municipal input was held in favor in the early twentieth century. It produced architecture like Union Station and amenities like public drinking fountains. It required city government to be more active, but this was necessary with increasing industrial urbanization. Progressives believed in the relationship between the social and physical environment and thought it was in their power to change either one for positive social benefits. The urban innards, even in their absence, could be powerful tools to shape society and consequently infrastructure was a weighty, contested issue. Through it all ran the river.

Suburbanites and Slumlords

Kansas City's urban growth from 1900 to 1930 was mainly to the south. The new suburban districts in the metropolitan area were primarily white. As the black population grew, it expanded its boundaries by pressing into white neighborhoods. By the 1920s tensions between black and white communities had increased and whites moved further southward. By the 1930s the intersection of Eighteenth and Vine Streets was the center of the largest black population in the Kansas Cities, as well as its cultural hub, known for jazz clubs and the *Call*, one of the nation's most important African American newspapers. It was also the home of the Monarchs, the dynastic Negro Leagues baseball team. By midcentury—about the time Jackie Robinson would be playing shortstop for the Monarchs—Kansas City neighborhoods were less integrated and segregation was more extreme.

Kansas City was a real estate developer's dream, noteworthy for planned developments and aesthetically pleasing subdivisions bounded by boulevards. J. C. Nichols was the most famous developer, responsible for The Plaza, the Country Club District, and thousands of homes and apartments elsewhere in the metropolitan area.[57] Nichols was influential not just in Kansas City but also in the national housing market, where he used private and government avenues to his advantage. He perfected the use of the restrictive covenant in his subdivisions and required neighborhood associations. These associations enforced the covenants in order to protect property values and they allowed residents to impose segregating restrictions in the name of "character," family, and "security." White neighbors policed each other, ensuring that no one sold homes to undesirables, like black people and Jews.[58] Neighborhoods were not mixed use; segregation of the places where people lived, worked, and played increased the need for automobile or railway transportation and ensured that industry and saloons would be kept at bay.[59]

Beginning in the early 1900s Nichols worked with George Kessler of the park board to create his developments. Often platting in unserviced areas, Nichols would build his own streets and sewers. He had a close relationship with city administrations and the public works department so that he was given the freedom to build and maintain and then be reimbursed by the city for his infrastructure investments.[60] He advertised the Country Club as "Restricted" and emphasized "city conveniences" and access to the park and boulevard system "in advance of the growth of the city":

> *1000 Acres Restricted*
> The Country Club District
>
> * Restricted to single residences only, ranging in value from $3,000 to $10,000 and up. Large lots and acre tracts with city conveniences.
> No monotonous repetition of the same design of houses.
>
> * Ample parkways and boulevards provided in advance of the growth of the city. Street improvements permanently and uniformly constructed. Straight south from the center of the city. Direct car service—25 minutes from 12th and Main.
>
> * Surrounds Country Club for coming quarter century.
>
> * J. C. Nichols
> 1213–14 Commerce Building Phones Main 4345[61]

Serving a prosperous class, Nichols's developments quickly attracted municipal services like water and sewers because of customer demand and ability to pay. Suburban housing like the Nichols developments "insulated" middle- and upper-class whites from the less-desirable social, industrial, and environmental realities of the Kansas Cities.[62] Wealth, it appeared, could afford good health. Wealth also afforded tastefully designed gasoline stations. In 1916 a filling station opened on Brookside Boulevard that matched the architecture of the neighborhood—due to proximity, it may have been the station at which the McDonnell family fueled up. Architect John Van Brunt designed it so that it would not look like a filling station: it was nicely landscaped and staffed by tidy attendants wearing white. Though this unique Standard Oil station was not far from the refinery in the bottoms, it disguised its industrial origins.[63]

With increased tension over space in the 1920s, progressive reformers wanted to keep the peace in the city and wondered how to resolve a "housing problem"—they did not regard racism as the problem. The Kansas City Inter-Racial Committee—whose intention was the promotion of "friendly relations between the races"—weighed in on the issue in 1922. They recommended an investigation of demographic trends, deciding where blacks and whites should live, and developing better black housing.[64] Black health and housing were oft-covered topics of the nonpartisan, civic-minded, reform organization the Citizens' League. The league's *Bulletin* included an essay by D. J. Haff, a reformer and businessman. He advocated black housing developments—a kind of country club exclusively for black residents "where the colored people may establish a complete and separate society." Haff volunteered himself as one of the "white capitalists" willing to finance such an endeavor.[65] Though they desired better housing and a healthier urban environment for African Americans, most white reformers of the 1920s accepted and promoted segregation.

For both the welfare board reformers and African American leaders, the best solution to the housing problem was homeownership. It was heralded as a key to the quandary of race and class. A handful of middle-class black Kansas Citians organized "The Pull Together Club," through which weekly subscriptions added up to thousands of dollars invested in property. Black newspapers advertised new homes and one pronounced that there was "Great Prosperity Among the Negroes in this Community, Especially in the South End of Town." A small but growing black middle class lived in neighborhoods nicknamed Negro Hyde Park, Vanderbilt's Lane, and Big Four Hundred. Importantly, the percentage of homeownership was rising among the middle class. The *Rising Son* newspaper noted:

> The white real estate men in the last few years have begun to deny to him [the prospective black homebuyer] any house one could really call

decent. They have driven him out of most all respectable neighborhoods and have even begun to set him back in alleys to live, and also are building old barns into houses for them to make their homes and rear their families. Some of our best Negroes have taken exception to this treatment and their great object now is to buy them[selves] a piece of property.[66]

Reformers commonly blamed unhealthy housing on landlords, property owners, or, as they were sometimes called, "slumlords." From the standpoint of profit, there was little incentive for property owners to invest in their properties. One letter to the editor signed "Slumlord"— possibly planted because it seemed to represent an unwelcome perspective—said he paid his city water fee and the problem was the tenants who did not access the good water.[67] Another landlord explained that he could not afford to invest in plumbing because his tenants were too poor to afford coal and therefore the pipes might freeze in the winter, risking his investment.[68]

The landlord J. T. Hartman wrote Mayor Albert Beach and Director of Health Ernest Cavaness to plead his case. He was living in Phoenix, Arizona, but owned property near Swope Parkway and, as part of a cleanup campaign, the city had served notice to his tenants, requiring that his rental houses undergo sanitary improvements. This was a financial burden, Hartman said, and he was requesting leniency from the city. He explained that he was a benevolent landlord; his property had three "cottages" that were "humble" and his tenant families were "good but poor people." Because their homes were down in the ravine, they could not connect to city water or sewers. Hartman assured the mayor that the people renting from him were healthy despite the "outside toilets" and the "low ground." The essential problem for Hartman was that the property value was too low to justify sanitary improvements and therefore compliance with the health orders.

Hartman played almost every card he could to appeal to the mayor's sympathy. He claimed his Republican affiliations, his church connections, his past relationship to the mayor and his wife, and the fact that he never complained about high taxes. He blamed "poison tongues" and "busy bodies"—likely the reformers—for getting the city to go after him and his property. Mayor Beach consulted the health department on behalf of Hartman and even asked Cavaness to "relieve him of this burden," but finally concluded that the landlord should be held responsible. Beach wrote Hartman that sanitation was "essential for the city's well being," and while Kansas City would grant an extension of time to remedy the situation, "eventually the conditions must be taken care of," even if the incurred expense meant he would lose his property. Hartman conceded: "If the sanitary orders are part of a citywide movement, I would not want to be one of those who are obstructing the good work," as opposed to the "special effort to annoy me or give me extra expenses" that he first assumed.[69] In this case, public health would prevail over private interest and the city health department and mayor would require that the housing of poor people be improved.

Class warfare played out over infrastructure. A white woman complained to the mayor that her part of town in the East Bottoms, a low-lying industrial area, was neglected by the city. Although residents "ought to have the sewer" and wanted the "unsanitary conditions of those toilets" remedied, improvements were stalled or blocked, she scrawled in a letter with enough grammatical errors to indicate that her education was limited. Her critique was one of class bias, saying that the real problem was that "those men that own so much property does not have to live down in the East Bottoms. No they reap the harvest off renters, drive big cars and live out South." Absentee landlords had no incentive to pay higher taxes to improve the infrastructure for tenants.[70]

City parks were a symbol of civilized refinement and a citizen's haven, but they could also be used as a weapon of resource control and racial and class prejudice. Not all Kansas City residents had equal access to the grand park and boulevard system for which Kansas City, Missouri, was so well known. The dense river wards had fewer city parks than the southern part of the city. In the 1920s black golfers brought a suit against the city for banning them from playing at the Swope Park Golf Club. The golfers failed to gain access to the public golf course because the court ruled that it was in the power of the city to use race to regulate its parks.[71] As part of a social uplift campaign in the 1920s, reformers took children from the North End to see Swope Park. Located in the south of the city, Swope Park was a long railcar ride from the North End.[72] The media coverage portrayed reformers positively, duteously exposing African American children who were isolated in the river ward slums to a wholesome experience.

Although it was a park used primarily by whites, Swope had a segregated picnic area for Mexican and African American patrons until 1954. An African American woman remembers her grandmother, as late as the 1980s, only taking her to "Watermelon Hill"—the nickname for the area of Swope Park that the city segregated for black parkgoers. Growing up with segregation, it was the only place in the park her grandmother had experienced, the woman said, and therefore she felt safer taking her granddaughter there. Moving into other public spaces carried risks for the grandmother, and this legacy of racism shaped the young woman's urban environmental experiences in Kansas City.[73]

Parks often were located on land less desirable for housing development; therefore, people in dire situations were often the ones evicted when parks were built. As it determined property value in the condemnation process of The Paseo, the Kansas City, Missouri, Park Board recorded notes such as: "house in dilapidated condition occupied by negroes," "no plumbing," "no furnace," "cold water supply," "tenant sick," and other information that devalued the property and indicated the socioeconomic status of the people who lived there.[74] When the city and commercial interests reported on the progress of the park system, they often illustrated progress with before-and-after photographs. The first photo would usually show a scraggly landscape that was the domain of black children, weeds, privies, and maybe a shack. The final photo would show how the homes and playgrounds of poor people had been turned into a genteel park landscape for white people to amble or take Sunday drives in. These public improvements defined social and economic progress for city planners and commercial interests and came at the expense of those who called the place home.

In the perception of some working-class whites, blacks unfairly benefited from parks. One woman complained that the city was giving special treatment to the black community because a park was being located near an African American neighborhood. Mrs. M. J. Loftus wrote the mayor complaining that, in comparison to the taxes she paid, the city neglected her. She suffered unsanitary conditions because, she said, the city manager would not make her neighborhood "as nice as he has for the negro at the Paseo."[75] Whatever the assumptions of such white residents, it must be noted that many black homes had been razed to create The Paseo.[76]

The logic of turning "low, cheap, undesirable property ... into beautiful and valuable boulevards," as one mayor put it, had social costs, as discussed above, but there were further complications in the feasibility of developing such land.[77] When engineer George E. Waring Jr. had surveyed Kansas City toward the end of the nineteenth century, he ascertained that "the low-lying districts along the river and the water-courses have the least value, but require the most expensive sewerage works, while the more valuable property, which is better able to pay taxes, lies in the higher portions which are easily drained and require less expensive sewers."[78]

The most challenging land for engineering infrastructure was where poor people lived, said Waring, which described the Kansas Cities' social and environmental landscape as well as relationships of risk. In 1929 Mayor Beach boasted of the park system he inherited: the "valleys, the lowlands, so unsightly in most cities, have been turned into beautiful drives and boulevards to serve the sightly property on the hills." Poor lands with poor people were likely places for parks while the hillside "upstreamer" enjoyed the increased property values and ease of building infrastructure.[79]

The control of land or zoning by the city was an important power, and residents used parks to protect their interests. Residents demanded that the city or park board build parks to replace unseemly developments or prevent unwanted neighbors. In some cases parks were used to keep black people out of white neighborhoods. If white residents learned of impending home sales to black buyers, they might request that the city tear down the homes and build a park. Right on the heels of the Western College ordeal, which resulted in the law that forbade a black Baptist school to be near white schools, the Linwood Improvement Association organized to prevent black residents from moving into the area.[80] The association recommended that the city condemn the land between Vine and Woodland Streets and tear down the subpar homes.[81] At times, the association used posters, at other times violence. Intimidation and the occasional bomb discouraged African Americans from encroaching on white neighborhoods.[82]

Reformers, city officials, and the media directed their attention to the core—the vice district, crime, slums, floods, the tensions of race, class, and sex, and so on. In contrast, the suburbs were treated as a kind of natural phenomenon growing quietly in the south. Mayor Beach bragged about exclusive districts like Country Club and Rockhill, saying that they were "*scientifically planned and protected* for individual homes," as if an absolute, natural form of housing had been discovered in Kansas City's elite neighborhoods, like an element found on the periodic table.[83] Despite the invisibility of privilege, the north and south, the poor and rich, the black and white, were related by environmental, economic, and social factors. The blame for the physical outcome of segregation and the social outcome of disadvantage is laid at three doorsteps: the real estate industry, federal policies, and the prejudiced perceptions of white Kansas Citians.[84]

The Kansas City elite exhibited the prejudices of race and class, though they were rarely acknowledged. This is significant because prejudice had real implications for health and wealth in the city. The records of the Kansas City Chamber of Commerce, formerly the Kansas City Commercial Club, reveal that these wealthy, white businessmen and industrialists believed in civic duty and felt they served the city's best interests. Though not elected, the clubmen felt, as wealthy men, that it was their responsibility to lead the city, and understood their role in the urban economy and landscape as a paternalistic one. In return, they expected the city to protect their wealth.[85] The men of the club were sincerely devoted to "mak[ing] Kansas City a great place to live in," but they could, at times, be self-serving from the perspective of others. When James McQueeny from the Loose-Wiles Biscuit company—creator of the ingeniously named Uneeda Biscuit—stepped down as president of the chamber, Mayor Beach sent a personal letter, thanking him for his service to the club and Kansas City. Surprisingly, Beach wrote the usually unstated when he said, "You have done more to break down the feeling of the masses against the Chamber and humanize it than any man in my experience."[86] This class tension can be understood by the historian after reading years of minutes and articles from the chamber, but it was rarely noted directly.

The West Bluffs, before. Reports by the Kansas City, Missouri, Park Board included before-and-after photographs to demonstrate its work. This photo of a home overlooking the bottoms was labeled "Kersey Coates Terrace North of Twelfth Street, Before Roadway was Graded." In the Progressive Era, the "before" shots often showed African Americans—either their dilapidated homes or children playing. Missouri Valley Special Collections, Kansas City Public Library, Kansas City, Missouri

The West Bluffs, after. This photo of the same location shows a white version of a genteel landscape, one without poverty or people of color. Missouri Valley Special Collections, Kansas City Public Library, Kansas City, Missouri

The Economics of Health

Many reformers believed that poverty was a main cause of illness.[87] Asa Martin, the sociologist, had traced the vicious cycle that caught black Kansas Citians. As long as discrimination existed in the workplace, black workers would be unable to afford better housing, which would continue to keep them in unhealthy conditions, which contributed to their overall degraded social situation. Many industries did not hire African Americans, and the jobs for which they were "qualified" were usually of higher risk. This was the case in the meatpacking industry, which employed about 10 percent of black men in the workforce.[88] A 1927 study found that one-third of Kansas City's largest employers did not even hire black employees, and many industries reported that they would do so only as a last resort.[89] The situation for black workers became dire with the economic depression.[90]

Black women workers became examples of how, as Martin said, "the health and prosperity of one race depends on the prosperity and health of the other." Black women were overwhelmingly employed as domestics, which meant that they worked for low wages in the homes and private spaces of white people. Emphasizing that illness crossed boundaries, Citizens' League reformers repeated this story:

> A colored woman had been laundering in her home, the soiled clothing of a white family. The children of the white family became ill of scarlet fever. The white woman told the colored woman that for the present she need not take the clothes to her home as it might give the fever to the colored children. "Day's no trouble about that, Missus," said the colored woman "because my children done had scarlet fever a month ago!"[91]

The campaign was supposed to illustrate how "the health of [one] part of the community affects us all," although this story seems mainly to hold poor black women responsible for the city's health. The thoughtful white woman is portrayed as the victim. Thus, the white readers of the *Citizens' League Bulletin* might conclude that black women, through their labor, were threats to the unsuspecting middle-class white home. Reformers who could not insulate their families were then motivated to clean up the "other" urban places.⁹² Although these campaigns reinforced race and class stereotypes, progressive reformers did, importantly, perceive the interconnections between society and environment via health.

The Kansas City Urban League, founded in 1920, wrote: "The job is the all important thing to an economically poor group." Black citizens had the "greatest need" and, even though they were willing to do anything for survival, found themselves "ordained to exist on the returns from jobs that pay the least."⁹³ If employed by the city, African Americans were most likely in the public works department. Nearly one-quarter of black men worked "on the streets" as laborers in the 1910s.⁹⁴ Specifically, they worked as street cleaners in the river wards, did street and sewer maintenance, and carried out sanitary inspections regarding garbage. Often, city jobs were in positions that serviced other black residents, such as at the black swimming pool or welfare workers within the black community.⁹⁵ Archival photographs show black men working to dig sewers, flush sewers, and drive garbage wagons. Clearly, African Americans were associated with the less-desirable jobs that constructed and maintained the urban innards. When the Beach administration fulfilled its campaign promises to appoint African Americans to city jobs in the 1920s, it managed to find places for them only as night janitors or for "common labor." Tired of the stereotype of blacks as "gang labor," black leaders pushed Beach to appoint educated black men to at least low clerical positions and therefore improve their status.⁹⁶

Executive Secretary E. S. Lewis of the Urban League reported in the *Citizens' League Bulletin* about the status of Kansas City's black labor force in 1930. He laid out the problem of a largely unskilled workforce: advancement was difficult because so few opportunities existed. Lewis appealed to his white readers to see African Americans as contributing members of the heart of America, saying that black workers constituted "a vital and integral part of the fabric of this community." Lewis connected black labor directly to the physical infrastructure of the city by speaking of the black bodies that "mould the iron for our new magnificent structures" and "crush and carry the rocks and mortar that go into the construction of our buildings." Lewis also connected the black labor force to the private home, where those bodies "help prepare the meat that comes daily to our table" and maintain "the cleanliness of our beautiful homes and institutions."[97] Black bodies, for better or worse, were central to the health and wealth of the whole.

The urban environment had a major influence on health. Nationally, from 1900 to 1920, African Americans had typhoid fever rates that were double that of white Americans, further evidence that housing conditions and access to water services were disparate.[98] Into the 1930s, the death rate among black Kansas Citians, including babies, was double that of white citizens, and the tuberculosis rates were anywhere from four to six times the white average. For reformers as well as city boosters, these statistics were unacceptable and were "accounted for by insanitary living conditions and by bad housing."[99]

Progressive philosophy and the "new" public health tended to look beyond the individual to the community. While reformers conceived of health issues in terms of interconnectedness and social networks, popular culture associated the genteel, boulevard-lined southern city with healthfulness whereas the congested, older, northern part of the city was stereotyped as the locus of disease. Medical knowledge and germ theory focused less on the individual body as a source for disease and instead saw bodies in interaction with each other and their environment. As the *Citizens' League Bulletin* put it: "The boulevards and the slums are alike affected by sources of infection." Founding member of the Kansas City Urban League and owner of a private hospital Dr. J. E. Perry echoed this on the heels of Negro Health Week: it was "axiomatic fact" that the poor health and housing of black populations had ramifications for the rest of the city.[100] A high rate of tuberculosis in black communities was of concern to everyone, African American doctors urged. Contagious diseases like tuberculosis or typhoid fever had complex causes that included but were not limited to personal habits and could be spread to anyone. Rather than blame the individual for poor sanitation, some reformers looked at the larger urban system to explain disease, which acknowledged the interrelationship of social, economic, and environmental issues.[101] Structural changes were needed.

Like black residents, Mexican Americans in Kansas City also continued to have increased health risk and inequitable access to the public healthcare system. Not allowed in the white hospital, Mexicans grudgingly used the black hospital where Dr. T. C. Unthank spoke Spanish.[102] In 1916 and again in 1928, the Mexican community, with help from white social workers, fought this segregation in health institutions and finally won.[103] For Mexicans, the most significant organization eventually took the name Guadalupe Center in 1926 and existed for decades under the guidance of a wealthy white woman, Dorothy Gallagher.[104] The center participated in Americanization campaigns, especially involving children, and provided social welfare to the community. The center also represented Mexicans before the city, helping get the first neighborhood park and baseball diamond in Westside, for example.

The Mexican American community did have access to garbage collection, a basic municipal service. The streets of Westside were littered, reinforcing the image of an unsanitary community. It was reported that the residents could not afford the proper garbage cans that the city required for garbage service. The social workers at the Guadalupe Center arranged in 1926 to get garbage cans and the community bought them at cost. The newspaper photo commemorating this shows a woman of Mexican descent holding a baby and with a young girl standing at her side. Three black men, probably sanitary workers for the city, are handing her a garbage pail.[105] It took intervention by white social workers to overcome institutional barriers; immigrants had to make compromises with the white reform community in order to get a basic service that many white middle-class residents felt they were entitled to.

Immigrants learned to take what benefits they could from interactions with white social workers, while at the same time preserving their own culture and autonomy.[106] Social welfare organizations like the Guadalupe Center and the Red Cross, which campaigned for better healthcare and "Americanization," meant well, but sometimes treated the objects of reform with cultural insensitivity. More significantly, reformers were blind to the economic situation of the communities they wanted to reform. A reformer might interpret scavenging and recycling, for example, as evidence of ignorant squalor and not a resourceful method of survival within the larger socioeconomic landscape. "The reformers were a lot more squeamish than the [immigrant] children or their families could afford to be," comments one historian.[107]

Segregated healthcare functioned well in the 1920s, according to the Woman's City Club. The organization had a mostly married, white membership whose addresses indicate that they lived at least fifty blocks south of the river—the part of town characterized by the developments of J. C. Nichols. Health and welfare were issues "we women consider our major concern," the City Club confidently explained. After lobbying to increase funds and establish facilities, the women continued to support the healthcare institutions by doing inspections or gathering toys and books for patients. They reported being impressed by the black hospital on a visit. During the "very hard times" of the 1930s, upon a visit to deliver clothes and magazines, both the Leeds hospital unit for black tuberculosis patients and the black hospital met the club members' approval. Healthcare remained a combination of public and private investments and when the city came up short the club would raise funds, as it did when it purchased a "paralysis tank" for black children with polio. During the Depression, the demands on club committees increased greatly in relation to social need. In response to need and pressure from reformers like the Woman's City Club, local, state, and then federal governments created a health and welfare system that replaced voluntary work, something these women viewed with some "regret" as well as "satisfaction."[108]

Overall, health and welfare reforms were moving from the private sphere to the public sphere. It was clear to public health officials, reformers, and citizens of all colors that Kansas City, Missouri's healthcare system was still inadequate. In 1922 the Kansas City Public Service Institute published a report on the health conditions in the city. The report, hoping to shock, announced: "Kansas City can no longer claim to be one of the healthiest cities in the country."[109] Kansas City's poor ranking proved that it was in need of a better-financed system, argued Walter Matscheck and other institute members. Kansas City was not reducing its death rate when compared to the successes of other cities. The report depicted the years around 1910 as comparatively some of the healthiest because of the city's position in the rankings. Typhoid fever was no longer the killer threat that pneumonia or heart disease were, but the city's typhoid fever death rate was still higher than the national average.[110] Health statistics were even worse when the black population was looked at separately.[111] The Kansas

City Women's Chamber of Commerce endorsed the report and rhetorically asked: "Is KC to stand alone as the one large city of the country unable to improve its public health?"[112] This report would go on to structure the health elements of the Ten-Year Plan proposal.

Kansas City reformers sought economic and political solutions. When the Women's Chamber of Commerce inquired, "THE CITY'S HEALTH—An Asset or a Liability?" it clearly hoped health could be turned into wealth.[113] National black leader Booker T. Washington campaigned on the message that health was at the root of success because "WITHOUT HEALTH AND LONG LIFE ALL ELSE FAILS"—including economic aims.[114] Through the 1930s, progressive reformers adopted economic language to advocate for health issues. "Health pays dividends," reformers said, meaning that spending money on healthcare paid off by prolonging life and therefore contributing to greater wealth. The insurance companies had calculated the "economic value of human lives," and reformers applied these data to their calls for better health.[115] By putting a monetary value on life and appealing to the rhetoric of economic development, reformers hoped to make the case for better sanitation, more education campaigns, and stronger regulations, all of which meant increased government involvement in public health. Health advocates like the Women's Chamber of Commerce and the Citizens' League adopted the rhetoric of economic boosters, subsuming healthcare within the economy. Increasingly, over the twentieth century, an economic model of health became the paradigm that industry, policymakers, and more and more public health officials subscribed to.

Just like R. E. McDonnell argued for the professionalization of water and public works departments, so health reformers sought political independence for the health department. A study by the American Public Health Association concluded that health should be in the hands of professionals and not politicians—much to the agreement of reformers.[116] In an acidic essay on the "health prospects" of the city, a local doctor spoke of the need to "fumigate" the health board, to bring it out of the Dark Ages, to upgrade its perspective, and to adopt more effective campaigns than advising individuals how to dress for cold weather.[117] The board should be professional and empowered, he argued, and should not displace its responsibility onto citizens. The Hospital and Health Board of Kansas City, Missouri, finally became nonpartisan in 1940, after the fall of the Pendergast machine.[118] The second solution was to invest more money in public health. Over and over, spanning the decades, the Citizens' League and McDonnell as a key member argued: "Good health can be purchased."[119]

As progressive as Kansas City was over the decades, it never led in the field of public health. Whether protecting drinking water, lowering typhoid fever rates, or advancing minority healthcare, Kansas City was slow to make improvements. The state of Missouri did not have a penchant for government activism at the state level, unlike Kansas, which was a public health leader. What Kansas City was able to accomplish was limited because many health issues went beyond the city, or even the state, as the flow of water proved. Spending on public health in the Kansas Cities continued to rank low in comparison to other American cities in the 1920s, although the importance of investing in public health was recognized.[120] That said, Kansas Citians committed to their park and boulevard system, which they saw as an element of the city's health. Today, in Kansas City, Missouri, the city spends less money on parks that are in the older, poorer, minority sections of the city. Sidewalks around parks are also lacking in low-income neighborhoods.[121] Inequality in health and infrastructure spending continues.

Live, Work, and Play

The Progressive Era saw a growth in concern for the "issues of where and how people live, work, and play," as historian Robert Gottlieb put it.[122] Urbanization and industrialization posed some critical environmental problems that simultaneously affected people and their environment. The people with less power—women, minorities, immigrants, and the poor of the Kansas Cities—lived, worked, and played in distinct locations identifiable by both social and environmental characteristics. Available sources make it possible to examine the black situation in the Kansas Cities descriptively to make the point that health and environment are closely connected. Black Kansas Citians were limited in their employment, healthcare, and housing options in ways that white citizens were not. Black and white reformers alike knew that the places where people lived, worked, and played were environments shaped by social and physical forces. Drinking water, sewers, garbage service, and parks were what Kansas Citians made of them; the discrimination that manifested in the urban innards was the product of the socioeconomic system.

Environmental justice, a term coined in the 1980s, is a grassroots reform movement in disadvantaged communities that makes the connections between race and class, and increased exposure to environmental hazards. The movement's critique has helped historians recognize similar issues in the past.[123] Studies and activists have shown that polluting industries have targeted poor and minority communities, assuming they would offer little resistance to the dangers of working for and living near these industries. Robert Bullard, writing actively of the movement, says: "Many African-American communities are subsidizing, with their health, the siting of risky industries other communities refuse to accept." Academic activists like Bullard write that the urban environmental and environmental justice movements are central to the success of a larger environmental ethic and put the onus on the environmental movement to recognize the diversity of the tasks before it, particularly to be more inclusive of race and class.[124] Class, by the numbers, has been a more predominant form of environmental injustice than race. That is, more poor whites are exposed to high-risk environments in the workplace and neighborhood. To choose race as the only factor is to miss the big picture of socioeconomic power in America. However, Latino/a and African American communities have suffered disproportional risks.[125]

The double-edged sword of poverty and racism was compounded in Kansas City's river wards by the "double jeopardy of poverty and pollution," as Bullard puts it.[126] Two environmental ways to see jeopardy in the river wards are through housing and health. We can use the space of parks, sidewalks, and drinking water taps to investigate race and class in Kansas City and in so doing add a new dimension to our understanding of the city's relationship to the river. Race and class influenced the way Kansas Citians related to their waterways. The environmental experience each citizen had was mediated by the infrastructure, and therefore, if race and class affected access to infrastructure, then they changed the way river and nature were experienced. Water and waste were a combination of the river and people. A drinking water flow line or a sewer served as a connection between Kansas Citians and the Missouri River, but water and waste also connected people to each other.

PART III

BASIN

HEALTH AND WEALTH

The laws of our land require us as individuals to use our property, or property in our possession, in such a way as not to interfere with the property of others; and the Congress of the United States, having exclusive control over our navigable rivers, should, in the name of commerce, control them in such a way as not to injure the property of the People.

Every cent of money invested in our natural resources is adding so much to the national wealth. It will be used in purchasing American material; it will be paid out to American laborers; it will go into circulation in American cities; and prosperity will follow in its wake, and leave us richer in national wealth.

    North Dakota governor John Burke addressing the
    1908 Conference of Governors

A sewer is a channel for human excreta [leading] to the sewage treatment plant or to a river; if to the latter, the stream becomes a channel to carry the poison to the human intestinal tract next below. . . . And what sanitation is trying to do is convert streams from extensions of intestinal tracts into channels of national good will.

    Editorial in engineering journal Municipal Sanitation, 1937

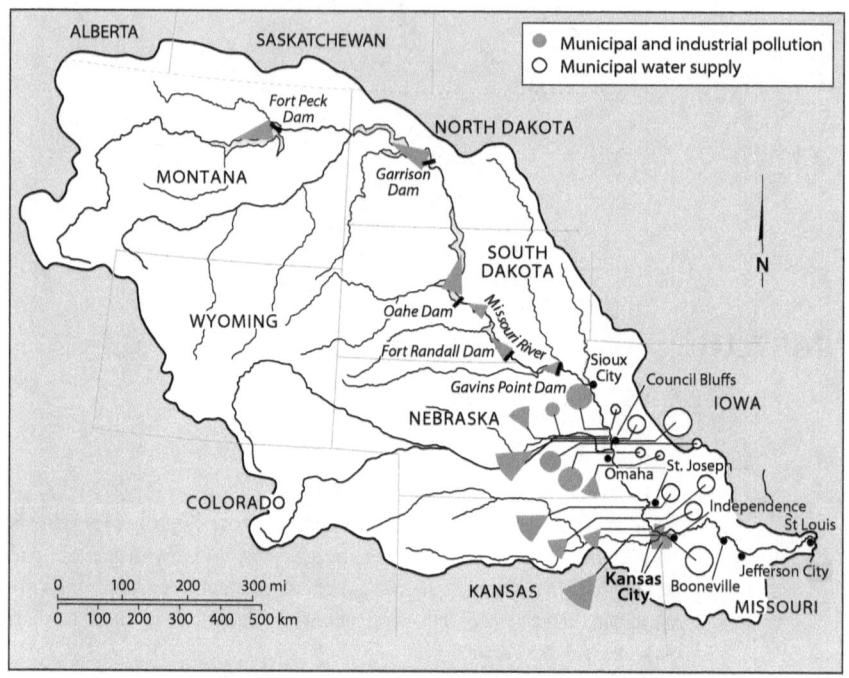

The Missouri Basin after Pick-Sloan. Cartography by Bill Nelson

A BROADER VISION

7

> *Thoroughly Washed and Screened Kaw River Sand*
> Frankenfeld Sand and Fuel Company, Kansas City, Kansas
>
> *All our sand is dredged from current of river and is sharp and clean.*
> Stewart-Peck Sand Co., Kansas City, Missouri
>
> *Clean Washed Channel Sand—A Specialty*
> Chas. Meierhoffer Company, Boonville, Missouri

Speaking at the 1885 Missouri River Convention in Kansas City, future congressional representative and avid navigation booster Champion S. Chase boastfully proclaimed that the Missouri River held the "healthiest water in the world." Tumbling and rolling down from the Rocky Mountains, the Missouri River collected sand and gold dust, and its "elegant muddy, golden color," Chase reasoned, signified "health from its source to its mouth."[1] In Chase's time, the Missouri River symbolized health. Like other economic boosters, Chase believed that health was the outcome of economic wealth. The health of the Missouri River basin could be measured in its freight tonnage, in its bushels of grain, heads of cattle, and jobber receipts. His healthy economic forecast rested on the industrial and agricultural development of the river and its basin. In 1906 Lawrence M. Jones, another navigation booster, addressed the Kansas City Business Men. He sold the idea that "the Missouri River runs through the richest country in the world," and cheap water transport would bring great wealth to the Kansas Cities if

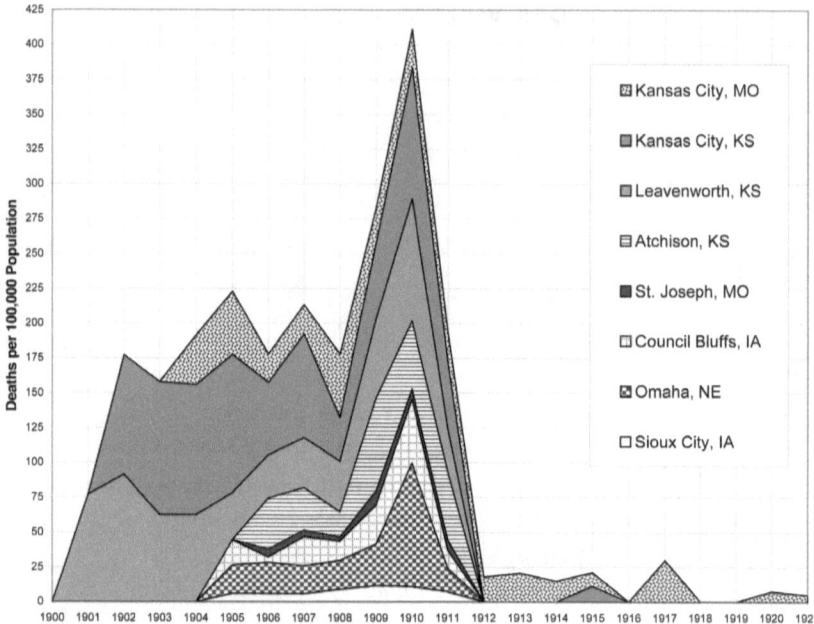

Typhoid fever deaths in cities on the lower Missouri River, 1900–1921. River pollution increased and illness followed suit until water treatment was introduced. The outbreak around 1910 resulted in calls for action. Graph by Pat Sachs

they harnessed the river, developed midwestern resources, and created jobs. "Make Kansas City a Water City" was the phrase Jones used, playing on the oft-quoted phrase "Let's make Kansas City a great place to live in," also adopted by social reformers.[2] The first responsibility of government, boosters believed, should be aiding the growth of economic wealth; good public health would grow out of this wealth. This was the Economic River.

In 1910, twenty-five years to the day after Chase stumped for the river's health, the Missouri River Sanitary Conference met up on the bluffs at the river's bend in Kansas City, Missouri. From within the chambers of the Kansas City Commercial Club, the bastion of river boosterism, the conferees discussed the growing problem of pollution in the Missouri River basin and its threat to public health. For sanitarians, health was measured not in barges, but in bacteria. These sanitarians—like R. E. McDonnell and Samuel Crumbine—felt that the public's good health formed the basis of social and economic wealth and while they did not oppose the economic development of the basin, they did wrestle with the complications and externalities of that development. Sanitarians argued that health should have equal importance if not precedence to economic growth and that the government was responsible for, or at least a key factor in, securing a healthy public and environment. In fact, for some sanitarians, the definition of wealth *was* good health. This was the Healthy River.

By 1910 the visions of the Economic River and the Healthy River were fleshed out, and their definitions of wealth and health were in competition. The use of the river was at the heart of this debate across the twentieth century because control over it would define the health and wealth of the cities and land alongside it. Population growth, urbanization, and industrialization created significant changes in the Missouri River basin, threatening health, and shifted the view of the river from pure to polluted.

By midcentury popular opinion regarded the Missouri River as a dangerous, disgusting cesspool of sewage and pollution. As if the smell were not warning enough, parents told their children not to go near the river because it was rife with meatpacking offal and the dismembered bodies of prostitutes that floated downstream from Kansas City.[3] These warnings were equal parts truth and rumor. Yes, the river did have organic waste and there was a suspected serial murderer in the 1980–1990s who dumped the body parts of seven women in the river, but the warnings also represented negative attitudes toward the city and urban river districts. This view of the river is strikingly different from that of nineteenth-century boosters.

Border War: Missouri v. Illinois

The state of Missouri sought and failed to invoke the power of the federal courts to deal with a pollution threat on its other great river, the Mississippi. Beginning at the turn of the century, St. Louis found itself on the receiving end of a spectacular engineering project. Chicago, the second-largest US city, began to divert sewage out of its natural watershed, which drained to Lake Michigan, and into the nearby Mississippi River watershed. Of the cities downstream, Saint Louis was the largest, and it was quite vocal in its disapproval of the Chicago sewage canal. The significance of this thirty-mile diversion canal is multiple. First, the state of

Missouri fought to stop the canal, and the ensuing case *State of Missouri v. State of Illinois and the Sanitary District of Chicago* was the first water pollution case heard by the US Supreme Court. Second, helping prosecute the case was Missouri's trust-busting attorney general, Herbert S. Hadley, a Kansas Citian. He was immensely influential in the state and was the epitome of progressivism. Third, this "border war" allows us to investigate typhoid fever and sewage on an interstate waterway at the turn of the century. This example of contested jurisdiction over rivers and public health became a touchstone for similar debates throughout the century.

Hadley, originally from Olathe, Kansas, moved to the big city after completing law school. He was appointed and then elected to legal positions in Kansas City and Jackson County. Active in the Commercial Club, he exhibited the political strains of national progressivism. Hadley was elected Missouri's attorney general under Governor Joseph Folk, making him a Republican in a Democrat's administration. As a prosecutor and attorney general, Hadley had no qualms about targeting powerful corporations and had a stupendous record for winning cases. He gained a national reputation for successfully using the state's antitrust laws to "bust" Standard Oil, among other monopolies.[4]

Hadley was driven by the "Missouri Idea"—a phrase used by Governor Folk to prioritize the consumer and citizen and to rein in corruption in business and government. Hadley carried forth his progressive activism when elected governor in 1908, serving from 1909 to 1913. He was named as a potential Republican presidential candidate in 1912; by this time, however, his health was in decline and he faded from public life, preferring to teach law. One historian identifies the Hadley of the early 1900s as radical, but the Hadley who became governor was a conservative progressive, interested in developing the institutional capacities of the state to regulate industries instead of handicapping them, mirroring the general trajectory of national politics.[5]

At the state level, Hadley was committed to conservation, and worked on a number of issues that dealt with water quality.[6] As attorney general, Hadley prosecuted the Supreme Court's first pollution case. Heard in 1906, the case against Chicago was actually begun as early as 1900, the year the canal opened. The city of Chicago was viewed as a corrupt monopoly—the

Herbert Hadley. Redpath Chautauqua Collection, University of Iowa Libraries, Iowa City, Iowa

polluter whose reckless use of waterways must be "busted" in order to protect the water consumers downstream.

The problem began with sewage. To improve public health, Chicago needed to divert its industrial and municipal waste away from Lake Michigan, which was also its drinking water source.[7] First conceived in the 1880s and under construction by the 1890s, the canal was an amazing engineering project that would force a river to run backward and send the wastes of 1.7 million people toward the Gulf of Mexico instead of the Great Lakes. But by redirecting waste to the Mississippi via the Des Plaines and Illinois Rivers, the health of downstream cities like Saint Louis would be put at risk. To protect public health, Saint Louis believed it needed to avert the diversion. The closer to completion the Sanitary and Ship Canal came, the more seriously Saint Louis began to take the threat of being downstream from such a populous city. Had Saint Louis started legal action earlier, it might have had a chance to affect the outcome of the project.[8]

Chicago was "feverishly eager" to divert its sewage and its officials remained confident that dilution would take care of the problem for those downstream.[9] In December of 1899, when some but not all of Chicago's sewage was already being deposited into the canal, Alton, Illinois, and Saint Louis had simultaneous typhoid fever outbreaks. The timing made Chicago seem guilty.[10] Saint Louis was constructing its new Bissell Point waterworks on the Mississippi and, with the big investment, hated to think that its water source was threatened. Meanwhile, the Saint Louis health commissioner, a Dr. Starkloff, remained adamant that Chicago's sewage was a "serious menace"; his official medical opinion was that dilution would not be adequate, especially during the winter.[11] The Missouri State Board of Health, which was rather weak, did not appear to be vocal in the debate. And despite requests from downstreamers, Congress did not appoint an investigative commission in this interstate matter.[12] Therefore, at the last minute, Saint Louis, led by prominent Missouri lawyers, went to the courts to enjoin the canal from being opened.[13]

In January of 1900, knowing that Missouri sought an injunction against the opening of the sewage canal, Chicago sanitary district officials raced to open the gates and send the water of questionable quality downstream toward Saint Louis. It was an exciting moment for the Chicago engineers and officials, but the story did not end there.[14] Saint Louis and Missouri continued their opposition, and six years later they would appear before the Supreme Court. Hadley claimed that Chicago knowingly victimized downstreamers and suggested that the situation chanced starting a war between the states.[15] War was the theoretical legal basis of the court case; the federal courts could step in to remediate if there were no other recourse to solve a serious conflict between states. There were no federal laws regarding water quality for Saint Louis to call upon. Because cities were limited entities within states, the court case ended up being between the states.[16]

Although no violence erupted, there were battalions of researchers, chemists, and engineers along the river banks from 1900 to 1906. Two cities, two states, and the US Public Health Service (USPHS) all monitored the situation, tested the water, and considered the data.[17] The stakes were high. Saint Louis did have a small increase in typhoid fever rates right after the canal opened, but cities closer to the pollution source did not suffer a higher number of cases. Still, Saint Louisans claimed the condition and smell of their drinking water was conspicuously different.[18]

Chicago and Illinois researchers claimed that Chicago's sewage was diluted by the time it entered the Mississippi; therefore, pollution had to come from other sources upstream like Peoria, Illinois, or from the Missouri River.[19] The Saint Louis waterworks were downstream from the Missouri-Mississippi confluence and were therefore exposed to the waters of both rivers. Isham Randolph, an engineer for Chicago, contacted General G. L. Gillespie of the Corps of Engineers in July of 1903 asking for more information about the supposed "series of investigations" the corps had made about water quality. The letter circulated through the chain of command and the response to Randolph was that, no, the corps had made no investigations about typhoid fever. Importantly, these letters document that, even as early as 1903, it was not out of the range of imagination that the Missouri River could be so polluted upstream as to cause illness downstream.[20] In fact, the Chicago defendants argued in court that Saint Louis's typhoid fever cases were the fault of bad Missouri River water.

The World's Fair, held in Saint Louis in 1904, became the goad to modernize water treatment. In March 1904 the city adopted a coagulation method of treatment using lime, created by Saint Louis engineer John Wixford.[21] In correlation, from 1904 onward, typhoid fever rates declined.[22]

In 1906 the US Supreme Court dismissed the case in favor of Chicago; Hadley and Saint Louis lost. The *Missouri v. Illinois* opinion, read by Justice Oliver Wendell Holmes, questioned whether the court had the responsibility to get involved in this interstate matter. Because the case had been in the making for a few years, the court had the benefit of examining the accumulated evidence—some eight thousand pages of testimony. The court concluded that the canal was not proven to "poison the water," as Hadley was paraphrased, and the justices ruled that Chicago's wastes did not adversely affect public health.[23] Six years after the canal opened, Saint Louisans were not all dead. If illness was a problem, said Justice Holmes, then Saint Louis could look to plenty of closer upstream cities. Besides, he reminded the court, Saint Louis dumped its own raw sewage into the Mississippi River, which seemed irresponsible, considering what it accused. The court ruling stated: "The evidence is very strong that it is necessary for St. Louis to take preventive measures, by filtration or otherwise, against the dangers of the plaintiff's own creation or from other sources than Illinois. What will protect against one will protect against another. The presence of causes of infection from the plaintiff's action makes the case weaker in principle."[24] That Saint Louis was also a polluter became one of the most important facts the court used in weighing the case. Let the city without sewage sins cast the first stone into the river.

Missouri v. Illinois was about more than just federal court power to abate nuisances between states. The court ruled that it did not have the responsibility to remedy all harm done between states, adding that there was no proof of harm. Despite the interesting situation of creating a nuisance where one had not previously been, the fact that Chicago had engineered a new river and was not using a "natural" waterway was not critical to the debate, surprisingly. The court opinion put it this way: "Some stress was laid on the proposition that Chicago is not on the natural watershed of the Mississippi, because of a rise of a few feet between the Desplaines [sic] and the Chicago rivers. We perceive no reason for distinction on this ground. The natural features relied upon are of the smallest."[25] Whereas the "natural" definition of a waterway had once been elemental, the courts now consented to let waterways be altered to support the urban industrial economy.

The future of American rivers was up for debate. As Justice Holmes suggestively said, "It is a question of the first magnitude whether the destiny of the great rivers is to be the sewers of the cities along their banks or to be protected against everything which threatens their purity." No one had the right to dump pollution, but neither should every harmful act be prevented. The court dismissed the case in large part because it did not take Saint Louis seriously. Holmes did not find the medical science convincing, saying, "We assume the now-prevailing scientific explanation of typhoid fever to be correct. But when we go beyond that assumption, everything is involved in doubt." At the turn of the century, when medical knowledge and science were still in a state of flux, relying upon the unseen was a bit of a stretch for the court. Saint Louis's case rested on the number of bacteria, which were invisible. The court recognized that the water-quality facts were disputed and that experts disagreed on the efficacy of dilution.[26] The apex of this debate would come in the next decade as American sanitarians wrestled with the definition of rivers as urban sinks or protected entities.

The legal option having failed, the short-term solution for Saint Louis was the adoption of better filtration, and a future option would be relocating its waterworks. The city began to think about the Meramec or Missouri River as a source.[27] Considering the Chicago sewage problem in 1900, the *Waterways Journal* had found the proper main stream enticingly clean, saying, "Its headwaters are about 3,000 miles away in the Rocky Mountains, and it is fed by innumerable streams as clear as crystal that flow from the snow clad peaks of many mountain ranges."[28] Pollution problems on the Missouri River had not yet reached a critical point and, to Saint Louis, these echoes of Champion S. Chase depicted it as a pristine river coming from the remote reaches and unpopulated expanses of the West. On the other hand, the Mississippi River, laden with waste from Chicago, the Twin Cities, and the Quad Cities, represented industry, urbanization, and immigration.[29]

Sanitary Seers and the Missouri River Sanitary Conference

When the Missouri River Sanitary Conference (MRSC) met at the Baltimore Hotel in Kansas City, Missouri, on December 29 in 1910, conferees dwelled on present and future health conditions in the Missouri River basin. Typhoid fever illnesses and deaths were an indicator of contamination by fecal coliform bacteria (*Salmonella typhi*) and every state in the lower basin wrestled with a high number of typhoid fever cases in the early 1900s. A punctuated outbreak occurred in the winter of 1909 and 1910.[30]

The MRSC conferees were concerned about public health issues and the river allowed them to transcend borders and see a larger set of environmental relationships than was common in the legal and political contexts of their day. The river was used freely by all the states for drinking water and sewage carriage but no entity existed to oversee water quality, which was regarded as a local, and sometimes state, but not a federal issue. With no restraints against waste disposal, cities along the Missouri found it cheap and easy to dump in the river. The conferees were keenly aware that the same advantages applied to those upstream. With no jurisdiction beyond their own backyards, cities and states found it difficult to prevent upstream pollution, and there was no incentive for them to protect communities downstream. At the forefront of a "movement"—as an observing engineer called the conference—for public environmental health, the conferees of the MRSC recommended that legislation regulating water quality be adopted by all states contiguous to the lower basin.[31]

On the MRSC guest list were the five lower basin states of South Dakota, Nebraska, Iowa, Kansas, and Missouri. According to correspondence about the conference, all of the invited states appointed representatives. Dr. Samuel Crumbine, of the Kansas State Board of Health, led the conference, along with sanitary engineer W. C. Hoad from the University of Kansas and Kansas governor W. R. Stubbs. In addition to impelling the conference, Kansas had eight delegates in attendance, of a total of nine attendees. The host state of Missouri had no delegates in attendance, though McDonnell and Governor Hadley were interested and supportive. One representative from Nebraska attended, but nobody from Iowa or South Dakota.

Typhoid fever topped the list of conference concerns. Attendees shared their cities' experiences dealing with typhoid—a bane of public health officials and the cause of stomach problems, fevers, and sometimes death. Dr. Ralph W. Connell, for example, talked about Omaha's typhoid fever outbreak from contaminated river water.[32] Of concern to the attendees was the fact that a large and growing population relied on the Missouri River for drinking water. From Sioux City to Kansas City alone, eight hundred thousand people were dependent on the river. Saint Louis's seven hundred thousand residents were not yet tapping the Missouri, but they would be within twenty years. By 1910 over half the population of the watershed was discharging raw sewage to the river, representing as many as two million people.[33] Kansas City's population of well over two hundred thousand was discharging forty million gallons of sewage daily directly into the river.[34] For all of these people, upstream and downstream alike, the Missouri was an essential resource.

Beyond municipal wastes like sewage and garbage, the "considerable burden" to the river included industrial wastes. Attendees discussed the large meatpacking enterprises discharging into the river—notably in Omaha, Saint Joseph, and the Kansas Cities. These packing plants were not considered the root of the problem, but the conference memorandum concluded that the packing wastes "help to form in the River a favorable environment for the multiplication of disease germs introduced from city sewers."[35] In 1910 public health officials' primary concern was sewage.

Having established the problem, the conferees next had to consider solutions, the most favored of which was interstate action. Crumbine, the impetus for the MRSC, was frustrated by the "old problem" of abating pollution on interstate waterways. They engaged in "considerable talk" of federal-level involvement but "the idea was finally abandoned" because no one believed the federal government "would be acceded to," as a conference summary put it. In the absence of federal willingness, the conferees felt that state legislatures should take leadership to pass laws.[36]

Of the sanitary conference, the doctor in his memoirs stated simply: "The agreement failed." Crumbine reported that due to the expense of requiring sewage treatment, other cities and states on the Missouri River balked.[37] "Shades of Neptune!" Crumbine commented sardonically in his memoirs:

No federal sanitary control! But wait a minute, Kansas has sanitary control over one-half of the Missouri River, the boundary line being exactly in the center of the stream. Of course, if the sewage and industrial waste discharged into the river from St. Joseph would only keep to the Missouri side of the river, all would be "fine and dandy," but the "Big Muddy" is not that kind of a stream.[38]

He had long been frustrated with jurisdiction and power because "time and time again" the problem of sanitation on interstate waters had been "brought to the attention" of the federal government but the federal level "definitely decided" it could not control waterways for this purpose.[39] Despite the powers of the USPHS and US Army Corps of Engineers, pollution was viewed as a local concern. But this political ideology of a limited role for federal government evolved at a time before rivers had become so polluted as to cause cities to play Russian roulette with their drinking water. The Progressive Era was a period of adjustment; new environmental, economic, and social realities required a shift in politics and law as well.

MRSC Context: Health versus Wealth

For the Kansas Cities and other municipalities downstream from sewage outfalls, the first decade of the twentieth century was already a time of looming crisis with no recourse to federal authority in sight. The related debates and discussions occurring nationally influenced Crumbine and the Missouri River Sanitary Conference.

First, concurrent with MRSC, sanitary professionals debated whether water filtration was sufficient or if sewage treatment was necessary. The debate highlighted the differences between the fields of engineers and physicians.[40] Engineers like the influential George C. Whipple believed that the natural process of dilution would suffice. Concentrate on water filtration, he counseled in 1909, "leaving the question of sewage disposal to be treated locally and principally from the standpoint of nuisance."[41] Engineers blamed physicians for whipping up public sentiment against dilution.[42] The public—with its "imperfect knowledge," as one engineer put it—mistakenly believed that sewage purification was necessary, whereas engineers could calculate the success of dilution.[43]

Physicians and public health officials like Crumbine, and the developing breed of sanitary engineers like McDonnell, on the other hand, felt that sewage treatment was a preventative priority. State boards of health wanted to manage water quality at both ends.[44] The debate between dilution and treatment was a question of the rights and responsibilities of downstreamers and upstreamers. Dilution would fall short for regions that required cooperative efforts between upstream and downstream parties—like interstate watersheds. The disparate economic costs involved—treating drinking water was cheaper—made health an economic question. Yet engineers, sanitarians, doctors, and city officials suspected or assumed that treatment would be required in the future. But until there was a crackdown on raw sewage, dilution was acceptable. The MRSC met amid this "dilution is the solution to pollution" debate.

Crumbine kept up with the sanitary activities of other states, reading health board reports and meeting fellow public health officials at conferences. He celebrated the victories of peers around the nation and decried the limited powers available to public health officials. Crumbine and Kansas were on the cusp of a movement among the states to control water quality in their respective watersheds. Surely these national debates, sanitary surveys, and conferences gave impetus to Crumbine in organizing the MRSC.

In 1908 a national conference of state boards of health was held, and it is likely that Crumbine attended. A national committee on water pollution condemned using waterways for sewage disposal. Meanwhile, federal-level sanitarians met at problem-solving conferences regarding national and international water pollution issues.[45] That same year, in Ohio the state board of health had a sanitary survey under way on the Mahoning River, an Ohio River tributary that received raw sewage and industrial wastes yet served as a drinking water source. A resultant "remarkable conference" brought all the Mahoning watershed stakeholders in Pennsylvania and Ohio together to "unite them in common cause." This basinwide conference, united in "mutual interest," was the first of its kind, the Ohio health board reported.[46] An editorial in the *Engineering Record* applauded their efforts to deal with water and sewage "collectively."[47]

Also in 1908, President Theodore Roosevelt convened the Conference of Governors in Washington, DC, an effort to build support for an integrated program of natural resource conservation. Some sanitarians probably found it useful to align themselves with conservation leaders like Gifford Pinchot and WJ McGee, who coordinated the conference, because it was under the umbrella of conservation that health and wealth could be united in the discussion of national waterways.[48] But in the Progressive Era, most political leaders supported harnessing and "improving" rivers for economic growth, saying little of public health.

Addressing the conservation-oriented Conference of Governors, Governor Folk focused on forestry and navigation. To loud applause, Folk spoke of the Missouri River as a potential "highway" for "carrying the products of farm and of mine and of factory across the seas into every land beneath the sun, and returning to pour into the lap of the Nation the golden stream of universal trade." He continued: "I think it is the duty of the Federal Government to improve these waterways and make them adequate for commerce . . . or give the States permission to do so."[49] The state of Missouri wanted either financial investment or the power to develop rivers themselves; a hands-off policy was not acceptable. Simultaneously, navigation boosters, adopting conservation rhetoric, organized the first convention of the Missouri Valley River Improvement Association, held in Sioux City. Their demands for river improvement were fueled by a Missouri navigation study released in 1908 by Colonel E. H. Schulz of the Corps of Engineers.[50]

Herbert Hadley, the trustbusting attorney general who took Chicago to court, was elected governor of Missouri in 1908. The national conservation movement inspired his protective politics and he tried to balance the visions of health and wealth as well as support the same kind of cooperation that Crumbine sought. Hadley won legislative authorization of a Missouri Waterway Commission composed of five men (some affiliated with the Kansas City Commercial Club) that addressed water quality in its first and only report. The commission decried poor sanitary practices because waterways were a public good that should benefit all. The recommended principle was to view water as belonging "jointly to the people of the State, whose rights therein are natural, inherent, inalienable, and indefeasible." As far as interstate waters were concerned, the water should also be enjoyed by upstreamers and downstreamers, but within the limits of federal laws permitted by the Constitution. Rather than strengthening the role of the federal government, the Missouri Waterway Commission recommended cooperative measures or arbitration among the states. Though it addressed the Healthy River, Hadley's commission devoted most of its report to the Economic River.[51]

The leading midwestern promoter of the Economic River was the Kansas City Commercial Club, the most influential business organization in whose chambers the Missouri River Sanitary Conference met in December 1910. It is the club's vision of the Missouri River as a navigable corridor of commerce that has most defined the Missouri River to the present day. Commercial boosters in the club believed barge navigation would connect the heart of America to the markets of the world. In its pursuit of the Economic River, it could have facilitated the Healthy River, too. But during the very week of the MRSC, the club was preoccupied with legislation in Washington, DC, that would secure funds for navigation. The club could have been useful in supporting the MRSC, as it certainly put its efforts into similar civic campaigns during the Progressive Era, but wealth trumped health for these men of privilege.[52]

Thus the 1910 MRSC represented a culmination of concern and a breaking point for American waterways and cities.[53] Government intervention was a popular solution for cities and states battling typhoid fever. Not long after the MRSC, the Kansas City Hospital and Health Board tested the Kaw and Missouri Rivers and ominously reported: "The time is not far distant when the government, states and large cities must appreciate the sanitary fact that the emptying of sewage into the rivers must cease, and our city, through its administrative officers, should begin at once to prepare to meet this big proposition in the way it should be handled."[54] Public health officials understood by 1910 that water quality regulation and sewage treatment would be expensive. By the 1910s sanitarians thought a moratorium on polluted waterways was reasonable and necessary.

Could advocates of the Healthy River have imagined it would be more than half a century before such regulations would be in place and begin to be enforced? To have been a success, the 1910 MRSC needed to reach out beyond public health officials. It needed to grow into a larger campaign, a movement that included more stakeholders and advocates. But the Healthy River had to compete with the rhetoric of the Economic River, and it came up short. Just as important, the boosters did not ally themselves with the sanitarians and did not embrace the Healthy River.

Limited Progress

In spite of Crumbine's judgment later in his life that the conference had failed, one of the goals of the MRSC that succeeded was a complete river survey of the five lower basin states. Crumbine needled the USPHS to conduct the survey and make a report as soon as possible. Frustrated that other states were not following through with the conference resolutions, Kansas told the US surgeon general that "the Kansas commission believes that the study should be undertaken independently," and to avoid default, "makes the request that you duly detail a competent man to join with the Kansas commission."[55] Kansas did not give up on federal involvement despite the other river states' reluctance to look beyond political boundaries. If it had not been for the persistence of the Kansas State Board of Health, the USPHS might not have "detailed" Allan McLaughlin to direct a report on Missouri River sanitary conditions.[56]

Coming from international public health work in Europe and the Great Lakes region, McLaughlin began his Missouri Basin work in 1912. That year the USPHS gained greater power from Congress to investigate stream pollution and subsequently initiated a number of studies.[57] Little money was available from the federal government, so the cost of the study was also borne by state health boards, cities, and universities. To gather the data, there was "cooperation" and a "public-spirited attitude" among the different levels of governance.[58] Iowa had one data collection point at Sioux City, the first one in the study area. The only collection points not located on the main stem were on the Kaw River, downstream from Lawrence. The Kansas and Missouri bacteriologists were busy; Kansas was responsible for seven survey stations and Missouri for nine.[59]

The resulting 1913 report, titled *Sewage Pollution of Interstate and International Waters with Special Reference to the Spread of Typhoid Fever: The Missouri River from Sioux City to Its Mouth*, concluded that sewage indeed was a problem.[60] In fact, the report unequivocally stated, "immediate control of pollution is necessary."[61] The USPHS report established the base data that would be referenced in nearly every report thereafter and framed the water-quality debate for the rest of the century.[62]

Kansas sanitarians, proud to instigate the MRSC and then the USPHS survey, assumed that the report's findings would be compelling and inspire action. "It is confidently hoped that those cities in other states now discharging their untreated sewage into the Missouri river may join with Kansas in constructing such sewage purification plants as will insure the comparative purity of the Missouri river," the Kansas health board stated.[63] But what McLaughlin's USPHS report recommended was not necessarily an interstate compact such as envisioned by the MRSC but "a central directing authority" that would be free from parochial viewpoints and equitable to all parties.[64] Follow-up MRSC gatherings failed to persuade other states to adopt common water-quality laws.[65] The states did not pursue the resolutions seriously, and the USPHS dropped the ball after completing its report—hence Crumbine's admission that the 1910 MRSC was a failure.[66] After the initial flurry of attention in the Progressive Era, the USPHS left the Missouri Basin to its interstate devices while the Great Lakes and Ohio River regions received more federal attention.

Crumbine had a larger vision of health that he wanted to extend to broader circles of relationships, as discussed in the chapter "Sister Cities." It was Crumbine and his colleagues who embarked upon outings down the Kaw River by boat, taking samples every few miles to gauge the bacteria levels and prove that dilution was insufficient. He was able to encompass broader communities and regions once he conceptualized everyone as both an upstreamer and a downstreamer. Kansas sanitarians were concerned about the Missouri River because they saw it as an extension of their own Kaw River—their "intestinal tract," as the engineering journal *Municipal Sanitation* referred to urban relationships with rivers.[67] Most of Kansas drained to the Kaw, which meant that Kansas City, a place for which Crumbine was responsible, was downstream from most of the state.

Iowa had not sent any delegates to the MRSC and the first mention the Iowa State Board of Health made of the conference was after the fact when it reviewed the USPHS's publication of the 1913 sanitary survey. At that point, the health board had plenty to say on the topic. Iowa credited Crumbine and counted itself among the lower basin states concerned about pollution of the Missouri River.[68] The "high degree of purification" necessary put the onus on water treatment plants. The use of hypochlorite and lime were effective when combined with sedimentation, but not foolproof, because removing 99 percent of bacteria was insufficient in times of "gross pollution." Such technological assessments led the Iowa health board to conclude: "This would naturally lead us to treat the sewage before allowing it to enter the streams, then follow this with sedimentation, filtration and treatment in order to prevent typhoid fever."[69] If the water were cleaner to begin with, there would be less risk at the waterworks.

The Missouri River, the Iowa health board declared, was "unfit for drinking" and required "very thorough treatment to render it safe." Although the board wanted more research beyond the "preliminary" 1913 findings, it stated with conviction that the river "probably exceeded even a liberal construction of the phrase 'permissible pollution.'" The Iowa board wanted "immediate control" because waterworks endured "too great a strain and responsibility."[70]

The health board of Iowa accepted that federal control was necessary. In 1914 it did not recommend an interstate agreement—as the MRSC had—but wanted "a central directing authority independent of local influences." The federal government should set minimum requirements for pollution and water quality, the board suggested. Each state would then have the freedom and "the right to exact more rigid requirements" and go above and beyond the federal minimum. The board imagined "perfect harmony" between federal and state entities to create a united front against water pollution.[71] The Iowa board's vision of federal intervention was closer to what would happen later in the century.

The Iowa board of health did not want to threaten legislators or cities with a ban. Rather, the board defined its work as the "logical position" between two extremes. The middle-of-the-road position held by most sanitarians, both public health officials and engineers, would allow some discharge and still protect public health. The first step in striking the balance was the sanitary survey. By studying conditions of the streams and urban environments, standards could be made that kept sewage discharge to safe levels.[72] Like progressive tendencies in general, the Iowa sanitarians wanted to do research, create a base of facts, and put decision-making in the hands of the professionals.

Not all public health officials addressed river pollution in the same holistic fashion as Crumbine. In 1915 public health representatives from the Missouri River Basin held their first meeting as the newly organized Missouri Valley Public Health Association (MVPHA). From Montana to Missouri, North Dakota to Oklahoma, the association was led by Paul Paquin, Kansas City, Missouri's director of public health. Paquin, a professional appointment during the Jost administration, was a flyswatter sanitarian who emphasized individual responsibility in his public health campaigns and philosophically blended health and wealth. Although only five years separated the two gatherings and they had similar objectives, Crumbine and the MRSC differed from Paquin and the MVPHA. By 1915 the innovative ideas of cooperative water quality management were gone, replaced by the more limited vision of the MVPHA.[73]

A newspaper recounted the MVPHA conference agenda, saying that conferees discussed how pollution from rural areas adversely affected the river and thus urban health. With the burden of pollution from sewage and industry, it seems absurd that farmers were singled out, but conferees thought pollution from farmers was "worse than sewage." The newspaper reported to its urban readers: "Our drinking water, it was said, is endangered more from the pollution of farms and small towns along the river bank than from the sewage of cities such as St. Joseph and Omaha." The MVPHA concluded that if farmers would clean up riverbanks, cities would have pure water.[74]

Perhaps the MVPHA conferees were making a basinwide environmental critique, looking at how water quality was influenced by soil erosion, animals on tributary banks, and the loss of vegetative cover like forests in exchange for grazing lands. But sources indicate that only one aspect of rural pollution was discussed. The "garbage" from farmers carried potentially disease-carrying flies down the river, endangering cities. Those are the flies that Paquin's educational campaigns urged swatting in order to avoid typhoid fever. The assertion that responsibility for public health rested on the farmers was specious when compared to the volume and content of urban and industrial wastes in the river. Kansas had proven that dilution of urban wastes was not sufficient. The 1913 USPHS survey, as well as data from the Iowa, Kansas, and Missouri boards of health, proved that bacteria levels downstream from cities were atrocious. McLaughlin had discussed rural pollution in his report, particularly animal wastes, but concluded that urban pollution was of greater concern, as well as easily controlled.[75]

The tactics of the MVPHA were to encourage fly swatting and farm cleanliness and not question the urban status quo. Addressing industrial and urban waste would be expensive and would challenge the Economic River. Absent were the MRSC's proposed interstate agreements to regulate water quality and its vision of the Healthy River. However, not all participants wanted to water down the campaign. McDonnell's address to the MVPHA on water and sewers was not about swatting flies. Reflecting the frustration of the MRSC, he said, "Sanitary reforms are slow in adoption and unlike political reform, they have few advocates." Critiquing the imbalance between health and wealth, McDonnell said, "Many people take great pride in the financial showing of their community and refer with pleasure to the high bank clearings, but there are few people who think very much about whether the death rate is high or low, and what may be done to better the conditions."[76]

In other river basins, there had been some tentative steps toward regionalism. By 1919, on the upper Mississippi River, four states entered into an interstate agreement to regulate the mussel industry.[77] Other places that considered regional decision-making around water quality were in New York, New Jersey, Rhode Island, and Massachusetts.[78] Urban areas needing to provide services and deal with environmental realities were adopting regionalism. Boston, for example, had a strong state board of health that looked beyond the city to the metropolitan area to control water quality.[79] Internationally, European countries like Great Britain also were beginning to make regional sewerage decisions, and engineers shared these international successes within their profession.[80] On the Missouri River, despite the early push, a broader vision was lacking. The lower-basin sanitarians could have been contenders—they could have been the heroes of the public health and environmental movement. Their efforts were notable but were drowned out by the determined lobbying of industrialists and navigation boosters.

The Quiet Years

The first two decades of the twentieth century were shrill with concern over typhoid fever and sewage pollution, and the 1950s and 1960s would also become a period of heightened concern and active abatement. In between these vocal periods of reform, sewage, garbage, and other pollution on the Missouri River did not raise a comparative stink, despite persisting pollution. Waterworks engineers and public health officials were aware of the problems on the river, but the period of agitation to improve water quality declined after World War I. Why, if the health threat was so dire in the 1910s, did nothing change?

One reason is that the adoption of new technological processes allowed communities to overcome the immediate problem of furnishing safe drinking water while continuing to contribute to the waste load. Before 1910 most cities had settled their water in basins and used coagulants, processes considered "primitive" by later standards. As a result of the 1913 USPHS report on typhoid fever, within a few years all ten river cities that McLaughlin studied began using chlorine and employing sedimentation and other filtration processes.[81] Drinking water was safer, but the Kansas Cities were still shamed by higher typhoid fever rates in national comparisons.[82] After Kansas City opened its new water treatment plant in 1928, it met the USPHS standards, however limited.[83] Yet while precautions advanced at the intake, pollution from sewage and other wastes continued to increase at the other end of the city's intestinal tract. While drinking water service was revenue producing, cities could not publicly justify the expense of sewage treatment, which only seemed to benefit downstreamers.[84]

Creative solutions to meet the growing problem of wastes in the 1920s exhibited the gamut from practical to wishful thinking. Conservationist WJ McGee advised the efficiency- and conservation-minded that "urban sewage should be converted and utilized as a source of municipal revenue."[85] Incineration and "sewage farming" were billed as miracle technologies.[86] The Commercial Club commented that "utilizing the waste matter now passing through the sewers" would increase the "manufacturing output" and also "public health would be conserved."[87] The attraction was that waste could be turned into profit, but like so many technological solutions for disposal, these plans never came to fruition or were infeasible.[88]

The stink multiplied. McDonnell cited statistics from 1900 to the 1930s that showed that bacteria counts taken near the intake at Kansas City had increased: "Our surface water supplies," McDonnell wrote, "are fast being protected by expensive purification works, but even these are now having difficulty in coping with the constantly increasing bacterial content of the rivers." For cities, it was "greatest of all engineering problems," he said, because "fully 90% of all sewage . . . is disposed of by dilution." Even before the New Deal, McDonnell thought the federal government had a positive role to play in building sewage plants. He also believed that state boards of health did too. "Their greatest handicap," he said of health boards, was "not having laws with teeth, and being without power to enforce recommendations."[89] He and Crumbine were on the same page.

Abatement did not proceed on the Missouri main stem during the quiet years, but tributaries did make some progress, notably on the Kaw River in Kansas.[90] In the early 1920s Topeka's sewage affected Lawrence, so the Kansas State Board of Health made an "elaborate sanitary survey" of the Kaw and recommended that, despite costs, large and small cities act in each other's interests by building treatment plants, especially in places like Topeka where flow was insufficient for dilution. In short, the state capital must protect Lawrence by treating its sewage.[91] Dr. Earle Brown of the board said that the expense "does not detract from the responsibility a city has to . . . not create a health hazard or a nuisance."[92] At the behest of the health board and aided by New Deal funds, Topeka opened a secondary treatment plant in the mid-1930s to protect the drinking water downstream.[93]

The Kansas board of health early recognized that wildlife health was related to water quality and human health. "Many of the streams of the state are so basely polluted as to have killed all the fish therein," a report

Table 7.1. Missouri River Bacteria Counts, Kansas City

Year	Number per cubic meter
1900	8,000
1910	12,000
1930	20,000
1931	100,000

read, finding the fish indicative of water quality.[94] Although fish kills and aquatic health had been the initial tipoff to problems, sanitarians aimed to protect the health of humans with the early Water and Sewage Law of 1907.[95] Earnest Boyce, who led the health board's Division of Sanitation at the time, said that state water law encompassed nonhuman life. The reasoning for this, Boyce said, was that the board "found itself constantly attempting to define the line of demarcation between pollution problems actually hazardous to the public health—and pollution problems of little public interest—but very destructive to aquatic life." Twenty years later, the first amendment to the Kansas water and sewage law was "extended to include pollution problems affecting the animal and aquatic life of the state."[96] Brown was pleased that the board could abate sewage "detrimental to animal and aquatic life as well as public health." Sewage dilution must not "interfere with the claims of other riparian users," he said.[97] In other words, the amendment acknowledged that the environment was more complex than the early law had allowed; the 1927 Water and Sewage Law connected human and ecological health.

In general, there were fewer reports that addressed water quality in the lower Missouri basin during the 1920s and 1930s. Of the reports made, a few pertained to wildlife, indicating that sanitarians were moving away from a human-centric focus on typhoid fever, were influenced by the conservation movement, and were seeing public health within a larger set of environmental relationships.[98]

Flirting with Federal Jurisdiction

The federal government took responsibility for the economy but not for public health. As both Crumbine and McDonnell had assessed, the lack of federal "teeth" in interstate sanitary matters was troublesome. To illustrate, in 1915 the engineer for the city of Liberty—just downstream from the Kansas Cities and Independence, Missouri—wrote the US Army Corps of Engineers asking about *federal* legislation regarding sewage in rivers. The Liberty engineer asked: "Has a law been passed prohibiting cities [from] emptying sewerage direct into government streams?" The response from the corps, of course, was that the Rivers and Harbors Act of 1899 ruled supreme in navigable waters, and it "prohibited the discharge of refuse matter into navigable streams, but excepted that flowing from streets and sewers and passing therefrom in a liquid state."[99] In a word, the answer was no. In practice, the corps sometimes employed the act to abate egregious pollution on behalf of navigation—like when the corps ordered Kansas City to stop using the river as its dump in the 1920s—but generous interpretations were not the norm. No federal law barred sewage disposal in the river; the Liberty engineer was unable to protect his water supply from upstream pollution and could freely send urban waste downstream.

In the 1930s federal agencies paused to consider their responsibility for river pollution but ultimately found the lack of regulation convenient. In 1935 US military bases prepared reports on their sewerage systems and considered whether or not their disposal infrastructure should be upgraded. From Fort Riley to the federal penitentiaries in Leavenworth, Kansas, to the boatyard at Gasconade, Missouri, several federal facilities used the Missouri River for sewage disposal. The reports of each institution detailed drinking water consumption and waste load contribution, down to the gallon. In each case, whether it was the Veterans Administration or the air base in Kansas City, the reports recommended that "no steps be taken" because federal facilities were a fraction of the total pollution load and because no one else was treating waste either. These reports document the prevailing thought of upstreamers.[100]

In reaction to extreme environmental and economic conditions like droughts, floods, and depression, the federal government played the most significant role in infrastructure development. First, navigation and flood control benefited. The barge channel (first authorized in 1912) was the beneficiary of New Deal largesse. With a new goal of a navigable river all the way up to Sioux City, the infusion of federal dollars furthered boosters' dreams of the Missouri as an artery of commerce and provided people with employment. Second, the federal funds expanded the urban innards—the pipes and conduits that connected people to the river. Midwestern communities built waterworks, sewers, and sewage treatment plants because federal agencies like the Public Works Administration made financing and matching grants readily available. The Missouri Department of Health happily reported: "It can truly be said that, although 1934 was a year of major depression, from the standpoint of improvements in existing public water supplies and construction of new water supplies, it was the best year the state ever had."[101] McDonnell's firm produced a report showing that more people gained access to better water and sewer facilities in the 1930s. Water consumption outpaced sewerage and waste treatment, which increased the waste load.[102] While the federal government expanded its economic jurisdiction with the New Deal, its sanitary jurisdiction remained limited.

Though sanitary activists on the Missouri and Ohio River started in the same place, the Ohio experienced noteworthy progress because sanitarians in the basin successfully organized interstate water quality control. Rooted in a historically aggressive Ohio State Board of Health, early water quality legislation and a good partnership with the USPHS, by 1935 planning for an interstate compact on the Ohio began. The governing interstate body, the Ohio River Valley Water Sanitation Commission (ORSANCO), was officially formed in 1948.[103] Interstate agreements and water quality control came to the Ohio before the federal government had the power or inclination to regulate waterways.

Critical to the success of interstate decision-making was the involvement and leadership of the Cincinnati business and political community, which advocated that cities and their voters take responsibility for water quality. The Cincinnati Chamber of Commerce led the campaign to limit pollution with a catchy slogan: "Let's take the dead horses out of the river," equating the volume of sewage waste in the river with the weight of horses.[104] It took years of work with an alliance of private and public forces, but the Ohio River came under sanitary control.

The Missouri had no such cheerleaders or allies. The Kansas City

Chamber of Commerce (formerly the Commercial Club) fixed its eyes on the Economic River. The Kansas City Citizens' League also would have been an ideal candidate to lead a campaign for the Healthy River in the 1920s and 1930s. However, the league's perspective on the local pollution situation was tempered: "The 'Big Muddy' can assimilate a lot of contamination" and "some day" downstreamers would "justly complain," but that day had not arrived yet.[105] McDonnell, one of the organization's leaders, had long espoused building treatment plants but his colleagues at the Citizens' League were more focused on political reform and the urban machine. No regional champion stepped forward on the Missouri River, which left a larger role for the federal government to play in the postwar period.

A final illustration of the failed Healthy River was an incident that took place in the 1930s. A typhoid fever death emphasized how a lack of sanitary responsibility at all levels eventually impacted the health of individuals. In Lexington, Missouri, below the Kansas Cities, the Corps of Engineers' Quarterboat "E" was doing some river work, maybe to lay woven mats or riprap along the banks. The hired firm, the Massman Construction Company, had a worker on the boat, Roy Higgs, die from typhoid fever. On Higgs's behalf, a suit was brought (by family, perhaps) against the construction company and thus the corps was involved. The insurance company and prosecutors began to ask questions about water quality and safety. The workboats used river water, but they treated it before consumption, said the Kansas City District Corps of Engineers. However, given the pollution, it was possible that treatment was inadequate or the chemical treatment tubes applied to the onboard water tanks were outdated.[106]

In response to Higgs's death and subsequent legal action, the Corps of Engineers had water samples sent to the Missouri State Board of Health. Those tests came back positive for contamination of the bacteria causing typhoid fever. The corps responded to a prosecuting attorney in 1933 that the federal government did not take responsibility for water quality testing. If the attorney wanted to locate evidence in the form of bacterial data, the corps suggested that the attorney contact a city waterworks, or Missouri's health board, or perhaps the USPHS. The lawsuit revealed that no rules or legal entities governed water quality and little knowledge about water quality existed, although the death was proof that the river posed a health threat.[107]

The river mythologist Cecil Griffith recalled how, after the death of Higgs, the Corps of Engineers disallowed drinking from the Missouri River and redoubled efforts to force river workers on the quarterboats to drink treated water. As it turned out, it was difficult to persuade "river rats" to drink treated water. Griffith claimed that he and other men who worked for the Corps in the 1930s "used to lie on our bellies on the sand bars and thoroughly enjoy a good long drink of the best-tasting water in the world, the old Missouri River—straight."[108] The corps announced that the river was unsanitary—"a vile and deadly poison"—and workers could be fired for drinking it. Griffith detailed methods of onboard purification, including using oatmeal, eggs, and finally chlorine. The chlorine came in capsules that had to be broken underwater and the person who did this wore a pair of filthy gloves for protection. Griffith humorously indicated that everyone was aware of the other (sometimes disgusting) daily tasks performed with these gloves.

Griffith's point in this story is that the process of effectively treating a tank of river water was less appealing than the original river, and "for a satisfying drink of water there remained the usual path out through the willows to the river—out of sight of the boss." River rats were unreformed. They still saw the river as naturally healthy. The men continued to get sick, so the corps required them to have typhoid fever shots. Griffith, perhaps jokingly, wondered if it was the chlorinated water and not the "unadulterated" river that made the river rats sick.[109]

The vacuum of sanitary power had ramifications for individual, city, state, and federal entities alike. There were no laws regulating water quality. Cities took advantage of the lack of jurisdiction over waterways and used the Missouri River as a sink, although cities also found themselves victimized by similar actions of those upstream. With so many different jurisdictions in competition, there was no accountability or incentive for any one entity to take responsibility to treat sewage. The only way to change the upstreamer and downstreamer dynamic was with mutual agreements between stakeholders, as the Missouri River Sanitary Conference had proposed. Alternatively, overarching power from the federal level could shift the dynamic. The federal government was clearly unempowered in the early twentieth century. Not even a directive to drink chlorinated water was effective among the men it employed! As different jurisdictions tussled and tugged for power and then shrugged when it came to responsibility, the river remained a health risk.

The Triumph of the Economic River

The scale of federal involvement on the Missouri River gathered steam. In the early twentieth century, Kansas City navigation boosters best represented the Economic River. Men like Champion S. Chase, clay sewer pipe magnate Walter S. Dickey, and the effervescent economic boosters in the Kansas City Commercial Club attached the future of the city, region, and basin to the ability to inexpensively run barges up and down the river. The federal government's involvement on the Missouri had begun as early as 1876, but its regulation of the waterway for commerce and navigation was limited in funding and scope. In fits and starts, lower basin advocates succeeded in securing incremental funds for channelizing the river to Kansas City. The Missouri River Improvement Association, dominated by Kansas City interests, lobbied for channelizing the river above Kansas City because its members envisioned themselves the recipients of commercial wealth across a swath of middle America.[110]

Floods impelled the engineering of the river. After the 1903 disaster, Congress renewed its interest in funding navigation as well as in flood control. Boosters like Kansas Citian Lawrence Jones maintained an "evangelical fervor," in the words of historian Robert Schneiders, for an appropriation that would match their visions.[111] The clamor from the lower basin resulted in the first significant commitment to a six-foot barge channel from Kansas City to the mouth in 1910, the work to be done by the Corps of Engineers beginning in 1912. A uniform channel depth required engineering the river to scour its own channel by building wing dams, levees, and stabilizing banks. The difficulty of controlling the fickle river meant that channel improvements were two steps forward and one step back, and by the 1920s the channel to Kansas City was still incomplete. Boosters remained vigilant until the 1940s, working hard from the local level to achieve a deeper nine-foot channel upstream to Sioux City.[112]

Floods on inland waterways produced national legislation like the 1928 and 1936 Flood Control Acts that shored up improvement campaigns on many American rivers. In 1935, after another impressive flood on the lower Missouri, Congress accepted the "308 Report," which identified the potential for economic development on the river. Along with a proposed Missouri Valley Authority, modeled on the Tennessee Valley Authority, the groundwork was laid for a plan that reflected the progressive, conservation-oriented, problem-solving optimism of the New Deal.[113] Upstream in Montana, Fort Peck—the world's largest earthen dam—was already underway as a New Deal project.[114] Montana novelist Ivan Doig described the "din" of dam construction in *Bucking the Sun* as "the opera shrieks of shale saws, the incessant comings and goings of locomotives and bulldozers and trucks, the falsetto of steam whistles, the attacks of jackhammers."[115] The dam promised to quiet flooding in the lower basin.

The vision of the Economic River evolved into what would become known as the Pick-Sloan Plan. Named for General Lewis A. Pick of the US Army Corps of Engineers and William Sloan of the US Bureau of Reclamation, the plan promised something for everyone. Whereas the 1920s and 1930s had generally been dry, the 1940s were wet, and high water compelled action. If it were not for a quick Dakotas snowmelt and severe flooding in 1943 and 1944, the Pick-Sloan Plan might not have appeared so attractive and been adopted so quickly in 1944 amid the war.

With congressional assurance of funding, the Pick-Sloan Plan represented the triumph of boosters' long-sought dream. In addition to Fort Peck, the plan called for five more mainstem dams in North Dakota and South Dakota, as well as hundreds of smaller dams on tributaries. These projects promised irrigation, hydroelectricity, flood control, navigation, and new landscapes for recreation throughout the basin. The semiarid upper basin would impound water during wet times and release water during dry times. Flow regulation would provide the flood-ridden lower basin with flood control and navigation, evening out the extremes of flood and drought. The lower river would be hemmed with levees, feathered with wing dikes, and channelized. The Pick-Sloan Plan sought to engineer the nature out of this unruly, dark river of peril, to borrow the descriptive words of river historians.[116]

Little discussed was how this massive basinwide engineering project would affect public health and the environment. McDonnell expressed concern that upstream agriculture negatively influenced water quality, saying, "The volume of water in the Missouri River is decreasing on account of the water from many of its tributaries being used for irrigation. At the same time the sewage that is poured into it is increasing. Any bacterial analysis shows the water not nearly so pure now as it was a few years ago."[117] Finally, in 1946, General Pick of the Corps of Engineers wrote an article for the *Journal of the American Water Works Association* that outlined the projected impact of the Pick-Sloan plan on public drinking water supplies.[118] Pick made big promises that the plan would "substantially improve the waters of the Missouri for domestic consumption." Each of the major cities would have benefits; the plan was intended to control many of the river's seasonal variations that threatened the functioning of waterworks. Cities would appreciate the regulated flow, which would ensure that their intakes were not flooded, left dry in drought, threatened by ice jams, or inundated by high bacterial counts and silt. In addition, the vagaries of the meandering Missouri would be quelled, giving more bank stability to waterworks facilities. For the Kansas Cities, a decreased silt load would make clarification easier and cheaper. Pick also predicted that money would be saved because the color and odor of the water would be more palatable, reducing the need for chemicals.[119] Americans who held faith in the ability to engineer nature for predictability and wealth development were pleased.

Less pleased were the farmers and tribes that lost land to dams and reservoirs—especially the bottomland. One-third of the residents in the five Lakota Sioux reservations along the Missouri were forced to relocate their homes.[120] The river was a massive resource critical to millions, but it was captured by the needs of a few. Worse yet, public dollars would be spent to engineer the river in a way that overlooked environmental and public health. The environmental injustice upstream on the Missouri was part and parcel of flood control, barge navigation, and the Economic

River. City waterworks directors and their urban customers were relegated to secondary status because they would have to adjust to the engineered river.[121]

The adoption of Pick-Sloan underscored that the federal government took responsibility for the Economic River, but not the Healthy River. The bonds of upstreamers and downstreamers would be altered by this massive federal face-lift of the Missouri River, bringing the basin into a new, yet still inseparable, economic and environmental relationship. The Missouri River joins the geographic east and west and Pick and Sloan bound the river with their contesting federal agencies. When the first shovel of dirt was turned in 1946, it was fitting that the basinwide plan's inauguration came on a "barren" spot along the river in the Kansas Cities, because the city's commercial boosters had lobbied so incessantly for this manifestation of the river.[122]

A VIEW FROM THE BLUFFS, 1951

8

On July 17, 1951, President Harry Truman, along with Bess and their daughter, Margaret, paused for photographs before boarding the presidential airplane *Independence*. President Truman waved his Panama hat and ducked inside. Their destination was the Midwest. The Truman family home was in Independence, Missouri, alongside the Missouri River and just downstream from the Kansas Cities, but this trip was prompted by an epic flood on the Kaw and Missouri Rivers. President Truman was going to take an aerial tour of the disaster, drive his motorcade through the Kansas Cities to witness damage in the bottoms, and promise midwesterners his support for recovery and prevention.

The Kaw River basin had received an average year's precipitation by early July, but the waters receded and flood danger appeared to be over. The collective sigh of relief was premature. Heavy rain fell and the river rose again. Upstream in Topeka, volunteers sandbagged the city drinking water plant and airplanes with loudspeakers encouraged evacuation while boats picked up the stranded.[1] Likely evacuees included Linda Brown and her family, who only five months prior had joined a class action lawsuit that challenged racial segregation in the public school system. The ballpark of the Topeka Owls—the racially integrated Chicago Cubs affiliate—went underwater. So did a motel, meat market, gas station, the Goodyear tire plant, and much of the capital city. Thousands of people were being displaced upstream but in the Kansas Cities the Kaw River still appeared to be under control. As the flood gathered volume on the Kaw River, it pushed a half to one million cubic feet of water past the Kansas Cities every second.[2] On July 12, storm drains and sewers began to "spout" as pressure backed them up, but authorities announced that levees were expected to hold and the industrial district was "safe."[3]

In the early hours of Friday, July 13, levees were topped and a flood emergency took hold of the Kansas Cities. Whistles, sirens, and loudspeakers announced danger to those in the bottoms. The Mexican American community in Argentine was one of the first in the path of the rushing water and residents and workers were evacuated. With levees breached in two places, the US Army Corps of Engineers admitted, "The fight is lost," and the National Guard was recalled from its protective labors. At ten in the morning, Mayor William E. Kemp and representatives from the Kansas City Chamber of Commerce and Corps of Engineers flew out of the Kansas Cities and up the Kaw River to survey the damage. Upon return an hour later, "the occupants of the plane gasped," a reporter wrote, "as a new and totally different view of Kansas City, Kansas came into view." Not only were Armourdale and Argentine obliterated by "muddy waste" but the industrial district had succumbed to a "solid brown sheet of raging water."[4] Large barrels bobbed around the Osage Paint and Varnish Company in Armourdale.[5] Businesses such as Procter & Gamble, Kansas City Structural Steel, and Sinclair Oil found "dirty water and stacks of debris" rising half a foot an hour. The planeload of dignitaries landed shortly before the airports closed. The numbers grew: fifteen thousand, then twenty thousand, and finally, by the end of the day, some twenty-five thousand people were homeless. A few hundred refused to go and were rescued from rooftops later.[6] Twelve thousand workers were displaced. The steeple of the Baptist church in Argentine, poking out of the floodwater, was a sentinel for the exiled community.[7] By evening, the mayor and city council declared a state of emergency.

The water situation became tenuous on Friday. The Kansas City, Kansas, water plant was precariously secured with sandbags. On the Missouri side, the Turkey Creek reservoir in the West Bottoms was threatened so the city closed one of its water supply tunnels that ran underneath the Missouri River. To accomplish this task, water department workers had to work in neck-deep water.[8] The Turkey Creek pump station shut down when floodwaters inundated the pumps. The station provided two-thirds of the water for much of the metropolitan region and, immediately, consumers were asked to conserve—if the reservoir went dry, so would city pipes. Industries needing water for operations announced that they would close during the water emergency. With only the smaller East Bottoms pump station in operation, maxed out at about forty million gallons per day, water pressure was low in elevated and outlying areas. Since water was needed for fighting fires, water department director Melvin Hatcher became exasperated that water pressure in mains was so low. "Despite repeated warnings to residents that every effort must be made to conserve water, consumption is continuing at [a] high level," he said.[9]

City manager L. P. Cookingham's suburban home had dry faucets, as did large portions of the city. For a few days, tanker trucks distributed water to advertised locations in both Kansas Cities for consumers to pick up. Everyone was advised to boil water for twenty minutes before drinking or cooking. Tests showed that the water was still safe, but boiling was a precaution reflecting the risk of contamination. Had the levees broken in North Kansas City, as they threatened to do, then water would have bypassed treatment, and the plant would have pumped impure flood waters into city mains.[10]

It continued to rain on Saturday and North Kansas City was evacuated. The Fairfax industrial district in Kansas flooded and General Motors and Phillips Refinery closed up shop. Fairfax, which was a major production center for the American defense industry, succumbed when a sewer exploded and gushing backwater eroded a levee.[11] Thousands of volunteers sandbagged and the airport levees were plugged with the bodies of junked automobiles. Electricity was sporadic in low-lying districts and telephone service was disrupted because the lines crossed the river. Trains were stalled and troops en route to the Korean War found themselves on a long layover in the heart of America.

The stockyards and packinghouses were again hit hard. As in 1903, Kansas City held a large share of the national meatpacking market.[12]

About ten thousand hogs and cattle drowned but just as many survived.[13] Some cattle were lassoed and pulled to higher ground. Hogs were moved into the upper floors of concrete packinghouse structures. Some 10,500 animals were thus made safe and workers returned over the next few days by boat and truck to care for them.[14] Some waterborne hogs, struggling to survive, provided a drama for onlookers. From West Terrace Park, spectators shouted encouragement to hogs that fought the current to find refuge. "That's the last car, buddy, you'd better grab on," one man advised a swimming hog. When the hog successfully climbed aboard a railcar roof, the crowd cheered.[15] The city warned the public not to salvage any live hogs because they might have come from a serum company—the Kansas Cities were the leading manufacturer of vaccines and drugs for hogs—and might be unsafe to eat.[16]

The waterworks of both Kansas Cities survived, though residents certainly were threatened with being severed from their lifelines to the Missouri River. The walls of the huge intake upstream withstood the pressure of high water, luckily, because inside were the pumps that drew the river into the plant for settling and treatment.[17] One week after the flood's peak, dropping water levels exposed the pumps at Turkey Creek and the water department started the coal-powered motors. Boiling was still urged and restrictions against watering lawns, washing cars, or filling swimming pools continued.[18] While the Kansas Cities did not lose drinking water, many cities upstream on the Kaw River did. In case of "water emergencies," the Kansas board of health maintained portable water treatment facilities and, for the three months following the flood, these mobile treatment plants supplied the thirty-seven cities in Kansas that lost water.[19]

The risk to public health in 1951 included water contaminated by sewage and industrial products. For sanitary purposes, washing clothes was encouraged and laundry facilities were made available for flood victims. Typhoid fever prevention was at the ready—the US Navy flew in 250,000 inoculations—though only those coming into contact with floodwaters were advised to receive them. One Kansas City, Kansas, health department gave 1,700 shots the first week. Despite the potential for typhoid fever, the incidence of illness was very low—no cases were reported in the Kansas Cities or Kansas, clearly a sign of better public health management in an

The 1951 floodwaters. The evening shadow of the photographer falls along the tracks by the Quindaro waterworks and Kansas City, Kansas, power plant. To the left, a flooded home and car can be seen, as well as the presumed homeowner in a boat. Burns & McDonnell Library

emergency.[20] Comparing the two floods, former head of the Kansas health board Samuel Crumbine, now aged, assessed that the "insanitary condition of 1903 had been eliminated," for which he thanked "modern sanitary science and scientific medicine."[21] In contrast to the 1903 flood, which had so dramatically affected the health and welfare of Kansas Citians, local, state, and federal public health officials coordinated to weather the 1951 flood without any disease outbreaks.[22]

There were other public health concerns. Rats fleeing floodwaters bit a few children and the public hospital reported treating twenty people for bites. One newspaper reported that forty-two extermination companies would be sent into the bottoms to catch typhus-infected rats.[23] Garbage disposal was curtailed and citizens were advised to cover waste tightly and bury it, if necessary. The thousands of hogs that consumed municipal garbage—not an uncommon system for cities—were drowned at their location in the river bottoms.[24] As soon as water levels dropped, as a substitute for the hogs, the city planned to dispose of garbage in the river, harkening back to the decades that Kansas City commonly used the Missouri River as its garbage receptacle.[25] The process of recovery involved more than waiting for water to recede; photographs after the flood show the piles of ruined goods and debris that thousands of homes and businesses had to deal with. One element that photos do not capture is the stench: the malodorous combination of water, mud, organic debris, and heat made the flood last a lot longer.

The pollution of the 1951 flood was riskier. Whereas one of the industrial spills in 1903 had been linseed oil, five decades later the major concerns were petrochemicals and agricultural chemicals like DDT. The region's oil refineries and suppliers, like Phillips and Sinclair, were alongside the rivers. Hundreds of fifty-gallon drums were afloat, contents unknown. Oil slicks made off for Saint Louis, oil tanker railcars rammed into barges, things exploded, and large refinery tanks were swept away, crashing into bridges. Barges and towboats lost their moorings and two boats came to rest aside the Hannibal Bridge. One petroleum-based fire along Southwest Boulevard raged uncontained for seventy-five hours.[26] The "peril from fumes" in the "oil- and gas-drenched" bottoms required caution, warned the *Times* newspaper.[27] It was forbidden to smoke or even carry matches in the industrial district because of flammable material in floodwaters.

Authority and institutional responsibility were more clear-cut in the 1951 flood. The city limited traffic and the National Guard stood armed, ready to enforce the restrictions of who could enter the disaster area. Thousands of people were coordinated in sandbag efforts. The National Guard and Corps of Engineers operated rescue boats and an emergency headquarters was established at the public works department. Whereas the flood response in 1903 had been decentralized and citizen organizations like the Commercial Club had spearheaded relief, the Red Cross and Salvation Army took responsibility in 1951. Unions aided cleanup. Unemployed workers were channeled into jobs aiding recovery. The challenges of the flood emergency and cleanup were successfully met, prompting Mayor Kemp to boast, "What a spirit!"[28]

Some watched the disaster on television or listened to the radio, while others drove to the bluffs to see. Photographers stood shoulder to shoulder with business owners, some of whom held field glasses and "were generally silent as they surveyed the sweep of the flood."[29] In a radio broadcast, John Thornberry described the stunning damage seen from the air. He gendered his description of the rivers, referring to the levees as protective corsets and saying, "The river was fat and broad—and angry, acting like a drunken harlot, destructive, vicious and immoral." His description brings to mind the prostitutes who were pushed "to the river" to control vice, and the prostitutes' bodies that children downstream feared were polluting the river. Thornberry assured his listeners that the future would be brighter because the Kansas Cities would bounce back, rebuilding to be stronger, by controlling "the mad woman." Control was the key because "this flood situation cannot repeat itself, ever again, if data and history can guide men to fashion stern controls for the now uncontrolled rivers."[30] Thornberry and other journalistic commentators chose language and descriptions that gendered nature as feminine and the control of nature as masculine.[31]

When President Truman flew over the soggy valley, he declared it "one of the worst disasters, I think, that the country has ever suffered from water. I am informed by General Pick that it is the worst."[32] General Lewis Pick and the Corps of Engineers lost no time promoting the Pick-Sloan Plan as the best response to the disaster because unless the basin continued to advance the plan, massive floods would repeat themselves. If dams and reservoirs upstream on the Kaw had already been built, this would not have happened, Pick said, exalting the ability of the corps to solve any problem.[33] Upstream on the tributary Kaw, the proposed Tuttle Creek Reservoir received rural opposition. One critic argued that river management was an urban coup because the "real purpose" of impounding water was to help "flush out Kansas City's sewage."[34] Had this contentious reservoir been in place, the Corps claimed, the damage could have been lessened. Colonel Pick took advantage of flood time, a time amenable to such confident claims, to present this as truth. Cities and politicians contributed to a swell of support for formerly contentious projects, particularly the Tuttle Creek Reservoir.

Under the title "Never Again!" the *Star* editorialized: "The long and costly period of soft glove boxing over Missouri valley flood control has been washed out with terrible devastation."[35] The debate over river development in the Kansas City region subsided and a clear winner emerged: more reservoirs, higher levees, and straighter channels. Four years later, *Life Magazine* proclaimed the "Big Muddy Tamed at Last." The article praises massive engineering projects in the basin that subdued the "wild" river.[36] The 1951 flood provided a cause célèbre to control the river; the hubris to achieve flood control and navigation would remake the watershed.

Although Pick and avid economic boosters were dominant in their campaign, in small circles the flooding of 1951 caused a reassessment of the engineering plan. It was already understood that flood control structures like levees restricted water and could *exacerbate* a flood. This had been discussed in regard to the Kansas Cities after the 1903 flood and on urban waterways like Brush and Turkey Creeks. A Kansas City newspaper in the 1930s had reported that a flood with the volume of 1903 would actually be six feet higher.[37] The 1951 flood had a lesser volume of water than 1903, yet it rose to a higher level and it did more damage. A corps historian has argued, however, that because Pick-Sloan was not fully implemented, the flood proved the weaknesses of not yet having upstream reservoirs and higher levees.[38] Importantly, most people—engineers and residents alike—believed the river could be controlled.[39] Ironically, all the flood prevention on the Kaw and Missouri caused restrictions that served to more quickly channel greater volumes of water toward the confluence and the Kansas Cities.[40]

Residents congratulated themselves in the following months that, in the face of one of the worst national disasters, "the Kansas City Spirit came to the fore . . . as it has *always* come forth in the past." As with previous fires and floods, "Kansas City has overcome its obstacle and is continuing to forge ahead into its second century as a modern, progressive American city . . . a city with a future as promising as the spirit it embodies."[41] Impressed by the character of the city in cleanup, Norman Rockwell joined forces with illustrator John Atherton and the Kansas City–based Hallmark Company to create the painting *The Kansas City Spirit*. Prints were sold to raise money for relief. The painting depicts the symbols of Kansas City: the art deco New Deal skyline, grain and cattle representing the hub of midwestern agricultural products, an airplane flying in the background as a reminder of the downtown airport and manufacture of B-25 bombers for World War II, and, finally, a man. He looks like an engineer. He's a no-nonsense professional, strong and clean-cut, and he's rolling up his sleeves, ready to get to work. In one hand he firmly holds a sheaf of blueprints, symbolizing Kansas City's planning tradition and "get it done" spirit of recovery.[42]

Rockwell and Atherton's painting was an optimistic view of the river. Another well-known artist painted a different perspective. In 1951 Kansas City artist Thomas Hart Benton painted *Flood Disaster*, depicting the Kansas City bottoms in the same flood. Benton portrays the suffering of the average person. His canvas has buildings, household goods, cars, and trees all caught in a muddy upheaval. A stricken family is returning to the scene of a destroyed home and their uprooted lives. Copies of this painting labeled "Homecoming—Kaw Valley 1951" were sent to members of Congress to encourage funding flood relief along with Benton's entreaty to "get out a new bill which will relieve the human side of this rotting catastrophe."[43] Whereas Rockwell and Atherton's painting depicted the flood as an economic opportunity, Benton hoped to appeal to legislators' compassion and to depict it as a social disaster.

The focus in 1951 was more on industrial loss and less on "refugees," despite the fact that a similar number of people were forced out of their homes and became unemployed. A significant working-class population had remained living in the bottoms after 1903 but many people did not return to the bottoms after 1951. The Colgan family was one such sign of the times. Mr. and Mrs. William Colgan had lived in Armourdale for over fifty years. Their two-story frame house survived all the floods, including 1903. They had rebuilt their lives in previous years but the couple, in their eighties, finally decided that the "old Kaw has them whipped" and they were not returning to their home with the stalwart cottonwood tree in the yard.[44]

Norman Rockwell and John Atherton, *The Kansas City Spirit*, 1951. Used with permission from the Hallmark Corporation, the Rockwell Family Agency, and the Atherton-Varchaver Family

Thomas Hart Benton, *Flood Disaster*, 1951. Art © T. H. Benton and R. P. Benton Testamentary Trusts/UMB Bank Trustee/Licensed by VAGA, New York, New York

The flood happened on the cusp of larger demographic and economic changes in America. The 1950s were an era of "slum clearance," public housing projects, and expressway building in the Kansas Cities. The displacement of impoverished people inadvertently aided in "clearance," and the bottoms were deemed a good place for public housing complexes, which continued the association of the bottoms with the poor. A pamphlet advertising low-rent homes for three thousand families in Armourdale assured "economy, flood-proof, fire-proof housing for flood-stricken families."[45] Because of segregation, poor and minority populations predominated in the urban core of the Kansas Cities while the white middle-class majority dominated the suburbs. This "white flight" happened in conjunction with the 1954 Supreme Court decision *Brown v. Board of Education*, the Topeka case named for the Brown family.[46] The ruling required integration of public schools, but many whites chose to move, using the coded phrases "good schools" and "safe neighborhoods." Just as the federal government was investing in the Economic River, so the collusion of real estate, local planning, and then federal policy was creating class

Public housing "improvements" after World War II. Public housing in the Kansas Cities proliferated in the river bottoms, including formerly industrial areas from Argentine to Fairfax that were flooded in 1951. High-rises, though criticized now, were a popular style. Wayne Miner Housing, built in an African American district near Eighteenth and Vine, is representative of such developments. Dorothea Eldridge Collection, Missouri Valley Special Collections, Kansas City Public Library, Kansas City, Missouri

and racial segregation in postwar America. The GI Bill, Federal Housing Administration, Homeowners Loan Corporation, and practices like "redlining" all encouraged white suburbanization, along with the 1956 Highway Act. The decline of the urban core included a decline in industrial production.[47] With fewer people, the bottoms were given over to an industrial landscape, and even that became a little ghostly, because some businesses did not return. The meatpacking industry was already frail; its death certificate was signed by the flood. In sum, the 1951 flood sped the process of urban decline and suburban segregation so that by the end of the twentieth century, the Kansas Cities had become one of the most racially segregated in America.

In the comparison to 1903, most striking may be that the story of the 1951 flood is repetitive. It affected homes, transportation, industry, and agriculture in similar ways. Basic needs and city services like the water supply were threatened. Both times there was raw sewage, but industrial pollutants were a greater danger to human health.[48] Public health was not as tenuous; fewer lives were lost and disease risk was reduced.[49] But the flood of 1951 was costlier, with about $1 billion in property damage in the basin. Shortly after his visit, President Truman requested a relief bill. Local relief was not enough, he said, especially when the national food supply of meat, wheat, and corn depended on this region. Truman also reminded Congress that funding flood prevention and flood disaster insurance was necessary.[50] Finally, in October Congress authorized millions for flood recovery. Many who voted for relief also sent Benton letters thanking him for his humanitarian artwork—except for the Iowa representative who sent a long letter explaining why he wouldn't vote to send a penny of funding to Kansas City. He assumed the infamous Pendergast machine would get hold of it.

Because of the Kansas Cities' importance as a transportation hub and one of the largest markets for major commodities, the national impact justified the financial outlay.[51] The basin above Kansas City would be hit by another record flood in 1952, devastating cities like Sioux City, Omaha, and Council Bluffs. The timing of these floods bolstered the investment in the Pick-Sloan Missouri Basin Plan.[52] Reservoirs upstream and taller levees provided a sense of security that encouraged valuable businesses and industries to continue locating in river districts. However, the cities would not be completely protected at the confluence.[53] In a critique that is becoming more common, such floods are an "unnatural disaster," meaning that human actions and policies turned a natural flood into a social and economic disaster.[54]

The 1951 flood marked the city's turn away from the river. In fact, the daily, intimate connection with the river continued; the river still ran through the city and its people and gurgled from fountains, but Kansas Citians in the years to come would barely be aware of their rivers as they sped over them on bridges. The advent of the Pick-Sloan Plan and the revamped river paralleled the urban disconnect; as the river was restricted and redirected in service of barge navigation, Kansas City lost interest in being a river city, its bottoms inhospitable to almost anything but public housing, industry, and outfalls.[55]

9

DOWNSTREAMERS AND POSTWAR POLLUTION IN A FEDERAL ERA

> Indeed one of the most alarming aspects of the chemical pollution of water is the fact that here—in the river or lake or reservoir, or for that matter in a glass of water served at your dinner table—are mingled chemicals that no responsible chemist would think of combining in his laboratory.
> Rachel Carson, Silent Spring, 1962

Melvin P. Hatcher had his hands full. As Kansas City's water director from 1943–1959, Hatcher got his department through the wartime labor and fuel shortages, and through the 1944 and 1951 floods. In the midst of baby-booming suburbanization, Hatcher converted the Missouri River into city water that quenched the thirst of the Kentucky bluegrass lawns and ran the appliances of postwar prosperity—like washing machines and air conditioners. When demand outpaced supply, new communities experienced low water pressure and the waterworks had to institute restrictions. Sometimes Hatcher's department sent out patrols in the middle of the night to catch people watering lawns. He oversaw millions of dollars in upgrades that increased pumping capacity and supported the growing metropolitan region.

The Kansas City, Missouri, waterworks consolidated water services as it expanded its infrastructural footprint to the sprawling metropolitan area. Hatcher's department had a nationally recognized method for assessing water rates, one that accounted for distance pumped and amount used. Although extending the urban innards was efficient, some customers bristled at the growing influence of Hatcher's department. One downstream water company accused Kansas City of polluting the Missouri River with its sewage, thus ruining it for downstreamers, and then offering to sell clean drinking water to those same communities at a high rate. In 1954 pollution caused the downstream city of Independence to build a new plant that drew on wells instead of the river. Pollution was a driving force in the decisions sanitary engineers like Hatcher made but, unfortunately, preventing pollution was not an important goal of federal river management.

As Hatcher was keenly aware, amid all the engineering attention, water quality problems resurfaced along the Missouri River. Hatcher and the Kansas City waterworks were the arbiter between the river and a growing population. Hatcher took his public responsibility seriously and he found himself greatly challenged to predict and manage the Missouri River. Hatcher's job—to facilitate safe drinking water in spite of postwar pollution—was not made easier by the presence of federal spending on the river. It was the responsibility of Hatcher and other public health officials to give voice to the Healthy River.

Facilitating River Health

In the 1950s the US Public Health Service (USPHS) renewed interest in the health of the Missouri River, holding hearings and issuing reports. Since the recommendations in the agency's first report in 1913, little had changed. The attention to water quality came on the heels of the 1948 Federal Water Pollution Control Act (FWPCA), which was the first major piece of federal legislation that addressed water pollution and was a precursor to the Clean Water Act. Thus, the FWPCA was the first law to give legal authority to the federal government to set a national water quality policy.[1] However, the FWPCA did not provide the legal teeth to enforce water quality. Under the act, federal entities created boards that urged different river users, like cities, states, and industry, to take responsibility and action. In the 1950s the USPHS worked with these stakeholders in "enforcement conferences" to develop plans for constructing treatment plants.[2] The agency could "recommend," it could propose programs and foster interstate plans, it could take legal action, but it simply could not enforce sanitary standards because the responsibilities to act in interstate matters of water pollution lay with the states.[3] Hence, enforcement was slow in the 1950s.

The FWPCA initiated a flood of studies about river health. In 1948 Congress authorized a water quality study on the stretch of river from Yankton, South Dakota, to the Kansas Cities. In 1950 the USPHS finished two studies in the Missouri basin—one on the South Platte River and the other on the Kaw—with the goal of using the findings for pollution abatement.[4] In 1952 the USPHS studied the impact of the existing and planned mainstem dams and reservoirs and concluded that there were both benefits and detriments. It found that dams could have positive ramifications for drinking water if flows were regulated for water quality. Beginning in 1952, Hatcher of the Kansas City waterworks, the state of Missouri, and the USPHS asked the Corps of Engineers to increase flow in order to dilute pollution.[5] Increased flow for navigation in the summer kept oxygen levels higher and increased the likelihood of dilution. Still, year-round releases from reservoirs did not provide enough dilution and bacteria counts rose.[6] Decreased sediment load could be a benefit of the dams, but taste and odor problems from algae were negative attributes.[7]

State and federal reports descriptively discussed the scum, grit, sludge, detritus, cake, influent, overflow, lagoons, solids putrefaction, and putrescibility of pollution and the river.[8] In 1953 a conference led by the USPHS and attended by the lower basin states recommended "a uniform program for pollution abatement along the mainstem of the Missouri River."[9] In conjunction with the ten river basin states, the USPHS officially adopted a "comprehensive" abatement program, and meetings set forth ways to achieve cooperative alignment.[10] Yet another report begun in 1955 punctuated the fact that the lower Missouri River was grossly contaminated with sewage. Such levels were thought to be harmful to people using the river for recreation, and the levels were so high that water treatment methods may not have been adequate.[11]

Amended in 1956, the FWPCA was given more teeth; it was now able to make requirements of interstate waters and of involved states without their consent.[12] Resuming under the newly created Department of Health,

Education and Welfare, the USPHS began a decade of tortuous hearings in the basin to achieve compliance. Conferences in 1957 and 1959 focused on lower basin river pollution in Saint Joseph, Kansas City, and Sioux City. The question of federal power and states' rights remained central to the debates over water quality standards and pollution abatement.[13]

Murray Stein of the USPHS, the chief of enforcement for water quality, was the ideal leader in the transition to pollution control. As a representative of the federal government, Stein was a negotiator and not a dictator. While some thought his methods soft, Stein defended his tactics, saying that abatement solutions should be arrived at through consensus. Stein criticized the work of scientists as "mental gymnastics"—achieving professional criteria without outcomes. He entreated scientists to engage in public service, to join the efforts to clean up waterways, and encouraged them to keep their narrow professional criteria in check, just like he expected of his stakeholders during conference proceedings.[14]

Chairperson Stein opened the 1957 conference on interstate river pollution, held in Saint Joseph, by recapping the relationship between state and federal entities. "Federal law recognizes that the states have the primary rights and responsibilities," said Stein, but the USPHS had responsibilities, too.[15] Stein then turned to Glen Hopkins, a USPHS engineer, and asked questions "largely for emphasis." According to the Constitution, legal precedence, and the law, the federal government had jurisdiction only over interstate waters. Stein asked Hopkins if the portion of the Missouri River that the conference studied could be defined as interstate waters. "Yes, sir," answered Hopkins. Stein continued: "As you know, pollution of interstate waters which endangers the health or welfare of persons in a State other than that in which the discharge originates is subject to abatement as provided under the Federal Act. In your opinion does the pollution you described fit this definition?" Hopkins answered, "Yes, sir." Stein moved on to the next presenter, content his point had been made to his constituency: the federal government had the responsibility to enforce sanitation on the river.[16]

Since the meetings and reports of the early 1950s achieved little, Hopkins added, "Obviously, the mere adoption of this program will not, in itself, reduce pollution or improve the usefulness of the waters in the

Missouri River Basin."[17] In other words, the USPHS could *facilitate* pollution abatement programs, but states would have to be trusted to follow through with their responsibilities. Kansas and Missouri had divergent responses that reflected their past attitudes toward the river. Kansas welcomed federal involvement and pursued abatement with gusto. Young environmental health pioneer Dwight Metzler organized the first Environmental Engineering Conference, cosponsored by the Kansas State Board of Health and the engineering school at the University of Kansas. Speakers reflected the complexity of "environment." Hatcher participated in these conferences in Lawrence (a Mizzou graduate able to overcome any animus toward Jayhawks!).[18]

While Kansas organized, Missouri state legislators in Jefferson City settled for drinking bottled water. The state capital's city water, drawn from the Missouri River, tasted unpleasant and was of questionable purity.[19] Missouri cities had well-established water quality problems and the state was often on the defensive at these conferences. In 1957 Clifford Summers of Missouri's division of health testified on his state's successes with its water pollution control agency.[20] The state had addressed pollution on its smaller waterways and 70 percent of identified pollution sources had already been abated, he claimed. However, when it came to the larger Missouri River, the state "has not progressed as rapidly," Summers admitted, "and while some preliminary planning has been accomplished, no abatement works on the mainstem have been completed."[21] Missouri was beginning to get organized at the state level to address its water quality issues, but had taken no initiative regarding the Missouri River, which was the most significant problem in scale.

At a 1957 conference led by the USPHS and held in Kansas City with a focus on the metropolitan area, Kansas City, Missouri's vibrant mayor H. Roe Bartle gave lengthy testimony on his city's situation. He reminded the conferees of Kansas City's fervent commitment to the Pick-Sloan Plan. As ardent supporters of economic development, the mayor said, we must therefore support the health of the stream. Mayor Bartle did not beat around the bush about the relationship between the city and the river. The city's municipal and industrial waste "results in a degree of contamination . . . far in excess of any conceivable limit of tolerance." If anyone wanted to know just how bad the situation was, the mayor recommended, "further confirmation" could easily be had with "a short observation of the outfalls of the Turkey Creek or Blue River sewer," indicating that the sight and smell of the two main sewage outfalls would enlighten the observer on just how poor the water quality was.[22] The mayor wasn't joking. Investigations of aquatic life showed that the taste of fish was affected, and fish caught below the Kansas Cities received the lowest scores for taste. Fish placed in cages in the river for study were killed within twenty-four hours by caustic substances that ate the flesh and left only bones. The river bottoms of the Blue River had no aquatic life and the Kaw River had only pollution-tolerant species like "sludge worms." These deleterious ecological effects continued for many miles downstream.[23]

Kansas City's Mayor Bartle outlined positive steps toward abatement, the most significant of which was that the city stopped dumping garbage to the river. The city had never been able to fund an incinerator, there was the setback of the garbage-eating hogs drowning in the flood, and then the state outlawed selling garbage-fed hogs because of a swine virus, making the system unprofitable. In 1953 the city planned to solve its garbage woes by contracting with a hauler to "grind the garbage and flush it" downriver.[24] Finally, the city began to use a sanitary landfill and stopped flushing garbage, with the exception of the waste from pens of the remaining garbage-eating hogs. The other accomplishment was a joint sewerage agreement between the city and a neighboring municipality—a sewer that ignored political boundaries. The mayor assured the hearing board that a feasible plan for abatement was now in the hands of the engineers.[25]

A sewage treatment plant was proposed for the Blue River, and the engineering firm Black & Veatch began preparing a study in the late 1950s. All sewage would go to a central location at the confluence of the Blue and Missouri. The sewage from the second major outlet at Turkey Creek would be pumped to the mouth of the Blue because Turkey Creek did not have enough room for its own treatment plant. At the time Mayor Bartle relayed this information, the study was in an early stage and costs were unknown, but the mayor did throw out cautionary remarks about financing the plant. Money might be a problem, he warned, because the city could not deficit spend and, in order to issue bonds, voters must approve them with a supermajority. He also concluded with a statement about timing, cautioning that Kansas City was not "in a position at this time to commit itself to a schedule for the attainment of this effort."[26] Kansas City, Missouri's leader, popular orator though he was, seemed to say to the USPHS: "when pigs fly," which was in contrast to the problem-solving attitude of Kansas.

Two years later the Kansas board of health proudly testified before the USPHS, harkening back to a history of battling pollution on the Missouri River. The board's testimony quoted Samuel Crumbine's 1909 letter to the mayor of Atchison about sewage in the river. Over the years, Kansas had cooperated with every study and abatement program in the region and had many successes at the state level, contributing to its reputation for public health innovation.[27]

After passage of the FWPCA, Kansas began surveying the Kaw and organized well-attended public meetings, which indicated "a strong interest on the part of citizens in cleaning up streams in Kansas," as a historian noted. Metzler took a watershed approach and the state facilitated adoption of similar laws among up and downstream communities, especially on the Kaw.[28] Kansas vowed in 1950 that every city must at least minimally treat sewage before dumping to a waterway and in 1953 its board of health adopted a policy requiring "primary treatment of all sewage, the elimination of toxic compounds from the waste discharges, notification of all cities involved . . . and review of the progress."[29] Eighty-one of the 129 river communities built their first treatment plants between 1950 and 1960 so that every discharger on the Kaw River treated wastes—either primary or secondary.[30] In almost all cases the voters approved bonds for treatment plants on the first ballot, indicating citizen willingness to pay for better public and environmental health.[31]

The interstate hearings diversified the range of those affected by pollution, including in less readily measured ways like fishing, recreation, and dangers to fish and wildlife. At the 1960 hearing, Stein asked people from a number of different perspectives to share observations and uses of the rivers. Such hearings were used as a "mechanism" to "develop the spectrum of public use desires" because federal policy assumed the river was a public resource that required input from stakeholders.[32] Among the speakers were a duck hunter who told of "big gobs of grease" and an associate city councilor in Kansas City who envisioned developing the waterfront. Benjamin Powers shared with the hearing board his vision of a marina and a restaurant with a window overlooking the river—the river that he then referred to as "disgusting" and a "cesspool." Powers told of people who boated and waterskied, regardless of water quality.[33] When asked to describe the river odors, Claude Relf, a train engineer who worked near the Kaw and Missouri confluence, replied, "I am not allowed to say in public." When prodded, he went on to confirm that he had seen piles of foam, contraceptives, garbage, gas bubbles, and the flushed waste from animals with pustules, and that the river sometimes ran red with blood from the packing plants.[34] Witnesses from the Corps of Engineers—a channel inspector and survey boat operator—talked about how smelly, greasy pollution clogged the cooling systems of the boats. One went on to say that he saw people swimming and fishing in the rivers. When asked to describe the odor, the witness said, "Well I couldn't tell you; just a terrible stink." He added that he saw "human manure" and offal downstream from the Kansas Cities.[35]

Saint Joseph, Missouri, was not the most polluting or populous city on the river, but it received a lot of attention at the USPHS conferences. Proximity to downstream cities was the most important reason; Saint Joseph was not far enough away for proper dilution. Moreover, three of the downstream cities—Leavenworth, Atchison, and Kansas City—were in the state of Kansas, which was active in pollution abatement, and there were ridiculously high bacteria counts near the Kansas City intake. In contrast, no major urban intake was downstream from the Kansas Cities; places like Lexington, Boonville, Jefferson City, or Saint Louis were either too small to be politically powerful or too far downstream to be severely affected. Because Kansas City's downstreamers were in the same state, Missouri was less likely to desire federal intervention to remedy a problem it had jurisdiction over.

The conference transcripts made evident that Saint Joseph, like all municipalities, had to deal with the complications of industry. Saint Joseph's packing industry contributed a larger volume of pollution to the river than did its municipal population. Downstreamers suffered bad tastes, odors, bobbing grease balls the size of footballs, and high bacterial counts.[36]

Commercial fishermen complained that the garbage, condoms, and animal entrails in the river interfered with netting and lowered fish value.[37] Municipal settling basins ended up with debris in them. One of Saint Joseph's largest polluters—a fertilizer plant—was not even hooked up to the sewer system.[38] Describing the danger, the mayor of Kansas City stated that "strong, organic industrial wastes" were complicated by inorganic metals, cyanides, and acids.[39] Cities had to decide how to manage industrial pollution so that discharge met new standards, but industries did not want to be regulated and resisted being connected to city sewer systems.[40] Industries could be required to "pretreat" their waste so that a city treatment plant could predictably deal with what came through the pipes, but with or without treatment, downstreamers had reason to worry. Industrial accidents in the 1960s in Sioux City, Omaha, and Kansas City raised concerns over unexpected events like fish kills that caused environmental damage.[41]

Just as the grease balls grew, so did the investigative discussions; whereas earlier USPHS reports were fewer than one hundred pages, by 1960 the hearing transcript was 511 pages.[42] In the 1960s, with federal facilitation, the four lower basin states began working cooperatively to lay out water quality criteria and take the first steps toward a management plan, but they did not agree on the standards because each state wanted to manage the river for a different purpose. Iowa and Nebraska wanted to manage the river with recreation in mind, arguing that water should always be clean enough to enable bodily contact. Despite this goal of keeping the river safe enough for people to touch it, the recommended safety limits for bacterial counts were frequently surpassed. A Missouri report said that there were "few locations on the river where these standards are not violated."[43]

Public Health in the Shadow of Pick-Sloan

Between the engineering, regulation, and management, the federal government came to the Missouri River on a massive scale in the postwar period. Pick-Sloan caused definitive environmental changes, but those monitoring the river were still learning how to evaluate them. Poor water quality affected a variety of river uses, but drinking water remained the focal point. Aside from its sheer importance, drinking water was discussed because many years of institutional data had accumulated. Most samples were gathered at intakes, which could be misleading; intakes were the least polluted places because they were immediately upstream from the outfalls. Hopkins said that the data from drinking water intakes "suggests that water plants have the most favorable locations under the circumstances," and that "damages to uses other than water supply may be considerably greater than can be measured by treatment problems posed at various municipal water plants."[44]

Twelve years after Pick's promise that basin projects would improve drinking water supply, Kansas City water director Hatcher offered his assessment. At a panel at the American Water Works Association conference in 1958, Hatcher reported that pollution levels in the river had worsened. Bacteria had continued to increase, although, Hatcher qualified, there was no proof that the dams were the cause. The amount of water released from the reservoirs was concerning to downstream cities, especially during winter when river flow was naturally low. Minimum water flows were regulated to provide just enough for city water supplies, although Missouri usually wanted higher flows. Despite dilution, in the seven days it took the water to travel from the lowest dam, Gavins Point, to Kansas City, organic wastes increased. In response, the USPHS had established minimum flow requirements on the river throughout the year in order to provide sufficient dilution. Regardless of regulated releases, cities still worried about keeping their intakes submerged in winter. Turning to benefits, Hatcher said that water hardness and turbidity decreased, because there were fewer minerals and less sediment. Kansas City reported an annual savings of about $84,000 in chemical purchases.[45] Hatcher described decision-making on the river as a "compromise."[46] Public health was not a priority but simply one competing need; waterworks operators had to remain vigilant to protect the quantity and quality of their drinking water supply.

Saint Louis County Water Company manager Herbert O. Hartung, presenting on the same panel, said that Saint Louis did not receive as many benefits as Kansas City because the positive or negative effects of the reservoirs were watered down. Although flow and turbidity were more uniform, Hartung said the reservoirs negatively affected taste and odor, including an increase in ammonia. The third panel presenter, Omaha chemist Joseph F. Erdei, reported that the minimal benefits of Pick-Sloan were counteracted by the industrialization of agriculture, which contributed to an increased algae growth during warm months because of irrigation, runoff, and the use of chemical fertilizers. Erdei also pointed out that the cost savings were negligible in light of what was originally promised. The impact of Pick-Sloan, the panel of waterworks engineers concluded, was a mixed bag for their thirsty basin residents.[47]

Evaluation of the health effects of Pick-Sloan continued. In 1960, at a USPHS hearing on interstate pollution in the Kansas City metropolitan area, the City of Saint Louis water commissioner testified that the data going across his desk made it obvious that the reservoir projects negatively affected water quality. Commissioner Conway Briscoe said that the city plant, now located at Howard Bend on the Missouri River, had suffered distinct taste and odor changes, necessitating more treatment, raising costs, and providing an inferior product to consumers. His opinion was that the lack of silt and sediment load in the river caused "a reduction in the absorptive powers of the river water," meaning that impurities no longer had anything to attach to, making it more difficult to settle, filter, and treat impurities in the city water supply.[48] In response, Briscoe's plant was more careful and they were forced to use more carbon for filtration.

Sensing that commissioner Briscoe felt a great injustice was being done to Saint Louis by the Pick-Sloan installations so far upstream, one of the members of the hearing board asked him if his blame was fair. Briscoe responded that these "minor negative features" of the reservoirs, like a lessened silt load, "should not be overlooked," and continued, "There is no quarrel with the project, but there is reason to feel that an equitable solution or an effective solution will have to be founded on a technically sound program and also one that involves all factors on an economically fair basis." In other words, Briscoe felt that drinking water, a vital public use of the river, was not prioritized in the larger basin plan.[49] The Economic River had taken precedence over the Healthy River.

A few years later, Hartung agreed that less turbidity in the Big Muddy made treatment more difficult.[50] The Missouri had once been described as "milk chocolate" because it went "ceaselessly about its age-old task of transporting the sunrise slope of the Rocky Mountains down to the Gulf of Mexico."[51] Before the mainstem dams were in place, the Kansas City waterworks daily removed at least six hundred tons of suspended matter—equal to filling thirty railcars with mud.[52] The corps estimated that, before Pick-Sloan engineering, 9,094 acres were naturally eroded and redeposited annually.[53] Hartung agreed with earlier reports, saying that all of them suspected silt and clay had aided clarification, and tastes and odors in the water had become more difficult to control.[54] Soon, the science would support the observations of public health officials. The "Big Muddy" was healthiest for public use when it was muddy.

At the 1966 annual conference of the Water Pollution Control Federation in Kansas City, Colonel R. W. Love of the Corps of Engineers engaged in post-Pick-Sloan boosterism when he said, "The Missouri River of today is a waterway that is being transformed from mud to gold, a river whose benefits are established facts rather than distant dreams."[55] Colonel Love's river resembled the healthy, muddy, golden river of earlier navigation booster Champion S. Chase. Refuting the arguments of drinking water suppliers, Love highlighted all the attributes that the dams had for drinking water, including reduced turbidity and increased predictability. Love's presentation repeated the claims of Pick two decades earlier. Nevertheless, Love had to admit that all the economic benefits were negated if "the ugly specter of pollution is allowed to go unchecked." Love supported the work of the federation to "make the public more and more aware of the need to clean up our lakes and rivers and to keep them clean."[56] Here, corps leadership was suggesting that environmental issues were the public's responsibility—and not something that the federally powerful, well-funded Corps of Engineers could do much about.

The New Pollution

Urban environmental historian Martin Melosi calls water supply in the postwar era a "time of unease," due in part to new types of environmental pollution.[57] A Missouri water quality agency reported, "Sampling on the Lower Missouri River during the 1970s has revealed the presence of ten heavy metals, three pesticides and two volatile organics on the priority list." The report mentioned high bacteria counts, but with sewage treatment plants coming online, concern shifted to toxic substances considered a "potential threat." Explained a Missouri official in 1990, "We've swapped one evil for another," referring to the increase of chemical and nonpoint pollution over the decades.[58] While cities struggled to raise the money to abate sewage, waterworks operators were dealing with the new pollution. Reports now spoke of both organic and inorganic wastes and listed metals, chemicals like DDT, and benzene, sulfate, mercury, fertilizers, and petroleum products as chief among the pollutants.[59] World War II and the years following were a watershed for the invention of industrial processes and thousands of new chemicals, many of which found their way into waterways. As the industrial economy and consumer affluence grew, so did waste, creating what Melosi has dubbed "effluent America."[60]

Pollution necessitated watchfulness. Water plants sometimes had to shut down when "slugs" of unknown contaminants made their way downriver, as the manager of the water company in Jefferson City testified. Chlorination needs skyrocketed and the intakes simply had to be closed until the slugs passed.[61] The Boonville water plant also closed "when the river gets so bad," as a water plant representative said, "that there are times when you can step out of your door anywhere in Boonville and it smells almost like walking up to an open septic tank." The only thing between the river and the good health of Boonville, he said, was the water plant's personnel and mechanical equipment. "I look upon my man in charge of our filtration plant as being more important to our community than any doctor," because if he or the equipment failed, health was at risk.[62]

Briscoe, of the Saint Louis city waterworks, when asked to describe the threat to the Saint Louis water supply, said: "I can't, because I am not capable of it." Evidence existed, he said, that the water contained "materials that are very subtle that we . . . are not capable of checking [and] that have at certain places very bad effects upon a community, certain chemicals and the like of that." Briscoe felt it was the duty of the state health department to deal with the ingredients of the river but, in their absence, his waterworks bore the responsibility.[63] Knowledge was incomplete, there was no government oversight, and, in the absence of the precautionary principle, water utility operators found themselves in a precarious position of ignorance.[64]

Hartung, who had worked for two Missouri River cities, explained: "The burden now being placed on the water utility operators for knowing at all times when and if objectionable quantities of pollutants such as synthetic organics or pesticides are present in the river water is overwhelming." Hartung's concern continued: "Pollutants can be wasted into the river either accidentally or intentionally without publicity or permission until discovered by someone who objects. Discovery usually results from damage to water supply quality." Hartung recommended a "pollution patrol and policing corps" to achieve strict enforcement of standards and violations.[65] In the 1960s little was known about the dangers of many of the chemicals, how to test for them, or if purification processes were comprehensive enough.

Twenty years later, engineer Paul Haney, of Black & Veatch, said of the change in risk and responsibility: "In the field of water supply, we have passed from the age of 'microbiology' into the age of 'microchemistry' and this age is getting more 'micro' and more 'chemical' all the time." The work done in the laboratory had changed; public works officials in lab coats could no longer assure the safety of public water supplies.[66] Sanitation no longer hinged on unseen bacteria but on a mystery brew of postwar chemicals.

In response to these new challenges, drinking water suppliers organized a basinwide nonprofit group. Formed in 1961, the Missouri River Public Water Suppliers Association (MRPWSA) sought "cooperative problem-solving." Its purposes were to conduct studies, build a base of data and share information—all with the common goal of "preservation of the Missouri River as a natural resource."[67] One task was to discover the origins of taste and odor problems, and members concluded that industrial processes were the most important contributor. The MRPWSA was a trendsetter and the Kaw basin followed suit with a similar organization.[68] These boundary-crossing organizations reflected the vision of the Healthy River that continued to resurface over the course of the century.

After reading Rachel Carson's 1962 groundbreaker, *Silent Spring*, the American public realized there were gaps in knowledge and a lack of political will to address these challenges. Technological advances like dilution, coagulation, and treatment had maxed out their potential and still pollution prevention was not prioritized. As the industrial economy developed, so did its shadow, environmentalism; historian Richard Andrews writes of postwar America that "demand rose for both material goods and for the environmental amenities threatened by their production."[69] The river provided a way to understand both the challenges and the solutions. Hartung, with a broad definition of environment in 1967, described the river as showing the vital signs of a community:

> Many of the materials used in industry and in community living which pollute our atmosphere or settle as dust on our streets, roads, and lands also ultimately pollute the river. One needs only to speculate a few moments about the final disposal of partially combusted fuels and lubricants, worn tires, and spilled chemicals in order to conclude that air pollution abatement and community housekeeping are also important to the preservation of our waterway.[70]

Hartung saw the Missouri River as a microcosm of the good and bad influences society had on its surroundings; an unhealthy economy made an unhealthy river. He used the phrase "community housekeeping" in much the same way Jane Addams or R. E. McDonnell used "municipal housekeeping." Both of these housekeeping concepts relied on responsible individuals who extended their responsibility to a broader, interconnected community and environment—though they needed political support along the way. The river, then, was an extension of the community and an example of how human and environmental health interrelated.

Crackdown and Treatment

A *Business Week* article in 1960 reported on an impending legal "crackdown" on the Missouri River. Under the amended FWPCA, the federal government was pressing Missouri River cities to stop dumping raw sewage and therefore to build waste treatment plants. The federal government gave legal attention to the Missouri basin, the article explained, because so many cities were so slow to abate. Plenty of American rivers suffered water quality problems, but what distinguished the Missouri River was that it served as drinking water for such a large number of people. The lower basin was gaining notoriety for its resistance to abatement and *Business Week* predicted a showdown between the urban status quo and federal regulations.[71]

Kansas was irritated with Missouri and the "sister cities," who were misbehaving and dragging their feet. The Kansas State Board of Health contacted the USPHS in 1960 to request that something be done about the "failure" of Kansas City, Missouri, to "fulfill the commitments which they made about the treatment of sewage."[72] The Kaw River and its tributaries, a watershed Kansas controlled, had no raw discharges anywhere in the state—a feat of national import—and its cities along the Missouri River were in the process of building treatment plants. Kansas City, Kansas, lagged behind the rest of the state, but it had begun building more sewers and making plans for treatment.[73] The Kansas side followed its bigger sister's lead—or "the slow rate of progress made by its sister city," as a Kansas official told the USPHS—in part because their infrastructure and economy were integrated. The "well-publicized lack of progress" on the Missouri side slowed down the entire region, much to the chagrin of Kansas.[74] Overall, the 1960s were a period of cooperation between the sister cities, and, to the annoyance of the Kansas board of health, the Kansas Cities cooperatively stalled pollution abatement.

While Kansas was vocal about its disapproval of Missouri and the Kansas Cities, even the Missouri Water Pollution Board (MWPB) admitted that Kansas City was disobedient, with progress at a "standstill." Kansas City had ignored the board's demands to reduce its pollution contribution to the Missouri River.[75] To keep from increasing the waste load, the MWPB threatened not to issue building permits to Kansas City or Raytown, which would limit urban development.[76] Frustrated, the MWPB hoped for federal intervention to rein in its renegade city.[77] Utterly slow to treat sewage and repeatedly missing deadlines, the Kansas Cities and the Missouri River were thus the potential targets in a court case testing the FWPCA by the early 1960s.[78] Important to the "crackdown" had been amendments to the FWPCA that gave federal agencies more power to set standards and force state compliance.

Wastewater treatment plants at the confluence. This 1960s picture of the West Bottoms taken from city hall shows open land, cleared after the 1951 flood, and the future location of the sewage treatment plants of both Kansas Cities. Dorothea Eldridge Collection, Missouri Valley Special Collections, Kansas City Public Library, Kansas City, Missouri

A major stumbling block to treatment facilities for Missouri cities was a state law making it difficult to pass bond measures. The state required a supermajority on general obligation bonds and it was difficult to get two-thirds of voters to approve multimillion dollar proposals. This law, written into the state constitution of 1875, was a backlash against burdensome railroad debts that saddled communities after the Civil War. This hesitancy to take on public debt has contributed to Missouri's "antitax" reputation.[79] Further complicating matters, Mayor Bartle chose to oppose one of Kansas City's bonds because it prioritized building sewers but was not comprehensive enough for sewage treatment.[80] The USPHS had ordered Kansas City to meet primary treatment requirements and required industry to join municipal sewer systems by 1962. None of the political entities in the Kansas City region made positive steps toward abatement within that timeline and the date by which contracts must be let was extended to 1963. The responsibility for cheerleading two-thirds of Kansas Citians to vote for sewerage bonds fell to the next mayor, Ilus "Ike" Davis. In the words of his colleague, Davis's "favorite topic seemed to be unexciting infrastructure such as sewers. Early on, it nearly cost him his mayorship."[81] North Kansas City was the first to build a primary treatment plant in 1964 but the rest of the metropolitan area still had not met the federal timeline. Federal grants helped sweeten Kansas City's bitter pill.[82]

Employing its growing powers, the USPHS both coaxed and scolded to get cities to plan for treatment. It ordered Saint Joseph to build a primary treatment plant by 1961, but when industries and voters did not support bonds to build the plant the target date was extended to 1963. Meanwhile, meatpacking and other industries planned for a separate primary treatment plant. The secretary of Health, Education and Welfare (the department in which the USPHS was housed in 1953) told Saint Joseph that it must "cease and desist" dumping raw waste, but the city and its industries did not comply. After the first steps in a federal court action were taken, Saint Joseph finally built a primary treatment facility. The case would be dropped in 1970, despite the fact that a percentage of the city's waste still bypassed treatment. Secondary treatment was still on the horizon.[83]

By 1965, the Kansas Cities still had not met their obligations for waste treatment and the date to comply was extended to 1967. Kansas City, Kansas, approved bonds and its first primary sewage treatment plant went into operation in 1968. However, its incomplete sewerage system caused the Department of the Interior (where Clean Water Act oversight was transferred) to issue a violation notification in 1970. In his State of the Union speech, President Richard Nixon mentioned the "Big Muddy" and taking responsibility for healthy waterways.[84] With the help of state and federal subsidies, most of metropolitan Kansas City achieved primary treatment by the 1970s. In addition, the Kansas City, Missouri, Blue River treatment plant had "agreements" with twenty-seven other municipalities whose pipes covered some five hundred square miles. The next stage in federal requirements was secondary treatment and Kansas City had to gear up for that.[85] The site of the plant at the base of the Blue River was the same spot Kansas City had considered earlier in the century.

Kansas City lagged furthest behind, but other river cities were not much better. Saint Louis's primary treatment plants did not go online until 1970.[86] Upstream, both Sioux City and Omaha struggled to accomplish their goals under the new federal water quality laws. At first Sioux City resisted sewage treatment, but, seeing the light after numerous meetings, it garnered accolades from the hearing board. The caveat was that meatpacking industries then moving to Sioux City did not meet new treatment standards. Also missing deadlines, Omaha finally had primary treatment by 1969, although it was "experiencing operational difficulties" because it could not adequately deal with industrial and packinghouse waste. Its goal for comprehensive secondary treatment would become 1978 while, across the river, Council Bluffs aimed for 1973. The Iowa, Kansas, and Missouri boards of health all issued edicts that secondary sewage treatment must be in place by December 31, 1975, which was the absolute final date required by the EPA as well.[87]

Murray Stein, who had presided over dozens of meetings on the Missouri and other waterways, began a 1965 hearing on interstate pollution by putting the work into perspective. In ten years, the Missouri River had seen much change since its pollution apex. Stein congratulated all involved; the battle for sewage treatment was nearly won and sanitary control would be achieved. While there was initial resistance, he recalled, success had come one city at a time. "Not only are we agreed on what should be done," said Stein, "but every city up and down the river, large and small, is either engaged in construction or a program of construction, and some of them have completed their works for pollution abatement. I think we have made tremendous progress."[88]

The Revolution

By the late 1960s the legal tide and popular sentiment had turned against pollution. Presenting at the Water Pollution Control Federation's conference in 1967, lawyer Mitchell Wendell summarized two main trends influencing water quality: federal power to regulate pollution, and public attitudes that did not accept polluted waterways. The thinking had changed so quickly that this was nothing short of a "revolution," he said, pointing out the differences between his generation and the sensibilities of a younger generation influenced by the growing environmental movement.[89]

Federal entities concerned with health and environment showed a spate of interest in the river in the 1950s and 1960s but had little power to follow through. Looking back from 1985, Haney said these were "termed the 'ping pong' years" because authority was not clearly consolidated. After decades of jurisdictional disputes that weakened abatement efforts, the power to control water quality on interstate rivers was more firmly in the gentle hands of Chairman Stein and federal entities by the late 1960s and 1970s. Perhaps because Americans have some historic discomfort with a powerful federal government, Stein treated the local and state-level officials with deference and did not wield an iron fist.

The federal government gained the legal strength to enforce environmental and public health through legislation like the 1969 National Environmental Policy Act, which led to the creation of the Environmental Protection Agency (EPA). To reduce the fragmenting of environmental laws, the FWPCA was amended in 1972 and a number of laws, including the Water Quality Act, Safe Drinking Water Act, and Clean Water Act, were brought together under the EPA. Success came not only through bureaucratic consolidation of responsibility but because the public so strongly supported abatement and protection. "Water quality protection," said Haney, "became newsworthy and advanced considerably on the public's priority list of things to be done."[90] The revolution at the federal level had strong support from the grassroots—indeed, public support for federal control and its outcomes was at the heart of the revolution.

In 1970 the first Earth Day almost spontaneously erupted with teach-ins and demonstrations on college campuses and in communities throughout the nation, an event symbolic of the popularity of environmental issues. There was continuity in the concern for places people lived, worked, and played, only now those basics were grounded in an array of federal laws.[91] To this day, the EPA sets standards but leaves the states responsible for meeting or exceeding them, reflecting the historic hesitancy to fully empower the federal government in matters of health. A broad-based support for clean water continues, though the public's working knowledge of policy is lacking.

The 1970s also marked the first, albeit slow, attempt to address nonpoint sources of pollution.[92] In contrast to point source pollution, which has an identifiable place of origin, nonpoint sources are more difficult to discern and control. Examples of nonpoint source pollution include rainstorm runoff from farms, cities, and roads, and leaching lawn-care chemicals. Even if point source pollution could be prevented, urban streams like the Blue River, owing to nonpoint pollution, were never expected to be healthy enough for human contact and were written off altogether.[93] In fact, in the 1960s and 1970s, the primary or secondary treatment of sewage was found to do little to change the pollution levels in the Missouri River around Kansas City because industrial and nonpoint source pollution had a greater impact on water quality. American rivers were so polluted that the water returned to the river from waste treatment plants was often cleaner than the rivers themselves.

Despite public support for the legal revolution of the 1970s, changes on the ground and in rivers were slow to come. The EPA assessed the years of multilevel work to clean the Missouri River, saying that resistance to abatement was strong and "rhetoric outweighed actions." Decades of wishful thinking and an admonition for the future were captured in the EPA's report title: *Everyone Can't Live Upstream*.[94] A polling favorability for clean rivers did not translate into less pollution and the report's title reminded readers that *everyone* was a downstreamer, suggesting that the problem was partly cultural. Perhaps conceptualizing one's place in the watershed and within a set of concentric relationships could help residents see themselves as part of the problem and the solution.

As has been evident, cities were not anxious to meet their new obligations. A combination of bottom-up support and top-down enforcement was the key to better sanitation on the Missouri River and, in the wake of failed efforts like the Missouri River Sanitary Conference of 1910 and a continuing lack of interstate cooperation, the federal government became the leader between the 1950s and 1980s. Pollution—its creation and prevention—became "a metaphor for the quality of daily life," as historian Hal Rothman put it, and government came to be seen as the protector of that quality.[95] Not until federal laws forced states and cities to abate pollution did Missouri River water quality improve. However, the concerns originated with public health officials at the local level.

It was the federal government and boosters that created the Economic River and it was the federal government that, with the support and demands of the public near the end of the century, was making the first strides toward the Healthy River. In the last decades of the twentieth century, both visions of the Missouri River were in play. Many of the changes among the public and federal agencies could be traced to the ecological turn in the sciences, which influenced research scientists, policymakers, and environmentalists.[96] Increasing concern over ecological health would lead to a questioning of the costs of the Economic River. Perhaps the best explanation of how to quantify a healthy river is through the concept of ecosystem services—the idea that natural processes have economic value to humans. The Healthy River, as defined by wildlife biologists, environmental organizations, and those pressing for ecological restoration, had the possibility of allying itself with historical precedents like the MRSC, Crumbine, MRPWSA, McDonnell, Hatcher, and Hartung. Plus, a burgeoning number of basin residents were anxious to rediscover the Missouri's "human face," as one journalist put it.[97]

The twentieth century saw a change in pollution on the Missouri whereby sewage became child's play compared to agricultural and industrial chemicals and nonpoint pollution. Protection of public health was far less tangible in the postwar era. Some of the earliest testing of treated water occurred in Kansas City in the 1970s and showed that pharmaceuticals (like birth control or heart medicine) and other drugs regularly flushed down the toilet (like caffeine) were present in the sewage. Once the water has been treated and returned to the river, the long-term effects of these chemicals are still unknown, though amphibians are showing effects from endocrine-disrupting compounds. Compared to the simplicity of sewage and typhoid fever, knowledge of pollution became vague and unspecific with possible long-term dangers. Writing in 1967, early in the federal era, Hartung summed it up nicely: "The primary and most important use of the Missouri River is for drinking water and other public water supply uses. The health and welfare of more than 3,000,000 people are unquestionably dependent on this water. No other use of the river should take precedence over this fundamental use."[98]

Regulation was necessary, and yet only provided limited success.[99] Most contentious was federal control over *interstate* water quality, which touched off a states' rights debate, led by industry.[100] Regulations, business argued, would stifle the economy and decrease profits. Public funds from cities (in the form of taxes, bond measures, and state and federal grants) covered the costs of ameliorating industrial effluent, in addition to their own wastes. The burden for pollution was on the public. From the perspective of twentieth-century public health officials and environmental engineers who found themselves on the frontlines—like Crumbine, McDonnell, Hatcher, and Hartung—cities and states failed to protect water quality and it was time that federal jurisdiction to protect public health be granted.

To this point, cities and states had been happy to accept federal dollars for economic development. The entire redevelopment of American rivers in the "big dam" era and all the engineering for flood control and navigation were done to expand and protect the American economy. But, argued sanitarians, shouldn't public health be part of economic wealth? Does good health not lead to wealth? Seeing concentric circles of relationships within a watershed makes it easier to see that no actor uses the river in isolation. It was easy for cities and states to see themselves as downstreamers but it was more difficult to recognize that everyone was an upstreamer who shared responsibility for the Healthy River.

10

CONCLUDING WITH A VIEW FROM THE RIVER

The lines are themselves only ideas.
   Donella Meadows, "Lines in the Mind, Not in the World" (1987)

On a crisp autumn day in 2012, on the eve of a Missouri River Relief cleanup at Kaw Point, Vicki Richmond stands in the middle of an industrial building in the West Bottoms. The building at 800 Woodswether Road is a former hog serum production facility—likely one that couldn't save its bacon in flood time.[1] The grid from the hog crates is still visible on the concrete floors. It is now the headquarters of her newly established Healthy Rivers Partnership and she is elbow-greasing it into a usable workspace for river activists, volunteers, and students. From a kitchen-table operation in the 1990s to this five-thousand-square-foot brick building, Richmond has seen a lot of trash. This self-described "trashologist" has pulled her share of tires out of the river and she has helped thousands of Kansas Citians get down alongside their waterways and haul out tons and tons of "nonnative species" like Styrofoam, kitchen appliances, car parts, wiring, and the ubiquitous plastic bottle. From where Richmond stands in the Healthy Rivers building, the river is only a stone's throw away, but it is not accessible because a floodwall obstructs it. This disconnect is symbolic of the relationship that Richmond works to change. This old building represents the evolution of the river bottoms, from manufacturing and warehousing dependent on the river to

Vicki Richmond organizing a Missouri River Relief and Healthy Rivers Partnership cleanup, based at Kaw Point Park in 2012. Photo by author

drug lair in an abandoned district, and now to nonprofit educational space and a home base for river cleanups as residents rediscover the river.

Richmond's first river cleanup was with her family on the Blue River in 1993. Inspired, she left her corporate job to pursue her vision of connecting Kansas Citians to their waterways. River cleanups are evidence of renewed interest in the river and, with successes in Kansas City, Richmond is now eyeing the basin. Concentric rings of awareness help Richmond and volunteers connect, from their backyard streams to the watershed. Upriver, cities like Omaha and Sioux City have adjusted their relationships to the Missouri, embracing it with infrastructure like pedestrian bridges, and designing green spaces that attract people for weddings, festivals, recreation, and soccer games. Around the world, postindustrial cities are revitalizing their riverscapes. Kansas City's evolving relationship—in love, out of love, and now flirting with renewed love—is a typical urban story except that Kansas City represents the extremes of affection and estrangement.

Love Faded

Kansas City started with respect for its providential location at the bend in the river and recognition that its waters were central to its identity, health, and wealth. However, after establishing and advertising itself as a "river city," Kansas City became forgetful. A developing cultural indifference was evident to an outsider by the late 1920s: when Lewis Freeman wrote in *National Geographic* about his travels in a small boat down the Missouri, he commented that Kansas City "clings partially to the river." The heart of America was then known as a rail center, but Freeman explained, almost apologetically, that Kansas City "has tried not to turn entirely from the river that served it so well."[2] A rift in the relationship was already visible to Freeman but the soggy break from the river was the Pick-Sloan Plan and the flood of 1951. After the flood, the bottoms were littered with abandoned businesses and homes; instead of making repairs, people moved. Increasingly, the city associated the river with its past. An image from the public library made in the 1950s splices two images of downtown Kansas City together (see photo on page 12). The view from the river is of a nineteenth-century city being built on the banks, dependent on and interactive with the river. The image of the modern 1950s skyline, however, hovers above, on the bluffs, detached from the river of old.

By midcentury many Kansas Citians already thought of the river bottoms as undesirable and of the river as dirty and unsafe. While suburbs boomed, declining investments reinforced the shift of economic and social power away from the river. Beginning early in the century, the average white, middle-class resident associated the river districts of both Kansas Cities with an immigrant working class, poverty, saloons, and the vice district. Those assumptions have stuck to the bottoms like the muddy silt left after a flood. Today's metropolitan residents (who may consider themselves separate from the city proper) still harbor these same notions about the river environment, but are not likely to realize the origins of their assumptions, and not likely to know how the social history of Kansas City figured into the coding of places like the bottoms. Kansas City has been marked by segregation, by the inequity of urban infrastructure, and by class distinctions. What stands between the average white Kansas

Citian and the river? Literally, the bulk of the African American and Latino population, households with below-average incomes, and an infrastructure lacking investment. A majority of African Americans lived in or near river districts and segregation has kept them there. Power shifted south from the river to the higher elevations of downtown, to J. C. Nichols's developments in the 1920s, then to the growing suburbs after World War II, and Johnson County, Kansas, since the 1970s. Though the geography of power has shifted, the river remains essential.[3]

The built environment has reinforced the city's alienation from the river. Under the Pick-Sloan Plan, the US Army Corps of Engineers built levees and walls for flood control, and channelized and built wing dikes for navigation—infrastructure that put barriers between people and the river. The Healthy Rivers building is next to the river, but Richmond must climb to the second floor or go out on the roof to see it. Many of the floodwalls in the Kansas Cities were built between the 1940s and 1950s, and are anywhere from eight to twenty-two feet tall—a very concrete obstacle. For the average person wishing to explore bankside, go fishing, or put in a boat, the hazards are multiple. There are railroad tracks, potholed roads, buildings in disrepair, dump sites, and (if you scale the wall) a rough and steep riprap bank. What isn't there is just as important: boat ramps, docks, fueling stations, sidewalks, public space, a place to get lunch, and places to live, work, and play. The obstructions and absences reinforce the stereotype of the bottoms and river as risky and unapproachable.

When author William Least Heat-Moon made his trek across America by water in the late 1990s, his view from the river echoed Lewis Freeman. Heat-Moon accused "Kansas City, born of the Missouri," of having "turned away from its great genetrix more than almost any other river city in America." He hardly exaggerated when he wrote, "Just to see the Missouri here, you have to cross a bridge at breakneck speed or take an elevator in a downtown skyscraper."[4] "A river of contradictions" is what the *Kansas City Star* called the Missouri River as residents blithely bridged their "lifeline." A man on his lunch break told the *Star* reporter, "I don't think too much about it. . . . It's dirty and industrial."[5] On his lower basin boat trip a few years later, river journalist Bill Lambrecht similarly commented: "Nowhere along the Missouri is the disconnection between the river and its people more acute than in Kansas City."⁶ Each of these views from the river showed increasing estrangement over the century.

The engineered reality, failed dreams, and illusive promises of the Economic River had turned the Missouri into the "empty river," devoid not just of people but of barges.⁷ The barge industry has been in decline since 1977 and research finds limited economic benefits from barge traffic downstream and especially upstream from Kansas City.⁸ Barge navigation is energy efficient if there are enough barges operating but, unlike the Mississippi, there are few on the lower Missouri.⁹ About one-third of the barges that pass Richmond at the Healthy Rivers Partnership are not engaged in commerce at all but are operated by the corps and are performing channel maintenance.

If You Build It, They Will Come

City planners had imagined parkland along the river since the early twentieth century, but those dreams never quite materialized. In 1965, the Kansas City Parks Department proposed Missouri River Drive, a six-mile ribbon of greenway along the south side of the river. This proposed park area was part of the levee district and was designated as open land, perfect for picnicking, tree plantings, athletic fields and other minimal infrastructure that would withstand flooding.¹⁰ A similar idea came to be when Riverfront Park in Kansas City opened in the 1970s. The Land and Water Conservation Fund aided the park with federal matching grants intended to increase public outdoor recreation land, but the park was not very accessible and never became popular. River Bluff Park, also near the river, received aid from the same federal fund in 1973.¹¹

In the wake of severe flooding in 1973, a growing constituency sought change. Growing research showed that the development associated with the Pick-Sloan Plan had negative effects on the lower river basin. The Missouri Department of Conservation produced an influential report in 1974 that charted changes in the river's channel and ecology in the twentieth century, most of which were set in motion by corps engineering. The narrower, deeper, and shortened Missouri lacked islands, sandbars, wetlands, cutoffs, and fluctuating water levels that supported a rich and diverse flora and fauna. The report argued that, beyond the negative environmental changes, human interaction with the river was directly affected; for example, the decline in the river's health meant less wildlife for fishing and hunting.[12] A generation influenced by ecology and environmentalism found itself dissatisfied with the model of the Economic River and sought to reconnect with the river through annual canoe flotillas. The corps helped support the annual canoe trips and began cooperating with the conservation department by notching wing dikes to restore riverine habitat. Meanwhile, within the city, smaller waterways kept residents on their toes. In 1977 Brush and Turkey Creeks flooded, killing people and causing millions of dollars of damage. The Plaza, the nation's first automobile shopping mall built by J. C. Nichols, sits alongside Brush Creek, and floodwaters left cars upside down in the streets.[13]

Emboldened by federal environmental legislation, public agencies took up the task of changing perceptions and reconnecting people to their environments. In an effort to raise awareness about the city-river connection, the Missouri Water Pollution Board in the late 1960s created a filmstrip for public exhibition called "How Our Town Saved the River." The board aimed to recruit public opinion and increase citizen participation in planning for sewage treatment facilities.[14] In the 1970s the Mid-America Regional Council (MARC), the regional planning entity of the Kansas City metropolitan area, similarly committed itself to changing attitudes toward waterways. Understanding that public participation was critical to water quality, MARC worked to persuade Kansas Citians that a cleaner river was a worthwhile financial investment. In 1978 MARC ran a public education campaign that included statements like "Please go near the river," challenging the common sentiment that the river was a dirty, unsafe place.[15] City newsletters encouraged volunteers to stencil "Dump No Waste—Drains to Stream" on sidewalks and curbsides to draw the connection between immediate infrastructure and the watershed.

In the aftermath of Pick-Sloan and its failed promises, the basin split bitterly over river management because there were more desires than water to fulfill them. The Missouri River Basin Association, founded in 1981 under the leadership of state governors, brought together the divergent interests of the basin states with the goal of problem solving, similar to predecessor organizations like the Missouri River Sanitary Conference. The onset of a drought in the late 1980s highlighted competing interests and also made clear that the lower basin held more clout; navigation and flood control were prioritized over recreation, power generation, and irrigation in the upper basin. The competing needs of so many users fueled debates during a time of scarcity. Upstream states sued the Corps of Engineers, pressing it to update river management to meet changed needs. Beginning in 1989, the Corps of Engineers began to revise its master manual—the document that guides management of the Missouri River—but the "fight" would be drawn out and contentious.[16] In the debate, waterworks operators found themselves just one of the interests jockeying for adequate water quality and quantity.

In 1993 the Missouri and Mississippi Rivers experienced another of their most significant flood events. The Missouri reasserted itself—an assertion measuring about seventeen feet above flood stage in Kansas City! The 1903 and 1951 floods had higher volumes of water, but the damage done by the 1993 flood was greatest owing to more constrictive engineering and a more vulnerable built environment. The Midwest has experienced several so-called one-hundred-year floods in the last twenty years. The fact that flood risk and damage is increasing, despite a century of flood control efforts, should give us pause. As in other years, the 1993 flood put people out on the river, sandbagging to protect levees and waterworks, and brought others to the bluffs to gawk.

The 1993 flood marked a significant turning point in attitudes and actions in the Missouri basin. It goaded people into rethinking their relationship with the river. Kansas City, among other cities visited by high water, stepped up efforts to increase "green infrastructure" and turn the "riverfront void" into profitable development. The historic uses of the riverbanks made creation of a park difficult. Urban archaeology revealed soil contamination from a gasworks industry and dump sites for demolished public housing projects; the hazards required expensive environmental cleanup.[17] This former industrial site, the "last great undeveloped waterfront property in America," as the *Star* put it, became Kansas City's new public interface with the river. Like the earlier Riverfront Park, the 1998 opening of Berkeley Waterfront Park seemed the culmination of a long dream; the park, however, did not become a popular destination.[18] Traveling upriver, Lambrecht found Berkeley Park vacant and remarked: "On the fabled Missouri River, the highway to American empire, I see no other boats or human beings during the entire trip."[19] Kansas City has been wasting its greatest resource.

Troy Gordon, an avid River Relief volunteer and founder of the Friends of Big Muddy, once answered the question of how to bring about change along the river. "Pray for rain," he said.[20] Floods have brought about the most significant policy shifts, then and now. Public land acreage increased after the 1993 flood because federal and state governments partnered to purchase inundated land from willing sellers along the lower river. In 1994 the Big Muddy National Fish and Wildlife Refuge was established, as were citizen organizations like Friends of the Kaw and Friends of Big Muddy. The US Fish and Wildlife Service (USFWS) and the Missouri Department of Conservation managed land on the floodplain for multiple uses: restoring wetlands, enhancing fish and wildlife habitat, and providing recreation. Taking agricultural land out of production would give the river more room and decrease flood risk. The emphasis has been on restoration to reconnect the river to its floodplain, which reflects a more comprehensive and complex definition of environmental health, encompassing the entire system, both natural and human.

But that same year, 1993, the corps revisited its long-term process of revising the master manual, which fed the antagonism between upstreamers and downstreamers. One commentator called the Missouri River "one of the most explosive environmental issues to face the middle of the country."[21] The upper basin, settled in behind the dams, preferred to manage the river for recreation and irrigation, while the lower basin interests wanted to maintain appropriate flows for navigation. The farm bureaus, diminishing barge industry, and the state of Missouri, however much in the minority, have proven to be the most forceful advocates of the restrained river and the status quo. These forces kept the river detached from its floodplain. The din heightened as years of less-than-average rainfall resulted in a drawdown of the reservoirs, exposing land not seen since the dams were closed.

Also in 2002, the National Research Council (NRC) released a significant report, *The Missouri River Ecosystem: Exploring the Prospects for Recovery*, which had been requested by the Corps of Engineers and the Environmental Protection Agency (EPA) in an effort to find a way forward through the morass.[22] River activists like Richmond had high hopes. The report called for different schemata of management on the river and generated much attention. "Degradation of the Missouri river ecosystem will continue unless some portion of the hydraulic and geomorphic processes that sustained the pre-regulation Missouri River and floodplain ecosystem are restored," warned the NRC.[23] The river's natural cycles of ebb and flow need both restoration and mimicry. To halt environmental degradation, the NRC members suggested that the basin consider "adaptive management." Using the concept of an environmental system, adaptive management deals integratively with natural resources. It is characterized by experimentation and flexibility in its implementation—a responsive dynamism that considers the needs of basin groups as well as the results of scientific studies. Integral to adaptive management, the NRC committee strongly recommended that decisions be more democratically made in a process involving a full array of "stakeholders"—those people, organizations, agencies, and scientists who have an interest in the river.

Health and wealth have remained central to the debate. The NRC findings were clear that the health of the river needed more attention. But how does a society measure or value river health? And where will the impetus for restoration and protection come from? The NRC report urged: "Ecosystem changes are not merely abstract, scientific measurements; they also represent the loss of valued goods and services to society." While society usually overlooks these goods and services, the NRC noted that "there is a growing recognition that the replacement costs of these services, assuming their replacement is even possible, would be very high."[24] Economists refer to these as externalities—costs and risks incurred by the whole.

Thus, the Healthy River has begun to mount a challenge to the seemingly megalithic Economic River. Using the Endangered Species Act, the USFWS, with support from environmental organizations, took the Corps of Engineers to court on behalf of threatened and endangered animal species native to the Missouri. At issue were releases of water from reservoirs that would mimic natural spring rises and are important to wildlife and habitat. In a 2004 settlement with the USFWS, the corps committed to develop a Missouri River Recovery Implementation Committee, the participatory process recommended in the NRC report.[25] Will the corps lean toward adaptive management, as the NRC and USFWS recommended, or will the river continue to be managed for narrow purposes? Everyone waited with bated breath: citizen activists, scientists, irrigating farmers, bottomland farmers, Native American tribes, environmentalists, piping plovers, pallid sturgeon, and waterworks directors all have an interest in the river.

The only ones who didn't wait were the so-called Asian carp, an invasive fish species that outcompetes native species and have colonized much of the Mississippi basin. These large fish are injurious to aquatic diversity, but they are best known for jumping out of the water when they hear the sound of motorboats—thus becoming injurious to people as well. I speak from experience, as I have been hit in the head by a carp—if you'd like visual proof of the hazard, search the internet. Every carp that flops out of the water is capturing nutrients and space from a diverse native ecosystem.

The Economic River has come at public cost and has produced questionable profits. Taxpayers annually pay millions to maintain the river for navigation and flood control, yet only a few companies run barges on the lower Missouri. Furthermore, taxpayers pay private landowners who successfully argue that they should be compensated for regulations that prevent development on the high-risk floodplain—called a "taking." Taxpayers also subsidize flood insurance and pay for disaster relief.[26] All of this adds up to billions spent on unsuccessfully keeping a river from doing what it naturally does—meander and flood. Additionally, the river has not been managed for public health or water quality. Preventing natural forces and cycles on the river has had costs for both human and environmental health.

Public perception of the Missouri River began shifting to see the Economic River as a relic. As the corps continued revision of its master manual, it opened the process up to public comment, raising the stakes for different interests and stimulating debate. The Missouri was once again bestowed with "most endangered river" status by American Rivers. The citizen advocacy group has ranked the Missouri and Kaw Rivers at or near the top of its list every year since 1994.[27] "As an artery of commerce, the Missouri River is a major disappointment," American Rivers averred.[28] Many agreed with the *Omaha World-Herald* that the Pick-Sloan Plan and the corps were dinosaurs of the midcentury. "It has become clear," the editorial stated, "that the limited way in which the corps historically has managed the Missouri, concentrating on navigation, flood control and power production, is no longer adequate for the times."[29] The river basin needed a "new Missouri Compromise," in the words of one contemporary reformer.[30]

Richmond points to the Lewis and Clark bicentennial commemoration as a singularly influential event. In anticipation of the expedition's bicentennial in 2004–2006, Kansas Citians joined others along the Missouri who looked anew at the riverscape. The Corps of Discovery's most significant passage was the Missouri River, and just as onlookers were drawn to the bluffs in flood time, planners and public officials assumed that the bicentennial would attract the public to the river. Local organizations, cities, states, and the federal government wanted to put on a big show for the commemoration, but they realized that encouraging people to discover history meant they had to rediscover the river.

In preparation for this rediscovery, the riverbanks needed a makeover, and a member of the Kansas City Port Authority predicted it would be the "decade of the riverfront."[31] Down on the old Kansas City wharf, plaques commemorating the most significant known floods in the Kansas Cities were set in place. Kaw Point, the location of the first waterworks, also became a Kansas City, Kansas, riverside park with a boat ramp and the bi-state Riverfront Heritage Trail opened.[32] Dan Sturdevant of the Lewis and Clark Trail Heritage Foundation credits the bicentennial as instrumental in reuniting people with the river.[33] This bicycle trail runs behind the Healthy Rivers Partnership building and Richmond has noticed increased traffic. To accommodate and encourage interaction with history and the riverscape, private groups and government entities at all levels cooperated to increase public lands, trails, and interpretive markers all along the Missouri. Slowly, the infrastructure that provides the public an interface with the river is being built. And, if you build it, the people will come.

The Spring Rise from the Grassroots

It is time to release the Missouri from its constraints, decommission the barge channel, address the flood risks of our built environment and climate change, consider working *with* the river and not against it, and define wealth as good health—both human and environmental. In the words of historian Robert K. Schneiders, it is time to "democratize" the river and consider it "everyone's river." The Economic River currently works for the few, but a dechannelized river would work for everyone. An unchanneled river will be wider and slower. This will reduce flood risk while increasing diversity of bird, animal, plant, and aquatic species, which will, in turn, attract recreation (swimming, boating, hunting, birdwatching, fishing, and walking trails). This healthier ecological regime will be better for the local economy as well. The federal government spends more to maintain the channel than it returns in wealth. Schneiders asserts that if the powerful few from the state of Missouri loosen control over the river, all states will benefit from the management shift. The tyranny of barge navigation, Schneiders writes, should yield to a "truly multiple-purpose river for everyone."[34] This vision of the river is an iteration of the Healthy River.

Since the bicentennial, the corps still manages the river similarly and, from a distance, it seems like not much has changed: the river is still in a navigation straitjacket and the city is still at odds with it. But look closer, to the grassroots. A groundswell of activists who seek a new vision for the river has emerged. River lovers are the coxswains of smaller boats that are challenging the megalithic barge. Activists are teaching a new generation about river culture, one kayak launch and river cleanup at a time. To an activist ten years is an eternity, but for comparison, look at the civil rights and environmental movements. Both began decades ago and are still in progress. Slowly but surely, activists are working toward the vision of a healthy river, first articulated over a hundred years ago. The Healthy River is managed for everyone, creating wealth through good health, and fulfilling goals of both the civil rights and environmental movements. There is no better city to create a new vision for the Missouri than Kansas City, the home of the boosters who dreamed up the commercial river that never came to fruition.

This broader consciousness of rivers began at the local level with activists like Richmond and Chad Pegracke. Starting as a teenager on the Mississippi River, Pegracke started picking up garbage while out on his boat and this has grown into an incredible volunteer-driven, educational nonprofit that connects residents to their waterways by organizing cleanups of trash. In 1998 Pegracke began Living Lands & Waters, and he and his crew (they are on a barge) have inspired similar cleanups and received dozens of awards and honors over the years. Pegracke called garbage a "gateway" issue because it increases interest in and concern for the river.[35] He is a poster child for what young, smart, concerned people do: roll up their sleeves and get to work. On the lower Missouri, Pegracke supported volunteer efforts that have taken on a life of their own. This was the start of Missouri River Relief, which organized its first cleanup in 2001 near Columbia, Missouri. I participated in this initial event; the excitement was palpable as river lovers and learners came together. We worked on boats and on the shore, and sometimes we got wet. Collectively, we were the Big Muddy. Since then, River Relief has organized well over a hundred cleanups with thousands of people, from Yankton, South Dakota, down to the confluence near Saint Louis. Dave Stous, a retired Burns & McDonnell engineer and avid supporter of River Relief, estimates that over twenty-five thousand people have been out on the Missouri doing cleanups.[36]

Richmond helped organize the first Missouri River Relief for the Kansas City reach in 2003. That September some 1,700 people, the majority of whom were youths, came to the banks of the Missouri in Kansas City to learn about the river and service its litter-strewn shores.[37] It was a chilly, wet day but spirits were not dampened. Perhaps the last time a crowd of that magnitude gathered was to gape at a flood, to greet a steamboat in the 1910s, or to welcome a barge to the city wharf in the 1930s.[38] Although these young Kansas Citians were connected to the Missouri through their city infrastructure—their toilets, taps, and tubs are filled with the river—for most it was their first time alongside it. The *Star* remarked that, as one student stood "near the mighty river cutting through Kansas City's heart," it was like standing "on a foreign shore."[39] Although Kansas Citians benefit from the river in their daily lives, most had never come into contact with the river in this manifestation. For the first time, they had a view from the river.

Like regrowth after a fertile spring rise, other activities and groups have sprung to life on and alongside the river. The Missouri River 340 Race is an extreme paddling race that starts in Kansas City and ends in Saint Charles, crossing the state of Missouri for 340 miles. Scott Mansker, also founder of an educational public television program called *River Miles*, started the MR340 in 2006 as an attempt to engage people in the river through recreation.[40] Now the race has taken on a life of its own, receiving worldwide attention because it is the longest nonstop paddling race in the world. Reflecting on this, Mansker said that kayaking the river was once a lonely outing but now it is common to see other paddlers. In 2007 a second Missouri River event began: Race for the Rivers: Bringing People to the River, which operates in the Saint Louis region and combines bicycling and paddling races. Riverside trails (like the Katy Trail State Park in Missouri and the Steamboat Trace in Nebraska) have increased access.

The river is an ideal teaching tool because it makes visible the intertwining of social and ecological systems, and the interconnectedness of upstreamers and downstreamers. The river is like "connective tissue" for communities, says Steve Schnarr of River Relief.[41] In commenting upon the popular interest in rivers, sociologist Karen O'Neill said it comes from an "impoverished language for discussing common interests in the U.S. today."[42] The river holds the promise of this for us. In the words of former Secretary of the Interior Bruce Babbitt, "water connects us all" and shows us to be interdependent. Babbitt trumpeted the work of Watershed Councils that developed common interest through waterways. Citizen groups are "protecting our natural heritage," Babbitt said, and are "re-connecting all the parts and making sure they function within the totality of one integrated watershed." Drawing on Aldo Leopold, Babbitt said that rivers are "the best examples of how everything is related, of how nature is more than the sum of its parts."[43] Many individuals and groups are finding common ground in the river.

The struggle of Kansas Citians to comprehend their long and sometimes troubled affair with the river resembles the general experience

Americans have had with nature. The cultural tendency has been to see people as separate from environment, and to see cities as unnatural. But it is essential to understand urban environments because these are how the vast majority of Americans (and half the world's population) experience daily life. As the next generations negotiate the way they see environment and their place in it, the river is sure to find itself in a new relationship. But the connection of city and river is doubly useful because it is not a static relationship but a dynamic one that includes connections upstream and downstream. Examining this history allows us to see the connections we have inherited—connections to both social and ecological systems.

Urban planners, environmentalists, historic preservationists, community leaders, and developers are supporting this rediscovery, though their motivations may differ. Threading the core, rivers are ideal locations for revitalizations because they are already at the center of cities, and potentially have high real estate value. Former manufacturing sites are converted to lofts and greenspace lines the river. Sometimes the developments are commercial, other times renewal includes public space. These riverside environments, formerly places of production, are converted to consumption and recreation. Currently, in Kansas City, the River Market district is a good example of this and there is something afoot in the West Bottoms as art spaces and nightlife pop up. Twenty years from now, things could look very different. As those cities that have already gone through this process can attest, revitalization must take into account the socioeconomic conditions of the city, strive to be accessible and democratic, and avoid gentrification. An urban river environment ought to represent all that we have learned from the past.

Amid this urban revitalization are attempts to redress the ecological impacts of industrialization. The Kansas Cities have long shirked their obligations to protect water quality under the Clean Water Act, but in 2010 Kansas City reached an unprecedented legal agreement with the EPA to control Combined Sewer Overflows (CSOs). Just as Kansas City was slow to treat its sewage, so it has failed to reduce its share of sewage overflows that occur in heavy rainstorms, sending an estimated seven billion gallons of raw sewage every year directly to the Kaw and Missouri Rivers. This pollution puts both human and riverine health at risk, especially in the parts of the city with the oldest infrastructure. Brush Creek, for example, has signs posted alongside the parkway warning runners, dog walkers, and strolling families not to come in contact with the water because of the high level of fecal bacteria, the result of sewage seeping into the creek.

In this historic agreement, the city pays a fine and agrees to build a green infrastructure that will reduce storm water and urban runoff in the future. This will also reduce flooding. It will be the costliest infrastructure investment the city has made since building the waterworks. Karl Brooks, the region's EPA administrator, called it a "landmark" decision that will put Kansas City at the forefront of green infrastructure leadership.[44] The firm helping Kansas City deal with this EPA decision and redesign its urban infrastructure to have a more responsible relationship to the watershed is Burns & McDonnell. Kansas City follows in the footsteps of other pioneering cities, but if successful, Kansas City could become the symbol of a revised urban infrastructure, a postindustrial city that acknowledges and understands the relationship between city and river. If Kansas City can fall in love with its river again, any city can.

✳ This book is more than an urban environmental history; it is also a story about the author. In my case, I fell in love with the river about fifteen years ago, when I moved to the Midwest. I participated in events that reintroduced people to the river and helped the public realize they were stakeholders. Those activists left their imprint on me. I took long bicycle trips along the Missouri River, to feel closer to it, to know it better, and to find the human side of the river. I have always intended to make visible the connection people have to the watershed, and try to tell a more social history of the river. The historical challenges of the Healthy River have echoes today, and not just in Kansas City.

Public drinking water systems in the United States are among the best in the world at providing broad access and inexpensive, well-regulated water quality. We are at a critical juncture. Threats to public drinking water systems are multiple and include aging infrastructure, scarcity, agricultural chemicals, climate change, zebra mussels, algae blooms, and lead. Think of the residents of Flint, Michigan—a city in which thousands of children were exposed to high lead levels through drinking water—or

Toledo, Ohio, where toxic blue-green algae (cyanobacteria) forced the city's water system to shut down. In the twenty-first century, American communities are boiling water and seeking shipments of bottled water as if we were an undeveloped nation. These threats undermine Americans' trust of the public system.

What are Americans to do—buy bottled water? No. Bottled water is even less regulated and more expensive. We must not let an essential resource like water become privatized. If trust in the public system wanes, then wealthy Americans will pay for safer water and leave the vulnerable at risk. The most democratic, energy-efficient, cost-saving, and safest way to protect public health is by investing in our public drinking water systems and better regulating our water quality. Now is the time to bolster our public health standards.

As fertilizer runoff from industrial agriculture continues to leach into our waterways, algae blooms are becoming the norm in the Great Lakes, the Gulf of Mexico (commonly called the "dead zone"), and in smaller bodies of water from Texas to Massachusetts. I now live in Des Moines, Iowa, and my city is fighting to improve the quality of our river water but we are up against a definition of wealth that is not inclusive of health. Des Moines knows what jeopardy looks like—during the 1993 flood the city lost drinking water for days. Now, waterworks engineers and chemists who are responsible for the health of half a million people remain vigilant to another issue. Des Moines spends over a million dollars annually to remove agricultural chemicals (especially nitrates) from the drinking water. Industrial agriculture, weak state laws, and a loophole in the Clean Water Act are responsible for this threat to public health. The problem is that farm runoff is considered nonpoint pollution, which is not regulated. The Des Moines Water Works resorted to a lawsuit against agricultural counties upstream. Cities need the Clean Water Act to define the drainage pipes from farm fields (called "tiling") as point sources. This would improve water quality in the Missouri-Mississippi watershed for downstreamers. Industrial agriculture has immense power and wealth and, just as policymakers engineered the Missouri River for these industries, so they allow for our water quality to be determined by them. The public's health is at risk. Waterworks engineers are finding it difficult to protect health and the public is paying the cost to purify the drinking water, whereas industrial agriculture reaps the benefits. Public cost and private gain—that might sound like an echo.

I encourage readers to discover their nearby stream, river, and watershed. From the grassroots, let your awareness grow. Look at it critically—is it an equitable body of water? Join others to raise awareness and remedy social and environmental concerns. Mostly, just fall in love with your river. Here are ways to ponder the river-city relationship and to "see" the river in Kansas City:

- Get a bird's-eye view of the bottoms from Ermine Case Jr. Park.
- See the *Muse of the Missouri* fountain.
- Visit the waterworks in North Kansas City and Quindaro.
- Picnic at Kaw Point Park.
- For an unconventional river exploration, paddle to the outlet of the West Bottoms treatment plant where water reenters the river.
- Walk along Brush Creek and admire the fountains of The Plaza.
- Follow a buried/covered creek using gravity and storm drains.
- Explore the Blue River trails.
- Participate in a Missouri River Relief cleanup.

NOTES

CHAPTER ONE: AN INTRODUCTION TO HEALTH AND WEALTH

1. See preface in William Cronon, *Nature's Metropolis: Chicago and the Great West* (New York: W. W. Norton, 1991); and Martin V. Melosi, "The Place of the City in Environmental History," *Environmental History Review* 17, no. 1 (1993): 1–23.

2. Cronon, *Nature's Metropolis*, 31–46.

3. Richard Wohl and Andrew Theodore Brown, "The Usable Past: A Study of Historical Traditions in Kansas City," in *The Pursuit of Local History: Readings on Theory and Practice*, ed. Carol Kammen (Walnut Creek, CA: Altamira Press, 1966), 145–163.

4. For an overview of the public policies that create segregation, see Richard Rothstein, "The Making of Ferguson: The Public Policies at the Root of its Troubles," Economic Policy Institute, Oct. 15, 2014, accessed Jan. 2016, http://www.epi.org/publication/making-ferguson/.

5. Jenny Price, "Thirteen Ways of Seeing Nature in L.A.," *Believer Magazine* (Apr. 2006); and Price, "Remaking American Environmentalism: On the Banks of the L.A. River," *Environmental History* 13, no. 3 (July 2008): 536–555.

6. Works of environmental history that address health include Robert Gottlieb, *Forcing the Spring: The Transformation of the American Environmental Movement* (Washington, DC: Island Press, 1993); Christopher C. Sellers, *Hazards of the Job: From Industrial Disease to Environmental Health Science* (Chapel Hill: University of North Carolina Press, 1997); and Sylvia Hood Washington, *Packing Them In: An Archaeology of Environmental Racism in Chicago, 1865–1954* (Lanham, MD: Lexington, 2005).

7. Melosi, "Place of the City," 18.

8. Lewis Mumford, *The City in History: Its Origins, Its Transformations, and Its Prospects* (New York: Harcourt, 1961), 563–567.

9. A number of fine environmental histories have been done for other cities, including Andrew Hurley, ed., *Common Fields: An Environmental History of St. Louis* (Saint Louis: Missouri Historical Society Press, 1997); Craig E. Colten, ed., *Transforming New Orleans and Its Environs: Centuries of Change* (Pittsburgh, PA: University of Pittsburgh Press, 2000); Char Miller, ed., *On the Border: An Environmental History of San Antonio* (Pittsburgh, PA: University of Pittsburgh Press, 2001); Joel A. Tarr, ed., *Devastation and Renewal: An Environmental History of Pittsburgh and Its Region* (Pittsburgh, PA: University of Pittsburgh Press, 2003); Craig E. Colten, *An Unnatural Metropolis: Wresting New Orleans from Nature* (Baton Rouge: Louisiana State University Press, 2005); Ari Kelman, *A River and Its City: The Nature of Landscape in New Orleans*

(Berkeley: University of California Press, 2003); Martin V. Melosi and Joseph A. Pratt, *Energy Metropolis: An Environmental History of Houston and the Gulf Coast* (Pittsburgh, PA: University of Pittsburgh Press, 2007); Michael F. Logan, *Desert Cities: The Environmental History of Phoenix and Tucson* (Pittsburgh, PA: University of Pittsburgh Press, 2006); and Matthew Klingle, *Emerald City: An Environmental History of Seattle* (New Haven, CT: Yale University Press, 2007). Finally, on the renewal of urban watersheds, see Paul Stanton Kibel, ed., *Rivertown: Rethinking Urban Rivers* (Cambridge, MA: MIT Press, 2007); and John R. Wennersten, *Anacostia: The Death and Life of an American River* (Baltimore, MD: Chesapeake Books, 2008).

10. Roy Ellis, *A Civic History of Kansas City, Missouri* (Springfield, MO: Elkins-Swyers, 1930); and Henry C. Haskell Jr. and Richard B. Fowler, *City of the Future: A Narrative History of Kansas City, 1850–1950* (Kansas City, MO: Frank Glenn, 1950); Charles N. Glaab, *Kansas City and the Railroads: Community Policy in the Growth of a Regional Metropolis* (Madison: State Historical Society of Wisconsin, 1962); Andrew Theodore Brown and Lyle W. Dorsett, *K.C.: A History of Kansas City, Missouri* (Boulder, CO: Pruett, 1978); and James Shortridge, *Kansas City and How it Grew, 1822–2011* (Lawrence: University Press of Kansas, 2012). Two works addressing social issues are Sherry Lamb Schirmer, *A City Divided: The Racial Landscape of Kansas City, 1900–1960* (Columbia: University of Missouri Press, 2002); and Kevin Fox Gotham, *Race, Real Estate, and Uneven Development: The Kansas City Experience, 1900–2000* (Albany: State University of New York Press, 2002).

11. Bruce Katz, "Kansas City: Region on the Rise" (presentation to Mid-America Regional Council, Brookings Institution Center on Urban and Metropolitan Policy, June 4, 2004), accessed Apr. 3, 2018, https://www.brookings.edu/wp-content/uploads/2016/06/20040604_KansasCity.pdf.

12. "Kansas City" by Richard Rodgers and Oscar Hammerstein II, copyright © 1943 by Williamson Music. Copyright renewed, international copyright secured. All rights reserved. Used by permission.

13. John Steinbeck, *Travels with Charley: In Search of America* (New York: Viking, 1962), 138.

14. Early works on the river include Stanley Vestal, *The Missouri* (Lincoln: University of Nebraska Press, 1945; repr., Lincoln: Bison Books, 1964); and Joseph Mills Hanson, *Conquest of the Missouri: Being the Story of the Life and Exploits of Captain Grant Marsh* (1909; repr., New York: Murray Hill Books, 1946). Academic works addressing the twentieth-century Missouri River tend to focus on political and economic issues, including Richard G. Baumhoff, *The Dammed Missouri Valley: One Sixth of Our Nation* (New York: Alfred Knopf, 1951); Henry C. Hart, *The Dark Missouri* (Madison: University of Wisconsin Press, 1957); John E. Thorson, *River of Promise, River of Peril: The Politics of Managing the Missouri River* (Lawrence: University Press

of Kansas, 1994); John R. Ferrell, *Big Dam Era: A Legislative and Institutional History of the Pick-Sloan Missouri Basin Program* (Omaha, NE: United States Army Corps of Engineers, 1993); Ferrell, *Soundings: One Hundred Years of the Missouri River Navigation Project* ([Kansas City]: United States Army Corps of Engineers, 1996); Bill Lambrecht, *Big Muddy Blues: True Tales and Twisted Politics along Lewis and Clark's Missouri River* (New York: St. Martin's Press, 2005); and Robert K. Schneiders, who was the first to bring the field of environmental history to bear in *Unruly River: Two Centuries of Change along the Missouri* (Lawrence: University Press of Kansas, 1999) and *Big Sky Rivers: The Yellowstone and Upper Missouri* (Lawrence: University Press of Kansas, 2003). Adding a social element to river politics is Michael Lawson, *Dammed Indians: The Pick-Sloan Plan and the Missouri River Sioux, 1944–1980* (Norman: University of Oklahoma Press, 1982).

15. Travel writings include John G. Neihardt, *The River and I* (New York: Macmillan, 1927); Lewis R. Freeman, "Trailing History Down the Big Muddy," *National Geographic Magazine*, July 1928, 73–120; Cecil Griffith, *The Missouri River: A River Rat's Guide to Missouri River History and Folklore*, ed. K. R. Canfield and R. L. Sutton (Leawood, KS: n.p., 1974); William Least Heat-Moon, *River-Horse: The Logbook of a Boat across America* (Boston: Houghton Mifflin, 1999); and John R. Ferrell, *Opposites* (Omaha, NE: Feather Works Books, 2000). People are more prominent in previous centuries—see Vestal, *Missouri*; Griffith, *Missouri River*; Stephen Ambrose, *Undaunted Courage: Meriwether Lewis, Thomas Jefferson, and the Opening of the American West* (New York: Simon & Schuster, 1996); James D. Harlan and James M. Denny, *Atlas of Lewis and Clark in Missouri* (Columbia: University of Missouri Press, 2003); and Brown and Wohl, "Usable Past."

16. Richard Lynn quoted by Jennifer Howe, "Mighty Missouri," *Kansas City Star*, Sept. 16, 1990, L1, 4.

PART I. CITY: URBAN INNARDS

Epigraph: Kate L. Cowick, *The Story of Kansas City* (booklet of articles published by the *Kansas City Kansan*, 1924–1926), 17, Kansas State Historical Society.

CHAPTER TWO: A VIEW FROM THE BLUFFS, 1903

1. *Liberator*, June 12, 1903, 1, State Historical Society of Missouri [hereafter SHSM].
2. *Kansas City Star* [hereafter *Star*], May 30, 1903, 1, SHSM.
3. "Special Meeting," May 31, 1903, vol. 17, Kansas City Chamber of Commerce Minutes, SHSM [hereafter CCM].

4. The Chicago & Alton Railway, *The Flood of 1903* (pamphlet, ca. 1903), Missouri Historical Society [hereafter MHS].

5. "The Dreary Hours in the History of Both Kansas Cities," *Reform*, June 6, 1903, SHSM (translated from the German-language newspaper).

6. *Star*, June 4, 1903, 9, SHSM.

7. "Mass Meeting," June 1, 1903, vol. 17, pp. 156–157, CCM.

8. *Star*, June 1, 1903, SHSM.

9. *Wichita Daily Eagle*, May 31 and June 2, 1903, Library of Congress.

10. "Mass Meeting," June 1, 1903, vol. 17, pp. 156–157, CCM.

11. *Liberator*, June 12, 1903, SHSM.

12. *Star*, June 8, 1903, 1, SHSM.

13. *Star*, June 9, 1903, 1, SHSM.

14. *Star*, June 6, 1903, 1; *Star*, June 2, 1903, 1; *Star*, June 8, 1903, 1, SHSM.

15. Different sources cite different population numbers. Daniel Serda says that 23,000 people lived in the West Bottoms (Serda, "A Blow to the Spirit: The Kaw River Flood of 1951 in Perspective" [paper presented at Midcontinent Perspectives lecture series, Midwest Research Institute, Kansas City, MO, Oct. 28, 1993], accessed June 30, 2006, html://www.umkc.edu/whmckc/PUBLICATIONS/MCP/MCPPDF/serda-10-28-93.pdf). The vertical file [hereafter abbreviated v.f.] "West Bottoms" at the Kansas City Public Library says that there were 20,000 worker and residential refugees. See also Haskell and Fowler, *City of the Future*; and Robert L. Branyan, *Taming the Mighty Missouri: A History of the Kansas City District Corps of Engineers, 1907–1971* (Kansas City, MO: US Army Corps of Engineers, 1974), 43. On June 1 and 4, the *Star* reported that 2,500 homes were flooded in the East Bottoms and that Kansas City, Kansas, had 20,000 refugees. The *Star* on June 1 and 2 also reported that many people were out of work—over 10,000, many of whom worked in meatpacking.

16. *Star*, June 7, 1903, SHSM.

17. Ellen G. Parkhurst, *Club Member* 1, no. 5 (1905): 17, Kansas State Historical Society [hereafter KSHS]. Being a member of this club, Parkhurst was probably white.

18. *Star*, June 6, 1903, SHSM.

19. *Star*, June 5 and 6, 1903, SHSM.

20. *Star*, June 4, 1903, 1; *Star*, June [n.d.], 1903, 1, SHSM.

21. *Star*, June 6, 1903, SHSM.

22. Letter from relief committee to Governor Bailey, "Flood of 1903," June 24, 1903, box 3, f. 1, Governor W. J. Bailey Papers, KSHS [hereafter WJB].

23. *Rising Son*, June 19, 1903, 2, SHSM.

24. *Star*, June 3, 1903, 1, SHSM.

25. *Kansas City Gazette*, June 1903 [dates unclear on microfilm], KSHS.

26. *Star*, June 8, 1903, 8, SHSM.

27. *Star*, June 12, 1903, 11, SHSM.

28. *Rising Son*, July 3, 1903, SHSM.

29. Samuel J. Crumbine, "A Few Highlights in the History of Sanitation," [draft, ca. 1953], 9, f. "Public Health Issues: Sanitation," series 1, subseries 2, Samuel Jay Crumbine Papers, Clendening History of Medicine Library, University of Kansas Medical Center [hereafter SJC].

30. *Star*, June 6, 1903, SHSM.

31. *Star*, June 7, 1903, SHSM.

32. Chicago & Alton Railway, *Flood of 1903* (pamphlet, ca. 1903), MHS.

33. Untitled clipping, *Kansas City Post* [hereafter *Post*], Sept. 14, 1907, v.f. "The Patch," 1, Wyandotte County Historical Society [hereafter WCHS].

34. *Star*, June 7, 1903.

35. *Star*, June 5, 1903.

36. *Reform*, June 6, 1903, SHSM.

37. V.f. "North Kansas City," 11–13, Kansas City Public Library [hereafter KCPL].

38. *Star*, June 5, 1903, SHSM; and Branyan, *Taming the Mighty Missouri*, 44.

39. *Reform*, June 6, 1903, SHSM; and "Mass Meeting," June 1, 1903, vol. 17, pp. 156–157, CCM.

40. "Flood Protection for the Kansas Cities," May 1947, box 3, f. 14, Central Industrial District Association Records, SHSM [hereafter CIDA].

41. Loren L. Taylor, *The Consolidated Ethnic History of Wyandotte County* (Kansas City, KS: Kansas City, Kansas, Ethnic Council, 2000), 85–86, WCHS; and v.f. "Argentine," WCHS.

42. *Star*, June 6, 1903, 3, SHSM.

43. *Star*, June 6, 1903, SHSM.

44. Neihardt, *River and I*, 2–3.

45. Chicago & Alton Railway, *Flood of 1903* (pamphlet, ca. 1903), MHS.

Chapter Three: Drinking the Water

1. John Walter, "Kansas City Engineering Firm Marks 100 Years of Service," *Jackson County Historical Society Journal* 38, no. 2 (1998): 6–7.

2. "Pen and Sunlight Sketches of Greater Kansas City," American Illustrating Company, ca. 1910, 101, KCPL; and Jack Cashill, *A Century of Excellence: Burns & McDonnell* (Kansas City, MO: Burns & McDonnell, 1998), 12, 18, 21.

3. Suellen Hoy, *Chasing Dirt: The American Pursuit of Cleanliness* (New York: Oxford University Press, 1995).

4. Nelson Manfred Blake, *Water for the Cities: A History of the Urban Water Supply Problem in the United States* (Syracuse, NY: Syracuse University Press, 1956), 9.

5. William E. Parrish, Charles T. Jones Jr., and Lawrence O. Christensen, *Missouri: The Heart of the Nation*, 2nd ed. (Wheeling, IL: Harlan Davidson, 1992), 224.

6. Joel A. Tarr, *The Search for the Ultimate Sink: Urban Pollution in Historical Perspective* (Akron, OH: University of Akron Press, 1996), 114.

7. Clark wrote this on June 28, 1804. Gary E. Moulton, ed., *The Journals of the Lewis and Clark Expedition* (Lincoln: University of Nebraska Press, 2001), 2:327.

8. Kansas City, Missouri, Water Services Department, Annual Report, 1998–1999 (1999); Ellis, *Civic History*, 130; and Blake, *Water for the Cities*, 77.

9. Brown and Dorsett, *K.C.*, 65; and Schneiders, *Unruly River*, 29.

10. Kansas City, Missouri, Water Services Department, Annual Report, 1998–1999.

11. T. D. Samuel Jr., "The Water Supply System of Kansas City, Missouri," *American Water Works Association Journal* 22, no. 9 (Sept. 1930): 1236.

12. Ellis, *Civic History*, 126–136; and Kansas City, Missouri, Water Services Department, Annual Report, 1998–1999.

13. Ellis, *Civic History*, 131–133, quote 129.

14. Ellis, *Civic History*, 133; and Kansas City, Missouri, Water Services Department, Annual Report, 1998–1999.

15. Martin V. Melosi, *The Sanitary City: Urban Infrastructure in America from Colonial Times to the Present* (Baltimore, MD: Johns Hopkins University Press, 2000), pp. 119–120.

16. Some water supply continued to be privatized, like bottled water and ice manufacture. See entry on the Consolidated Water Company, which brought bottled water from Excelsior Springs, in "Pen & Sunlight Sketches of Greater Kansas City," American Illustrating Company, ca. 1910, 152, KCPL.

17. R. E. McDonnell, "Where Cities Get Their Water Supplies," *Citizens' League Bulletin*, Nov. 24, 1928, 409, KCPL.

18. Cashill, *Century of Excellence*, 26, 34, 45; and Burns & McDonnell, *100 Reasons Why 100 Cities Approve Municipal Ownership of their Public Utilities*, n.d. (pamphlet, ca. 1920), Burns & McDonnell Library [hereafter B&M].

19. "City Water Works," Jan. 1, 1899, scrapbook vol. 15, p. 7, Kansas City Commercial Club Scrapbooks, SHSM [hereafter CCS].

20. "High Rates for City Water," *Star*, Jan. 5, 1905, scrapbook vol. 26, p. 5, CCS; and "A Municipal Failure," *Kansas City Journal*, Aug. 19, 1905, scrapbook vol. 27, p. 87, CCS.

21. Fuller & Maitland to Board of Fire and Water Commissioners, June 5, 1924, 6, f. 105, Albert I. Beach Papers, SHSM [hereafter ABP].

22. National Board of Fire Underwriters, Committee on Fire Prevention and

Engineering Standards, *Report on the City of Kansas City, Mo.*, no. 52 (Feb. 1924), 2, ABP.

23. Mark Twain, *Life on the Mississippi* (1896; repr., New York: Harper and Brothers, 1981), 112.

24. Fire Underwriters, *Report on the City of Kansas City*, 4.

25. Fire Underwriters, *Report on the City of Kansas City*, 4–5, 8.

26. Scrapbook clippings from 1905, vol. 27, p. 80, and vol. 28, p. 32, CCS; and Sept. 1, 1906, scrapbook vol. 31, p. 47, CCS.

27. Fire Underwriters, *Report on the City of Kansas City*, 4–5, 8.

28. Fire Underwriters, *Report on the City of Kansas City*, 5–7.

29. "About B.P.U.," Kansas City, Kansas, Board of Public Utilities, accessed Apr. 4, 2018, https://www.bpu.com/about.aspx; and Joseph H. McDowell, "Building a City: A Detailed History of Kansas City, Kansas" (Kansas City, KS: *Kansas City, Kansan*, 1970[?]), 13; Kansas City, Missouri, Water Services Department, *Annual Report*, 1998–1999; D. M. Bone, ed., *Annual Review of Greater Kansas City* (Kansas City: Bishop, 1908), 71, KCPL; and Dwight F. Metzler, *Kansas Public Water Supplies—A Century of Progress*, Kansas Water Environment Association, accessed Mar. 31, 2018, http://www.kwea.net/images/About/documents/ks-public-water-supplies.pdf.

30. Kansas State Board of Health, *Sixth Biennial Report of the Kansas State Board of Health, 1911–1912* (Topeka: Kansas State Printing Office, 1912), 51; Allan J. McLaughlin, *Sewage Pollution of Interstate and International Waters with Special Reference to the Spread of Typhoid Fever: The Missouri River from Sioux City to Its Mouth* (Washington, DC: US Government Printing Office, 1913); and Paul D. Haney, "The Missouri River—A Vital Resource" (paper presented at 58th Annual Conference of the Water Pollution Control Federation, Kansas City, MO, 1985), KCPL.

31. "To Aid a Sister City," Jan. 25, 1914, scrapbook vol. 1, p. 231, Henry Jost Papers, SHSM [hereafter HJP].

32. "Drinking Water Pure in Kansas City, Mo.," Aug. 23, 1913, scrapbook vol. 1, p. 120, HJP.

33. "Talk of Two Cities Using Same Intake," Nov. 2, 1913, scrapbook vol. 1, p. 172, HJP; "Connect Pumping Plants of 2 Cities," *Post*, June 5, 1914, scrapbook vol. 5, p. 7, HJP; and Samuel, "Water Supply System," 1237.

34. "River Channel May Change," Sept. 9, 1915, and "See Need of Haste to Prevent River from Leaving Intake Dry," Sept. 10, 1915, scrapbook vol. 2, n.p., HJP; clipping, *Post*, Sept. 29, 1908, scrapbook vol. 39, p. 58, CCS; and National Board of Fire Underwriters, Committee on Fire Prevention and Engineering Standards, *Report on the City of Kansas City, Mo.*, no. 52 (Feb. 1924), 2, ABP.

35. "City to Pay Taxes in Kansas," *Journal*, Dec. 22, 1913, box 4, scrapbook vol. 2, p. 11, HJP.

36. "Must Help Run K. C., K," Oct. 21, 1914, scrapbook vol. 2, p. 41, HJP.

37. "Mayor is against Park Bond Issues," Mar. 4, 1914, scrapbook vol. 1, p. 254, HJP.

38. "A Municipal Failure," *Journal*, Aug. 19, 1905, scrapbook vol. 27, p. 87, CCS.

39. Clipping, *Post*, Sept. 29, 1908, scrapbook vol. 39, p. 58, CCS.

40. "A House without Open Porches," *Star*, Apr. 24, 1914, B1, KCPL.

41. Schirmer, *City Divided*, 17–18.

42. William J. Novak, *The People's Welfare: Law and Regulation in Nineteenth-Century America* (Chapel Hill: University of North Carolina Press, 1996), 191.

43. John Duffy, *The Sanitarians: A History of American Public Health* (Urbana: University of Illinois Press, 1990), 4.

44. Cashill, *Century of Excellence*, 40.

45. Unknown author quoted by S. J. Crumbine, "Public Support in Public Health" [ca. 1920] 6, f. "Public health issues," series 1, subseries 2, SJC; William J. Novak, "Private Wealth and Public Health: A Critique of Richard Epstein's Defense of the 'Old' Public Health," *Perspectives in Biology and Medicine* 46, no. 3 (2003): 176–198; and Duffy, *Sanitarians*, 188.

46. "R.E. McDonnell," entry dated 1939, in Citizens' Historical Association, "Bibliographical Data of Kansas Citians," (1938–), 283–86, KCPL.

47. "R.E. McDonnell," entry dated 1939, 3–4, in Citizens' Historical Association, "Bibliographical Data of Kansas Citians," (1938–), 283–86, KCPL.

48. "Civics by Radio," *Citizens' League Bulletin*, Feb. 16, 1929, 459; and *Citizens' League Bulletin*, May 12, 1928, 320, KCPL.

49. Anne Whiston Spirn, *The Granite Garden: Urban Nature and Human Design* (New York: Basic Books, 1984); and William H. Wilson, *The City Beautiful Movement in Kansas City* (Columbia: University of Missouri Press, 1964).

50. William H. Wilson, *The City Beautiful Movement* (Baltimore, MD: Johns Hopkins University Press, 1989); and Brown and Dorsett, *K.C.*, esp. chap. 4.

51. Quoted in Vestal, *Missouri*, 125–126.

52. Lawrence H. Larsen and Nancy J. Hulston, *Pendergast!* (Columbia: University of Missouri Press, 1997); William M. Reddig, *Tom's Town: Kansas City and the Pendergast Legend* (Philadelphia: Lippincott, 1947; repr., Columbia: University of Missouri Press, 1986); and Brown and Dorsett, *K.C.*

53. Brown and Dorsett, *K.C.*, 158–159.

54. Kansas City [Missouri] Chamber of Commerce, *Where These Rocky Bluffs Meet: Including the Story of the Kansas City Ten-Year Plan* (Kansas City, MO: Chamber of Commerce, 1938), 240, KCPL.

55. Schneiders, *Unruly River*, see esp. chap. 6.

56. Daphne Spain, *How Women Saved the City* (Minneapolis: University of

Minnesota Press, 2001); Paula Baker, "Domestication of Politics: Women and American Political Society, 1790–1920," *American Historical Review* 89, no. 3 (1984): 620–647; William H. Chafe, "Women's History and Political History: Some Thoughts on Progressivism and the New Deal," in *Visible Women: New Essays on American Activism*, ed. Nancy Hewitt and Suzanne Lebsock (Urbana: University of Illinois Press, 1993); Kathryn Kish Sklar, *Florence Kelley and the Nation's Work: The Rise of Women's Political Culture, 1830–1900* (New Haven, CT: Yale University Press, 1995); and Sellers, *Hazards of the Job*.

57. "City Ownership Finds Champions in Suffragists," Nov. 29, 1913, scrapbook vol. 4, pp. 81–82, HJP.

58. Peggy Masters, "History of the Woman's City Club and Other City Clubs, 1890–1929" [ca. 1992], f. "History," box 31, Kansas City Women's Chamber of Commerce Records, SHSM [hereafter WCC]. See Chafe, "Women's History and Political History."

59. In 1908 the entity was called the Board of Pardons and Paroles and was led by reformer and philanthropist William Volker. For information on social reform, see Brown and Dorsett, *K.C.*, esp. chap. 5; and Ellis, *Civic History*, 182–185.

60. My assumption is that this was paid work. In the 1920s Mayor Beach's office indicated that women's organizations were sometimes asked to volunteer time toward such causes, which they apparently had a history of willingly and enthusiastically doing.

61. Hoy, *Chasing Dirt*, esp. chap. 4, "The American Way."

62. Kendra Smith-Howard, *Pure and Modern Milk: An Environmental History since 1900* (New York: Oxford University Press, 2014), esp. chap. 1.

63. "One Bath for Every 162," *Star*, Apr. 12, 1912, box 4, scrapbook vol. 3, p. 3, HJP.

64. Clipping, *Star*, Apr. 7, 1907, scrapbook vol. 34, p. 117, and vol. 33, p. 98, CCS.

65. Christine Meisner Rosen, *The Limits of Power: Great Fires and the Process of City Growth in America* (Cambridge: Cambridge University Press, 1986).

66. "Water Pressure," *Kansas City World*, Apr. 4, 1900, scrapbook vol. 17, p. 169 and vol. 17, p. 180, CCS.

67. Clipping, Apr. 4, 1900, scrapbook vol. 17, p. 180, CCS.

68. C. A. Burton, Ways and Means Committee of the Commercial Club, "One Hundred Ways to Make an Ideal City," Jan. 28, 1913, vol. 32, CCS.

69. "High Rates for City Water," *Star*, Jan. 5, 1905, scrapbook vol. 26, p. 5, CCS; and "A Municipal Failure," *Journal*, Aug. 19, 1905, scrapbook vol. 27, p. 87, CCS. "Insurance Rates Should Be Lower," *Journal*, Jan. 27, 1908, scrapbook vol. 38, pp. 18–19, CCS; and Club Meeting, Mar. 3 and 4, 1903, vol. 17, p. 49, CCM.

70. National Board of Fire Underwriters, Committee on Fire Prevention and Engineering Standards, *Report on the City of Kansas City, Mo.*, no. 52 (Feb. 1924), ABP.

71. Sanborn Fire Insurance Maps, Kansas City, 1909, Special Collections, University of Missouri, Columbia; and C. A. Burton, Ways and Means Committee of the Commercial Club, "One Hundred Ways to Make an Ideal City," Jan. 28, 1913, vol. 32, CCS.

72. For an idea about the extent to which packing plants altered their environments with water systems and ditches see the Saint Joseph Stockyards Company Records, SHSM.

73. Special Meeting of the Commercial Club Board of Directors, July 9, 1909, vol. 25, p. 2, CCM.

74. R. E. McDonnell, "Romance in Engineering," *Stanford Illustrated Review*, May 1927, 394–395, B&M.

75. Nancy Tomes, *The Gospel of Germs: Men, Women, and the Microbe in American Life* (Cambridge, MA: Harvard University Press, 1998), 2.

76. Duffy, *Sanitarians*, 153–154; and Kansas State Board of Health, *Sixth Biennial Report*, 15.

77. "Report of the State Bacteriologist and Chemist," *Annual Report of the Missouri State Board of Health*, Dec. 11, 1902, 77–79.

78. McLaughlin, *Sewage Pollution*, 37; and Paul D. Haney, "The Missouri River—A Vital Resource" (paper presented at 58th Annual Conference of the Water Pollution Control Federation, Kansas City, MO, 1985), KCPL.

79. Black & Veatch Engineers, graph, "Total Cases of Typhoid Fever, Kansas City, Kans." [ca. 1920], f. 25, Black & Veatch Engineers/Architects Records, SHSM [hereafter B&V]; and Melosi, *Sanitary City: Urban Infrastructure*, 145.

80. "A Municipal Failure," *Journal*, Aug. 19, 1905, scrapbook vol. 27, p. 87, CCS.

81. Letter to the editor from R. E. McDonnell, "High Rate for City Water," Jan. 5, 1905, scrapbook vol. 26, p. 5, CCS. Also "Water Filled with Bacteria," *World*, Dec. 5, 1905, scrapbook vol. 26, n.p., CCS.

82. McLaughlin, *Sewage Pollution*, 40–41; Kansas City, Missouri, Chamber of Commerce, *Health and Hospital Survey* (Kansas City, MO: Lechtman Printing, 1931), 152, KSHS; and Melosi, *Sanitary City: Urban Infrastructure*, 144–145, 147; and National Board of Fire Underwriters, Committee on Fire Prevention and Engineering Standards, *Report on the City of Kansas City, Mo.*, no. 52 (Feb. 1924), 4, ABP.

83. Tarr, *Search for the Ultimate Sink*, see chap. 7–8 generally and 169–173, 192, specifically.

84. Fire and Water Board of Commissioners meeting minutes, June 1924, f. 100, ABP.

85. Letter to Mayor Beach, "City Planning Commission" [ca. 1927], f. 63, ABP.

86. R. E. McDonnell, "Where Cities Get Their Water Supplies," *Citizens' League Bulletin*, Nov. 24, 1928, 409–410, KCPL.

87. Judith Walzer Leavitt, *The Healthiest City: Milwaukee and the Politics of Health Reform* (Princeton, NJ: Princeton University Press, 1982).

88. Tomes, *Gospel of Germs*, 5, 47, 177. The young Swope's death was originally thought to be a case of typhoid fever.

89. "Negro Health Concerns All," *Citizens' League Bulletin*, Apr. 18, 1931, 315, KCPL.

90. Tomes, *Gospel of Germs*, esp. chap. 8; and Tera W. Hunter, *To 'Joy My Freedom: Southern Black Women's Lives and Labors after the Civil War* (Cambridge, MA: Harvard University Press, 1997), esp. chap. 9, on tuberculosis.

91. Dr. J. E. Perry, "Local Disease is a Menace to All Parts of the City," *Citizens' League Bulletin*, May 11, 1935, 489, KCPL; and "Unhealthful Conditions," *Citizens' League Bulletin*, June 4, 1927, 156, KCPL.

92. George C. Whipple, "The Policy of Water Filtration," *Engineering Record* 60, no. 26 (1909): 718–719; "Typhoid Fever Carriers," 60, no. 7 (1909): 175; and "National and State Control over Sewage Disposal," 60, no. 26 (1909): 704.

93. First Annual Report of the Board of Hospital and Health of Kansas City, Missouri (Apr. 1908–Apr. 1909), 39, box 63, Department of Health and Senior Services, Communicable Disease and Environmental Public Health [hereafter DH], Missouri State Archives [hereafter MSA].

94. McLaughlin, *Sewage Pollution*, 41; and Black & Veatch Engineers, graph, "Total Cases of Typhoid Fever, Kansas City, Kans." [ca. 1920], f. 25, B&V.

95. McLaughlin, *Sewage Pollution*, 40–41; Kansas City, Missouri, Chamber of Commerce, *Health and Hospital Survey*, 152, KSHS; and Melosi, *Sanitary City: Urban Infrastructure*, 144–145, 147. No exact date for full-time chlorine use has been found, but current waterworks engineers in Kansas City are confident that it was in the 1910s. Kansas City Board of Hospital and Health records state that the city was sterilizing by 1911 but does not specify the process.

96. Sixth Annual Report of the Board of Hospital and Health of Kansas City, Missouri (Apr. 1913–Apr. 1914), 108–110, box 63, DH, MSA.

97. John McCool, "Present at the Creation," *KU History: The University of Kansas* (blog), accessed Apr. 3, 2015, http://kuhistory.com/articles/present-at-the-creation/.

98. George Hoxie, "Why the Typhoid Mortality in Kansas City?," *Citizens' League Bulletin*, Oct. 14, 1922, 1–2, KCPL; "Water is Worst Carrier of Germs," Sept. 30, 1915, scrapbook vol. 2, n.p., HJP; and Citizens' League letterhead, f. 107, ABP.

99. R. E. McDonnell, "Where Cities Get Their Water Supplies," *Citizens' League Bulletin*, Nov. 24, 1928, 409, KCPL.

100. Kansas City Hospital and Health Board weekly report, "Kansas City's Health," Jan. 30, 1926, 2, f. 264, ABP.

101. R. E. McDonnell, "The Bane of Politics in the Water Department and the Remedies," *Water Works Engineering* 84, no. 10 (1931): 659.

102. "Meeting of the Kansas Water and Sewage Works Association," *Water Works and Sewerage* 83, no. 6 (1936): 221; Robert E. McDonnell, "The Engineer Looks at Management," *Journal of the American Waterworks Association* 32, no. 6 (1940); R. E. McDonnell, "Money Used for Water Works Brings Satisfactory Rewards," *Water Works Engineering* 81, no. 17 (1928): 1195, which was based on a paper given at the Missouri Conference on Water Purification in 1928; and McDonnell, "Are Water Rates Adequate for the Service Now Demanded?," *Public Service Management* 44, no. 6 (1928): 213. McDonnell thought the courts were not aggressive enough, but by 1928 he could point to thirteen cases where the courts established the responsibility of a city to provide dependably clean water.

103. "Memorandum upon Application for Temporary Injunction," May 6, 1924, 9–17, f. 104, ABP.

104. National Board of Fire Underwriters, Committee on Fire Prevention and Engineering Standards, *Report on the City of Kansas City, Mo.*, no. 52 (Feb. 1924), 10–12, 32–33, and quote on 35, ABP.

105. Fire Underwriters, *Report on the City of Kansas City*, 3–8.

106. Brown and Dorsett, *K.C.*, esp. chap. 5.

107. See, for example, Fuller & Maitland letter to Board of Fire and Water Commissioners, June 5, 1924, 1–10, f. 105, ABP.

108. "Water Board Meets Tonight," *Star*, Apr. 13, 1922, SHSM; "Water Body in Tomorrow," *Star*, Apr. 16, 1922, SHSM; and "Charter Vote is Near," *Star*, Nov. 12, 1922, SHSM.

109. Alexander Maitland, "Why Kansas City Should Vote for Water Bonds," *Citizens' League Bulletin*, Mar. 4, 1922, 4, KCPL.

110. Maitland, "Why Kansas City Should Vote for Water Bonds," 4.

111. "What the American Medical Association Thinks of Kansas City's Water Supply," *Citizens' League Bulletin*, Mar. 11, 1922, 6, KCPL.

112. R. E. McDonnell, "How Good Health Can Be Purchased," *Citizens' League Bulletin*, Apr. 1, 1922, 4, KCPL; and Black & Veatch Engineers/Architects, graphs [ca. 1920], f. 25, B&V.

113. Maitland, "Why Kansas City Should Vote for Water Bonds," 3–4.

114. Louis Rothschild, "No City Can Expand beyond Its Water Supply," *Citizens' League Bulletin*, Mar. 25, 1922, 1, KCPL.

115. Mary M. Miller, "Shall We Have Soft Water in Kansas City?," *Citizens' League*

Bulletin, Mar. 3, 1928, 277–278, KCPL; and "Water Bonds Needed," Citizens' League Bulletin, Mar. 21, 1931, 299, KCPL.

116. Clipping, "Rally to Soft Water," Star [ca. 1928], Georgia McDonnell scrapbook, B&M.

117. "Charter Vote is Near," Star, Nov. 12, 1922, SHSM. For the number of the ballots cast for and against, see "The Water Bonds," Star, Apr. 5, 1922, SHSM.

118. "Alternative Fight is On," Star, Nov. 14, 1922, 2, SHSM.

119. "Back to Old Board," Star, Dec. 7, 1922, SHSM. For legal discussion see "Water Board Out," Star, Dec. 6, 1922, SHSM.

120. Cartoon appeared on the front page of the Star, Dec. 22, 1922, SHSM.

121. "Citizens Back into Action," Star, Dec. 7, 1922, SHSM.

122. Brown and Dorsett, K.C., chap. 5 and especially 144–152; and "Beach to Office Today," Times, Apr. 21, 1924, 1, SHSM.

123. "Water Board 'Hops to It,'" Kansas City Times, Apr. 25, 1924, 1, SHSM.

124. Contract between Kansas City and Fuller & Maitland engineers, June 1923, sec. 1, p. 1, f. 104, ABP.

125. Letter from Barnett to Bryant, July 21, 1924, 4, f. 105, ABP; Letter from Spencer to Beach, Apr. 4, 1925, f. 107, ABP; letter from City of Saint Louis to Kansas City, Missouri, Dec. 8, 1924, ABP; letter from Kingsley to Nolte, Dec. 13, 1924, ABP; and letter from Lee to Gilmore, Aug. 6, 1924, 1–14, f. 106, ABP.

126. "Beach to Office Today," Times, Apr. 21, 1924, 1, SHSM; and "Bosses Lose Kansas City," Weekly Kansas City Star, Apr. 16, 1924, 1, SHSM.

127. "Highest Salaries," Star, Apr. 13, 1924, 2A, SHSM.

128. "Water Suit Decision Today," Star, Apr. 13, 1924, 1, SHSM.

129. "Water Suit Decision Today," Star, Apr. 13, 1924, 1, SHSM. Weaver, Bullene, Woodward, Shepard, and Dean were the plaintiffs.

130. "Kills Water Deal," Star, Apr. 14, 1924, 1, SHSM.

131. "Court Halts Spending Orgy," Weekly Kansas City Star, Apr. 16, 1924, 1, SHSM.

132. US District Court of Western Missouri papers, filed Apr. 12, 1924, 11, f. 104, ABP. Letter from Ash to Fire and Water Commissioners, Aug. 13, 1924, 1, f. 106, ABP. As early as 1923, Fuller and Maitland reported as much. "Kills Water Deal," Star, Apr. 14, 1924, 1, SHSM.

133. "Kills Water Deal," Star, Apr. 14, 1924, 1, SHSM.

134. Information regarding the committee found in "Memorandum upon Application for Temporary Injunction," May 6, 1924, 9–17, f. 104, ABP.

135. "Signs Three Spite Bills," Star, Apr. 23, 1924, 1, SHSM; "Audit by Popular Gifts," Star, Apr. 24, 1924, 1, SHSM; and the front page of Times, Apr. 25, 1924,

SHSM, contains a list of names. Contributors included William Volker, W. S. Dickey, Conrad Mann, J. W. Perry, and Louis Rothschild. The audit included more than just the waterworks.

136. With the declarations of unconstitutionality, the board remained unreformed. However, Beach could appoint board members as if the reforms were in place. The word *nonpartisan* fell out of use and instead *bipartisan* was used to refer to the board. Occasionally the board was still called "nonpartisan," but political affiliations were listed. Edwards and Bryant had been part of the "ousted" and unconstitutional nonpartisan commission. Edwards was on the aforementioned committee that worked to influence the direction of the new waterworks. Buchholz had served on the park board. "New Water Board," Star, Apr. 24, 1924, SHSM.

137. "Engineers Oppose City," Star, Apr. 16, 1924, 1, SHSM.

138. US District Court of Western Missouri papers, filed Apr. 12, 1924, quote p. 4, f. 104, ABP. The injunction was dismissed May 6, 1924, by Judge Van Valkenburgh.

139. Fire and Water Board of Commissioners Minutes, June 12, 1924, and June 24, 1924, 7, f. 100, ABP.

140. Quote from Josiah Strong is printed on the letterhead of Citizens' League of Kansas City, from Nat Spencer to Beach, Apr. 30, 1925, f. 107, ABP.

141. "A New Deal in Water," Times, Apr. 25, 1924, 1, SHSM.

142. Letter from Kiersted to Board of Fire and Water Commissioners, June 10, 1924, 1–3, f. 105, ABP; and on the subject of ice, see letter from [Fuller & Maitland to Fire and Water] titled "Notes for discussion," Aug. 8, 1924, 7, f. 106, ABP.

143. Letter from E. E. Harper to I. N. Watson, May 7, 1924, 1, f. 105, ABP.

144. Kiersted to Fire and Water Commissioners, June 10, 1924, 1–3, f. 105, ABP; and Wynkoop Kiersted, "Estimate Cost of Building Water Supply Works at Quindaro, Kansas City, Missouri," and "Estimate of Constructing Water Supply Works and East Bottoms Pumping Station," both dated June 10, 1924, f. 105, ABP.

145. Edwards to Fire and Water, Aug. 13, 1924, f. 106, ABP; and quote from Advisory Committee letter signed by Beach to Fire and Water, Aug. 13, 1924, 1–2, f. 106, ABP.

146. On the water supply relationship between the cities, see letter from Bruce Barnett to Bryant of Fire and Water Board, July 21, 1924, 2–4, f. 105, ABP; and on Ash's position see his letter to Fire and Water Commissioners, Aug. 13, 1924, 1, f. 106, ABP.

147. Fuller & Maitland to Board of Fire and Water Commissioners, "Memorandum," June 20, 1924, 3–5, f. 105, ABP.

148. Cartoon by Gale Stockwell, *Citizens' League Bulletin*, Mar. 17, 1923, 4, KCPL.

149. For information on the election, charter, and new city operations, see Ellis, *Civic History*, 62–67. See also Brown and Dorsett, K.C., chap. 5.

150. Kansas City Board of Fire and Water Commissioners minutes from 1924 to 1926 contain extensive information on all aspects of the old and new waterworks, and for specific examples about the tunnels see Fire and Water Board minutes, Jan. 15, 1925, p. 2, f. 101; Sept. 8, 1925, p. 2; Oct. 15, 1925, p. 2; Oct. 27, 1925, p. 2; Nov. 16, 1925, p. 1; Nov. 10, 1925, pp. 1–2; Mar. 25, 1926, p. 3, f. 102, ABP.

151. Board of Fire and Water Commissioners minutes, Feb. 25, 1926, 4–6, f. 102, ABP.

152. Kansas City, Missouri Water Services Department, *Annual Report, 1998–1999*.

153. Debris screening, presedimentation, sedimentation, coagulation, filtration, and chlorination would all be part of the treatment process. "Kansas City Annual Meeting," *Journal of American Public Health* 28, no. 8 (Aug. 1938): 990.

154. Robert E. McDonnell, "How Water Works Can Give Real Service," *Water Works Engineering* 82, no. 24 (1929): 1666.

155. McDonnell, "Are Water Rates Adequate," 213.

156. For a discussion of professionals as members of society, see John P. Herron, *Science and the Social Good: Nature, Culture, and Community, 1865–1965* (New York: Oxford University Press, 2010).

157. Board of Fire and Water Commissioners minutes, Dec. 29, 1925, 3, f. 102, ABP. In 1926 Kansas City, Kansas, asked if it could buy some land to enlarge Quindaro. Apparently, when it took over Quindaro it upgraded water treatment.

158. *American City* 46, no. 6 (1932): 53–58. The Kansas side, which today supplies a number of other cities, also relies on riverine wells.

159. Burns & McDonnell, *The Purification of Water* (pamphlet, 1927), 29, B&M; and Martin Melosi, *The Sanitary City: Environmental Services in Urban America from Colonial Times to the Present*, abridged ed. (Pittsburgh, PA: University Press of Pittsburgh, 2008), 94–96.

160. American Public Health Association, "Annual Meeting," 1938, 990, KCPL.

161. Heat-Moon, *River-Horse*, 228.

Chapter Four: Sewers: The Waste Stream

Epigraph: Paul Hohl, "City beneath a City," *Kansas City Magazine*, Sept. 1979, 68, Missouri Valley Room, KCPL.

1. Richard Rhodes, *A Hole in the World: An American Boyhood* (New York: Simon & Schuster, 1990), 137–140.

2. 1946 report, p. 3, f. 15, B&V; and Kansas City, Missouri Water Services Department, *Annual Report, 1998–1999*, 1999.

3. Octave Chanute, "The Sewerage of Kansas City," *Kansas City Review of Science and Industry* 7, no. 9 (1884): 519–527; and George E. Waring Jr., ed., *Report on the*

Social Statistics of Cities (Washington, DC: US Government Printing Office, 1887), 2:557.

4. Tarr, Search for the Ultimate Sink, 117; and Martin V. Melosi, The Sanitary City: Urban Infrastructure in America from Colonial Times to the Present (Baltimore, MD: Johns Hopkins University Press, 2000), 92–98, 153–160, 238.

5. Chanute, "Sewerage of Kansas City," 520, 521.

6. Hoy, Chasing Dirt, 70–72; Melosi, Sanitary City: Urban Infrastructure; and Tomes, Gospel of Germs.

7. Chanute, "Sewerage of Kansas City."

8. Duffy, Sanitarians; and Melosi, Sanitary City: Urban Infrastructure, 111–112.

9. Medora Crane, "Water Plant Purification Expert Made His Home in Liberty," Dec. 2, 1987, v.f. "Liberty, Mo.," KCPL; and Wynkoop Kiersted, Prevailing Theories and Practices relating to Sewage Disposal (New York: John Wiley & Sons, 1894), f. 27, Wynkoop Kiersted Papers, SHSM [hereafter WKP].

10. Kiersted, Prevailing Theories and Practices, 172, f. 27, WKP.

11. Paul Hohl, "City beneath a City," Kansas City Magazine, Sept. 1979, 68, Missouri Valley Room, KCPL. Westport's separate system was converted after annexation.

12. Melosi, Sanitary City: Urban Infrastructure, 163–164.

13. Waring, Report on the Social Statistics of Cities, 2:557. Though still representative, the statistic of half is from the 1880s.

14. "No Grades, No Water; New Sections Suffer," Aug. 16, 1913, scrapbook vol. 1, p. 118, HJP.

15. "Sewers Menace the City," Post, Nov. 18, 1915, 1–2, CCS.

16. Ellis, Civic History, 172.

17. "Manufacture of Sewer Pipe," Jan. 1898, scrapbook vol. 13, p. 66, CCS.

18. "Dickey in Base Plot to Regain Pipe Control," and "How Dickey and Nelson Tried to Rebuild Sewer Monopoly," Post, June 21, 1914; and "Nelson and Dickey in a Combine for Mutual Welfare," Post, June 25, 1914, all scrapbook vol. 5, p. 9, HJP.

19. Brown and Dorsett, K.C., 198–199.

20. "A Day in Kansas City, Kansas," Club Member 4, no. 4 (1907): 12, KSHS.

21. Clipping, Journal, Jan. 22, 1907, scrapbook vol. 33, p. 99, CCS.

22. Ellis, Civic History, 69.

23. Waring, Report on the Social Statistics of Cities, 2:557.

24. See file on "Beardsley Road and Bluff Street," n.d., box 8, f. 39, CIDA.

25. Ellis, Civic History, 172–173.

26. Star, Mar. 12, 1922, box 1, f. 9, CIDA.

27. "Memorandum on Conference with Kansas City, Kansas Officials relative to Improvements of Kansas Side of the C.I.D.," Aug. 20, 1954, box 8, f. 33, CIDA.

28. "Sewer Study," 1972, 2–3, box 5, f. 15, CIDA.

29. *Arn v. City of Kansas City*, Federal Reporter, no. 14, 1882, 236–238.

30. "Odor of Sewers Draws Ire of Home Owner," Mar. 22, 1915, scrapbook vol. 5, p. 84, HJP.

31. Letter to the editor from L. M. Miller, "Want an Open Sewer Closed," June 20, 1905, scrapbook vol. 27, p. 28, CCS.

32. Quoted in "Those Filthy Sewers," *Journal*, Mar. 26, 1902, scrapbook vol. 21, pp. 103–104, CCS; and minutes, Nov. 1897, vol. 11, p. 163, CCM.

33. "War on Unflushed Sewers," *Star*, Mar. 26 1902, scrapbook vol. 21, p. 104, CCS.

34. Letter from George Waring, report of the Committee on Municipal Legislation, Feb. 1904, vol. 18, pp. 48–49, CCM.

35. Report of the Committee on Municipal Legislation, Commercial Club minutes, Feb. 1904, vol. 18, pp. 49–50, CCM; and "Kansas City Cleans its Sewers," *American City* 63, no. 11 (1948): 11.

36. Emphasis mine. "Big Bond Issue," *Journal*, Jan. 1, 1905, scrapbook vol. 26, p. 1, CCS.

37. "Flushing of Sewers is Ordered by Mayor," Mar. 26, 1915, scrapbook vol. 5, p. 86, HJP.

38. "Sewer Odors Can Be Stopped at Small Cost with 2 Extra Men, City Engineer Declares," Mar. 25, 1915, scrapbook vol. 5, p. 89, HJP.

39. Jost quoted in "Flushing of Sewers is Ordered by Mayor," Mar. 26, 1915, scrapbook vol. 5, p. 86, HJP.

40. "Kansas City Cleans Its Sewers," 11.

41. "Sewers and Drainage Problems" [ca. 1949], box 8, f. 33, CIDA.

42. Letters between Coerver, Shumway, Waddell, Beach, and Hughes, 1924–1926, f. 90, ABP; and Board of Public Work minutes, July 27, 1924, f. 246, ABP.

43. "Caring for Sewage a Knotty Problem," Nov. 29, 1908, scrapbook vol. 40, p. 109, CCS.

44. "Mayor Is against Park Bond Issues," Mar. 4, 1914, scrapbook vol. 1, p. 254, HJP.

45. "Sewage into Rivers," *Citizens' League Bulletin*, Oct. 9, 1926, 20, KCPL.

46. "Caring for Sewage a Knotty Problem," Nov. 29, 1908, scrapbook vol. 40, p. 109, CCS.

47. Wilson, *City Beautiful Movement in Kansas City*.

48. Board of Park Commissioners, "Park System of Kansas City," 1914, 44, Kansas City, Missouri, Parks and Recreation Archives [hereafter KCPR].

49. Commercial Club minutes, Nov. 1897, vol. 11, pp. 163–164, CCM; and "Big Bond Issue," *Journal*, Jan. 1, 1905, scrapbook vol. 26, p. 1, CCS.

50. McLaughlin, *Sewage Pollution*, 40, map 7.

51. Superintendent W. H. Dunn, "Condensed Resume of Park Problems," 1913, 23–24, f. "Park System, General Information," Board of Park Commissioners, Resolutions and Ordinances (roll PA-72), KCPR.

52. McLaughlin, *Sewage Pollution*, 38, map 4; and "Caring for Sewage a Knotty Problem," Nov. 29, 1908, scrapbook vol. 40, p. 109, CCS.

53. A. H. Pierce et al. opposed bonds in 1906 and were quoted in "Fight Sewer Park," *Journal*, Aug. 18, 1906, scrapbook vol. 31, pp. 28–29, CCS.

54. Kansas City Board of Public Welfare Research Bureau, *Social Prospectus of Kansas City, Missouri* (Kansas City, MO: Kansas City Board of Public Welfare, 1912), 121; and Brown and Dorsett, *K.C.*, 153–158. The Board of Public Welfare reports can be found in more than one location, including SHSM, KCPL, and the University of Missouri libraries.

55. Kansas City [Missouri] Board of Park Commissioners, "Special Report from the Blue Valley Parkway," Mar. 1912, f. 60, Kansas City's Ten-Year Plan Records, SHSM [hereafter TYP].

56. See, for example, Karl Jacoby, *Crimes against Nature: Squatters, Poachers, Thieves, and the Hidden History of American Conservation* (Berkeley: University of California Press, 2001).

57. *Report of the Board of Park and Boulevard Commissioners of Kansas City, Missouri* (Kansas City, MO: 1893), reprinted in Charles N. Glaab, *The American City: A Documentary History* (Homewood, IL: Dorsey Press, 1963), 257–263.

58. *Report of the Board of Park and Boulevard Commissioners of Kansas City, Missouri, 1893*, in Glaab, *American City*, 257.

59. Ellis, *Civic History*, 94.

60. Board of Park Commissioners, "Special Report from the Blue Valley Parkway," Mar. 1912, 28–30, f. 60, TYP.

61. Board of Park Commissioners, "Special Report from the Blue Valley Parkway," Mar. 1912, 24, f. 60, TYP; and Superintendent W. H. Dunn, 1913 report, f. "Park System, General Information," Board of Park Commissioners, Resolutions and Ordinances (roll PA-72), KCPR.

62. Superintendent W. H. Dunn, "Condensed Resume of Park Problems," 1913, 30–33, 37, f. "Park System, General Information," Board of Park Commissioners, Resolutions and Ordinances (roll PA-72), KCPR.

63. See essays in William Cronon, ed., *Uncommon Ground: Rethinking the Human Place in Nature* (New York: W. W. Norton, 1995).

64. Blake McKelvey, *The Urbanization of America, 1860–1915* (New Brunswick, NJ: Rutgers University Press, 1963); Brown and Dorsett, *K.C.*, 169; and Ellis, *Civic History*, 96.

65. "Sewers Menace Health of City," Post, Nov. 18, 1915, 1–2, CCS.

66. Speaker's Committee, "What K.C. Is to Get for Ten Year Plan," 1930, f. 66, TYP.

67. "Boil Water of Wells near Blue River; They May All Be Dangerous," Sept. 19, 1915, scrapbook vol. 2, HJP.

68. Board of Park Commissioners, "Report of the Board of Park Commissioners of Kansas City, Mo. 1914," 9, SHSM.

69. Letter from Pearse to Beach, July 17, 1925, f. 264, ABP.

70. "Kansas City's Health," 1925 conference summary, f. 264, ABP.

71. "Pollution of the Blue to Stop," July 16, 1913, scrapbook vol. 1, p. 95, HJP.

72. Frank Bartonek, "Armco-K.C., 75-Years Old and Still Growing," *Greater Kansas City Monthly*, Mar. 1976, 18–19, 28, box 46, f. 31, American Institute of Architects [hereafter AIA], Kansas City Chapter Records, SHSM; and A. H. Sheffield, "Blue Valley District an Important Asset," *Citizens' League Bulletin*, May 17, 1930, 1, v.f. "Citizens' League," KCPL.

73. C. A. Burton, "Protect the East Bottoms," *Citizens' League Bulletin*, Nov. 26, 1921, KCPL; clippings and photos, v.f. "East Bottoms," KCPL; and Kansas City Board of Public Welfare Research Bureau, *Social Prospectus of Kansas City*, 27, 113, 120.

74. Kansas City Board of Public Welfare Research Bureau, *Social Prospectus of Kansas City*, 27, 113, 120.

75. Letter from Loftus to Beach, June 17, 1927, f. 251, ABP.

76. Quote from A. H. Sheffield, "Blue Valley District an Important Asset," *Citizens' League Bulletin*, May 17, 1930, 1, v.f. "Citizens' League," KCPL; Superintendent W. H. Dunn, "Condensed Resume of Park Problems," 1913, 29, f. "Park System, General Information," Board of Park Commissioners, Resolutions and Ordinances (roll PA-72), KCPR; and Kansas City Board of Public Welfare Research Bureau, *Social Prospectus of Kansas City*, 27, 113, 120.

77. C. A. Burton, "Protect the East Bottoms," *Citizens' League Bulletin*, Nov. 26, 1921, KCPL; and clippings and photos, v.f. "East Bottoms," KCPL.

78. D. M. Bone, ed., *Annual Review of Greater Kansas City* (Kansas City, MO: Bishop Press, 1908), 14, KCPL.

79. Kansas City Board of Public Welfare Research Bureau, *Social Prospectus of Kansas City*, 121–123.

80. "City Playground Will Be Ruined by Sewer into Blue," Post, Jan. 20, 1914, scrapbook vol. 4, p. 104, HJP; and "Sewer into Blue Would Ruin It as Pleasure Stream," Post, Jan. 27, 1914, scrapbook vol. 4, p. 108, HJP.

81. Kansas City Board of Public Welfare Research Bureau, *Social Prospectus of Kansas City*, 70–71.

82. Kansas City [Missouri] Chamber of Commerce, *Where These Rocky Bluffs Meet: Including the Story of the Kansas City Ten-Year Plan* (Kansas City, MO: Chamber of Commerce, 1938), KCPL.

83. Kansas City [Missouri] Chamber of Commerce, *Where These Rocky Bluffs Meet*, 69, 87; and Ellis, *Civic History*, 180. In June 1915 and November 1925, large bond measures for hospitals passed.

84. "Sewers Menace Health of City," *Post*, Nov. 18, 1915, 1, CCS.

85. "Kansas City's Health," 1925 conference summary, f. 264, ABP.

86. See *Star* clipping and letters, Nov. 1924, f. 265, ABP; Mayor Beach statements [ca. 1925], f. 265; and "Mayor's Statement on Sewer Engineering Bill" [ca. 1928], f. 236, ABP.

87. "Kansas City's Health," 1925 conference summary, f. 264, ABP.

88. Hospital and Health Board meeting, 1925, f. 264, ABP; and "Kansas City's Health," 1925 conference summary, 2, f. 264, ABP.

89. Letter from Stoeltzing to Beach, June 13, 1927, f. 251, ABP.

90. Letter to Beach, f. 137, ABP.

91. *Third Annual Report of the Hospital and Health Board of Kansas City, Missouri, 1910–1911*, 30–31, box 63, DH, MSA.

92. "Sewers Menace Health of City," *Post*, Nov. 18, 1915, 1–2, CCS.

93. Kansas City Chamber of Commerce, *Kansas City Health and Hospital Survey*, 156.

94. "Impure Water Causes High Death Rate Here," *Journal*, May 4, 1912, box 4, scrapbook vol. 1, p. 22, HJP; and letter from McDonnell to DeLano, Feb. 23, 1925, f. 221, ABP.

95. Kansas City Chamber of Commerce, *Kansas City Health and Hospital Survey*, 152.

96. C. A. Burton, Ways and Means Committee of the Commercial Club, "One Hundred Ways to Make an Ideal City," Jan. 28, 1913, vol. 32, CCS; Kansas City Chamber of Commerce, *Kansas City Health and Hospital Survey*, 152; and Civic Department of the Kansas City Chamber of Commerce, Public Health and Welfare Committee, "Health Conservation Contest Report of Kansas City, Missouri for 1933," Feb. 1934, KCPL.

97. Kansas City Chamber of Commerce, *Kansas City Health and Hospital Survey*, 155–156.

98. Melosi, *Sanitary City: Urban Infrastructure*, 235–236.

99. McDonnell to Delano, Feb. 23, 1925, f. 221, ABP.

100. Kansas City Public Service Institute, "Jackson County Sewer District Number 1," May 1930, 1–5; and *Citizens' League Bulletin*, June 6, 1931, KCPL.

101. See letters dated 1925–1928, f. "Brush Creek Parkway" and "Condemna-

tions," Board of Park Commissioners, Correspondence & Report Files (roll PA-28-29), KCPR; and Kansas City [Missouri] Chamber of Commerce, *Where These Rocky Bluffs Meet: Including the Story of the Kansas City Ten-Year Plan* (Kansas City, MO: Chamber of Commerce, 1938), KCPL.

102. Kansas City [Missouri] Chamber of Commerce, *Where These Rocky Bluffs Meet*, 202.

103. "News and Ideas for Commercial and Civic Organizations," *American City* 24, no. 5 (1921): 513, 515.

104. Brown and Dorsett, *K.C.*, esp. chap. 6; and Larsen and Hulston, *Pendergast!*, esp. chaps. 3 and 4.

105. Kansas City [Missouri] Chamber of Commerce, *Where These Rocky Bluffs Meet*, 88, 94.

106. See clippings, f. 9, TYP; Brown and Dorsett, *K.C.*, 196; and Kansas City [Missouri] Chamber of Commerce, *Where These Rocky Bluffs Meet*, 97.

107. Larsen and Hulston, *Pendergast!*, 89.

108. Kansas City [Missouri] Chamber of Commerce, *Where These Rocky Bluffs Meet*, 235–236.

109. Kansas City [Missouri] Chamber of Commerce, *Where These Rocky Bluffs Meet*, 192; and Brown and Dorsett, *K.C.*, 196.

110. As quoted in Kansas City [Missouri] Chamber of Commerce, *Where These Rocky Bluffs Meet*, 173.

111. Larsen and Hulston, *Pendergast!*, 90.

112. Parrish, Jones, and Christensen, *Missouri*, 310–311.

113. Brown and Dorsett, *K.C.*, 199–200.

114. For clippings, see v.f. "New Deal" parts I and II, KCPL.

115. List of bonds ("No. 9"), *Journal-Post*, May 24, 1931, f. 9, TYP; and Kansas City [Missouri] Chamber of Commerce, *Where These Rocky Bluffs Meet*, 137–138.

116. Kansas City [Missouri] Chamber of Commerce, *Where These Rocky Bluffs Meet*, 145–147, 202, 218.

117. "A New Channel," *Times*, Feb. 29, 1940, box 6, f. 1648-535; and "To Stretch Blue River," *Times*, Aug. 29, 1939, box 6, f. 1648-414, Kansas City, MO, District Office, General Administrative Files, 1930–1943, Records of the Office of the Chief of Engineers, US Army Corps of Engineers, National Archives and Records Administration–Central Plains Region [hereafter USACE, NARA-CPR].

118. "The Future of the Blue," *Times*, Mar. 16, 1937, box 4, f. 1648-107, Kansas City District, General Administrative Files, 1930–1943, USACE.

119. Kansas City [Missouri] Chamber of Commerce, *Where These Rocky Bluffs Meet*, 137–138.

120. "Relief—Missouri," *Times*, Nov. 16, 1936, v.f. "New Deal" part I, KCPL;

and Kansas City [Missouri] Chamber of Commerce, *Where These Rocky Bluffs Meet*, 147.

121. Kansas City [Missouri] Chamber of Commerce, *Where These Rocky Bluffs Meet*, 208–209, 214.

122. Kansas City [Missouri] Chamber of Commerce, *Where These Rocky Bluffs Meet*, 194.

123. Kansas City [Missouri] Chamber of Commerce, *Where These Rocky Bluffs Meet*, 198–199; Larsen and Hulston, *Pendergast!*, 88–89; and Brown and Dorsett, *K.C.*, 198–200.

124. "Relief—Missouri," *Times*, Apr. 28, 1936, v.f. "New Deal" part I, KCPL.

125. Letter from "JCN" to Arthur Hardgrave, Sept. 12, 1928, p. 2, f. "Brush Creek Parkway," Correspondence and Report Files (roll PA-28), Board of Park Commissioners, KCPR. "JCN" is probably J. C. Nichols, whose developments bordered the creek.

126. Paul Hohl, "City beneath a City," *Kansas City Magazine*, Sept. 1979, 69, Missouri Valley Room, KCPL.

127. Ken Thornton of the US Army Corps of Engineers, quoted in Hohl, "City beneath a City," 82.

128. "A Portfolio of New Deal Construction," *Fortune* 14, no. 5 (1936): 76, 80–81; and Neil Maher, "A New Deal Body Politic: Landscape, Labor, and the Civilian Conservation Corps," *Environmental History* 7, no. 3 (July 2002): 435–461.

129. "Renting Sewer Service," *Citizens' League Bulletin*, Nov. 12, 1932, 40; and Melosi, *Sanitary City: Urban Infrastructure*, 240–241.

130. Kansas City [Missouri] Chamber of Commerce, *Where These Rocky Bluffs Meet*, 202, 218.

131. Metro Planning Commission, "1990 Sketch Plan," 1968, 28–31, 38–39, box 6, f. 24, AIA.

132. Postcard and text to Star article, May 22, 1971, accessed June 27, 2001, http://www.kclibrary.org/sc/post/rivers/20000161.htm.

133. Water Pollution Control, Department of Environmental Quality [hereafter WPC, DEQ], "Water Quality Management Basin Plan for Lower Missouri River Basin," June 1976, 55, Missouri Department of Natural Resources [hereafter DNR], MSA.

134. United States Environmental Protection Agency [hereafter EPA], Office of Water Quality, Region VII, *Everyone Can't Live Upstream: A Contemporary History of Water Quality Problems on the Missouri River, Sioux City, Iowa to Hermann, Missouri* (Kansas City, MO: US Environmental Protection Agency, 1971), 163–164.

135. WPC, DEQ, "Water Quality Management Basin Plan for Lower Missouri

River Basin," June 1976, 55, 57, 68–70, quotes to 57, DNR, MSA; and Karen Dillon, "Water Study Finds Polluted Streams," Star, Sept. 11, 2006.

136. In theory, clean water is returned to the river after secondary treatment. There are fewer bacteria and nutrients, for example. However, recent studies have begun to show that chemicals might become dangerously airborne in the incineration process or that pharmaceuticals remain in the water after treatment. One of the earliest studies to prove the latter involved the Blue River waste treatment plant in the 1970s. Leah Eisenstadt, "Drugs in the Water: How Our Medicine Cabinets are Contaminating Nature," Triplepoint, 2005, accessed Apr. 4, 2018, http://www.bu.edu/sjmag/scimag2005/features/drugsinwater.htm.

PART II. REGION: BORDERS, BONDS, AND BODIES

Epigraph: John Muir, My First Summer in the Sierra (Boston: Houghton Mifflin, 1944), 157.

CHAPTER FIVE: SISTER CITIES

Epigraph: Gov. Edward W. Hoch, "Address by Edward W. Hoch, Governor of Kansas," in Proceedings of a Conference of Governors in the White House, May 13–15, 1908 (Washington, DC: Government Printing Office, 1909), available online in The Evolution of the Conservation Movement, 1850–1920, American Memory Collection, Library of Congress, accessed Apr. 4, 2018, http://memory.loc.gov/cgi-bin/query/r?ammem/consrv:@field(DOCID+@lit(amrvgvg16div83)).

1. Kate L. Cowick, The Story of Kansas City (booklet of articles published by the Kansas City Kansan, 1924–1926), 3, 14, KSHS.

2. Susan Strasser, Waste and Want: A Social History of Trash (New York: Metropolitan Books, 1999), 32.

3. Kansas City, Kansas, Chamber of Commerce, "Tour of Kansas City, Kansas" [1920s], 3 pp., f. Wyandotte County Manuscripts, KSHS.

4. Greater Kansas City Official Yearbook, 1904–1905, 75, KCPL.

5. Charles Brokaw, "Pulling Together in Greater Kansas City," Citizens' League Bulletin, June 16, 1929, 341–342, KCPL.

6. Carl Dehoney, "What Kansas City Owes to Kansas," Kansas City Spirit 1, no. 3 (1908): 3–4, KSHS.

7. Dehoney, "What Kansas City Owes to Kansas," 3–4.

8. Quote to L. R. Ash, "Comprehensive Development Plan Needed," Citizens' League Bulletin, June 2, 1928, 329–330, KCPL; for boosterism, see Wohl and Brown, "Usable Past"; and Cronon, Nature's Metropolis, 31–46, 396n45.

9. Brokaw, "Pulling Together in Greater Kansas City," 341–342.

10. Brokaw discussed by Dehoney, "What Kansas City Owes to Kansas," 4.

11. Kansas City Board of Public Welfare Research Bureau, *Social Prospectus of Kansas City, Missouri*, 52–53.

12. Bone, *Annual Review of Greater Kansas City*; and "Kansas City Most American of Cities," *Star*, Jan. 30, 1908, scrapbook vol. 38, pp. 23–24, CCS.

13. A. H. Sheffield, "Blue Valley District an Important Asset," *Citizens' League Bulletin*, May 17, 1930, KCPL.

14. Mrs. Ray Conlin in report of the Civics Committee, Annual Meeting Minutes, May 13, 1940, p. 5, box 1, f. 13, WCC. Decades later, in the 1970s, when cultural diversity had taken on a new meaning, the Kansas City, Kansas, mayor described the city as "cosmopolitan" and mentioned the diversity of residents with Slavic, Mexican, and Russian origins. See Joseph H. McDowell, *Building a City: A Detailed History of Kansas City, Kansas* (Kansas City, KS: Kansas City Kansan [ca. 1970]), 11–12.

15. Amahia Mallea, "Progressive Black Kansas City" (paper presented at Missouri Conference on History, Kansas City, MO, 2002).

16. Margaret Landis, "The Winding Valley and the Craggy Hillside: A History of the City of Rosedale, Kansas," 1976, Kansas Collection, Kansas City, Kansas, Public Library, accessed Apr. 4, 2018, https://www.kckpl.org/kansas/documents/winding-valley-rosedale-history.pdf. For geologic history, see "How Pre-Historic Turkey Creek Saved Kansas City Millions," in v.f. "Turkey Creek," KCPL.

17. Because sources often adhere to political boundaries, telling the story of a watershed or region can be difficult, but environmental historians are producing works that transgress boundaries: Donald Pisani, *Plumbing the Truckee: Water, Diversion and the Creation of Community along the Truckee River, Nevada* (ProQuest, 2007); Dan Flores, *Horizontal Yellow: Nature and History in the Near Southwest* (Albuquerque: University of New Mexico Press, 1999); and Schneiders, *Big Sky Rivers*.

18. Ellis, *Civic History*, 59.

19. Kansas City [Missouri] Chamber of Commerce, *Where These Rocky Bluffs Meet: Including the Story of the Kansas City Ten-Year Plan* (Kansas City, MO: Chamber of Commerce, 1938), 48–49, KCPL.

20. Waring, *Report on the Social Statistics of Cities*, 2:557.

21. W. B. Storey, *Committee on Surveys and Plans for Controlling Floods on the Kaw River* (Missouri River Commission report, 1905), p. 5, f. 281, box 12, Sioux City District, General Administration, USACE, NARA-CPR; and Kansas City engineer Pike and Hiram Chittenden of the US Army Corps Engineers discussion, Dec. 1904 to Jan. 1905, 4 pp., f. 364, box 23, Sioux City District, General Administration, USACE, NARA-CPR.

22. Kansas State Board of Health, *Sixth Biennial Report*, 55.

23. McDowell, *Building a City*, 63. Incidentally, because of its orphaned geographical status, Greystone Heights would still not be completely sewered in the 1960s.

24. Kansas State Board of Health, *Third Biennial Report of the Kansas State Board of Health, 1905–1906* (Topeka: Kansas State Printing Office, 1907).

25. "Railroads to Pay Cost of O.K. Sewer," *World*, Oct. 26, 1906; "To Stop Nuisance," *Journal*, Oct. 26, 1906; and "One Long Depot Stride," *Times*, Oct. 26, 1906, all scrapbook vol. 32, pp. 62–63, CCS; "To Build O.K. Sewer," *Post*, Feb. 11, 1907; "No O.K. Sewer Bonds," *World*, Feb. 11, 1907, both scrapbook vol. 33, p. 153, CCS; and "Big Bond Issue," *Journal*, Jan. 1, 1905, scrapbook vol. 26, p. 1, CCS.

26. "No O.K. Sewer Bonds," *World*, Feb. 11, 1907, scrapbook vol. 33, p. 153, CCS.

27. "Ask Turkey Creek Dredging," Apr. 14, 1915, scrapbook vol. 2, pp. 140, 149, 160, HJP, SHSM.

28. "Mayor Smiles at Flood Suit Hints," Sept. 17, 1914, vol. 2, p. 35, HJP.

29. Ways and Means Committee of the Kansas City Commercial Club, Dec. 20, 1912, vol. 32, pp. 16–18, CCM.

30. "Say It Is Too Large," Feb. 12, 1913, vol. 1, p. 18, HJP.

31. Ed Shutt, "Businessman, Civic Leader Served as Mayor of 2 Communities," *Wyandotte County Star*, Nov. 18, 1987, v.f. "Green," WCHS. Note that in the Kansas State Board of Health records, the mayor is incorrectly referred to as T. W. Green.

32. "High Bridge Approach Checked up to K.C., K," and "Discussion by Club Now," Feb. 15, 1914, scrapbook vol. 1, p. 24, HJP.

33. "Says Kansas Blocks Turkey Creek Change," June 21, 1913, scrapbook vol. 1, p. 83, HJP.

34. "Flood Plan is Blocked," Feb. 11, 1914, scrapbook vol. 1, p. 243, HJP.

35. Kansas State Board of Health, minutes, June and July 1913, in *Seventh Biennial Report of the Kansas State Board of Health, 1912–1914* (Topeka: Kansas State Printing Office, 1914), 42–43.

36. Kansas State Board of Health, Minutes of the Annual Meeting of the State Board of Health, Topeka, Kansas, June 30, 1913, in *Seventh Biennial Report*, 41–44.

37. "Flood Protection Plan is Blocked," Feb. 11, 1914, scrapbook vol. 1, p. 243, HJP.

38. "Mayor Smiles at Flood Suit Hints," Sept. 17, 1914, scrapbook vol. 2, p. 35, HJP.

39. "High Bridge Approach Checked up to K.C., K," and "Discussion by Club Now," Feb. 15, 1914, scrapbook vol. 1, p. 24, HJP.

40. "May Open Way for Sewer Improvements," *Journal*, Mar. 1, 1914, scrapbook vol. 1, p. 252, HJP.

41. "Another Big Issue of Bonds Possible," *Journal*, Sept. 17, 1912, box 4, scrapbook vol. 2, p. 33; and see clippings, box 4, vol. 2, pp. 83, 166, 172–173, HJP.

42. "Turkey Creek Matter Up," Mar. 12, 1915, scrapbook vol. 2, p. 116, HJP.

43. Clipping, *Journal*, Nov. 1912, box 4, scrapbook vol. 2, p. 83, HJP. The engineer said, "As the work is to be paid for in special tax bills against all the taxable property in the drainage area, and there was some doubt as to the validity of the tax bills on account of a portion of the work being done in Kansas, a friendly suit was brought to test the validity of the entire proceedings." Alfred D. Ludlow, "Turkey Creek Sewer One of the Largest Yet Built," *Engineering News-Record* 87, no. 20 (1921): 817.

44. Ludlow, "Turkey Creek Sewer One of the Largest Yet Built," 815.

45. "City Has New Plan for Protecting West Bottom from Flood," May 14, 1915, scrapbook vol. 2, p. 159, HJP; "Diversion of Turkey Creek into Kaw River to Be Discussed Again," Jan. 25, 1916, scrapbook vol. 3, p. 29, HJP; and "Tunnel Plans Are Complete," Mar. 9, 1916, scrapbook vol. 3, p. 92, HJP.

46. Ludlow, "Turkey Creek Sewer One of the Largest Yet Built," 815–817.

47. "Bosses Lose Kansas City," *Weekly Kansas City Star*, Apr. 16, 1924, 1, SHSM.

48. Kansas City Board of Public Works minutes, June 17, 1924, f. 246; see also letters Aug. 1924–Feb. 1925, f. 265, ABP; and Brown and Dorsett, *K.C.*

49. Kansas City Board of Public Works minutes, Aug. 15, 1924, f. 246, ABP.

50. Letter from H. B. Walker to City Counselor Gilmore, May 28, 1925, f. 265, ABP. The 1923 legal suit pertained to the diversion tunnel and was called *Davidson Construction v. City of Rosedale, et al.* In the same file, see letter from Gilmore to comptroller Kimball.

51. See 1924–1925 documents, f. 265, ABP.

52. Letter to Beach, "The Bridge and Bond Issue" [ca. 1928], f. 77, ABP.

53. Kansas City Board of Public Works minutes, June 17, 1924, f. 246; see also letters Aug. 1924–Feb. 1925, f. 265, ABP.

54. Paul Hohl, "City beneath a City," *Kansas City Magazine*, Sept. 1979, 68, Missouri Valley Room, KCPL; and Ludlow, "Turkey Creek Sewer One of the Largest Yet Built," 815–817.

55. "Findings of Fact, no. 5" in proceedings, US Department of Health, Education and Welfare, Progress Evaluation Meeting in the Matter of Pollution of the Interstate Waters of the Missouri River and Tributary Waters, Kansas City's Metropolitan Area, April 21, 1965, in Predecessor Water Pollution Control Organizations, River Basin Files, Missouri River Basin, Kansas City, Missouri Area, Environmental Protection Agency [hereafter EPA], NARA-CPR.

56. "How Pre-Historic Turkey Creek Saved Kansas City Millions," v.f. "Turkey Creek," KCPL.

57. Lucas Wetzel, "Unheralded Turkey Creek May Yet Have Its Day, At Least with the Outdoor Crowd," *Pitch*, Feb. 9, 2016, http://www.pitch.com/news/article/20560705/unheralded-turkey-creek-may-yet-have-its-day-at-least-with-the-outdoor-crowd.

58. Carl Dehoney, "What Kansas City Owes to Kansas," *Kansas City Spirit* 1, no. 3 (1908): 3–4, KSHS; and "Kansas As an Inventor of Health Laws," *Star*, Feb. 9, 1918, Kansas State Board of Health clippings vol. 1, p. 67, KSHS.

59. "The Topeka Commercial Club," *Club Member* 7, no. 3 (1908): 13, KSHS.

60. Samuel J. Crumbine, *Frontier Doctor: The Autobiography of a Pioneer on the Frontier of Public Health* (New York: Dorrance, 1948), 114.

61. See Duffy's chapters "Growth of Boards of Health" and "Health at the Close of the Nineteenth Century" in *Sanitarians*; and Harriet S. Pfister, *Kansas State Board of Health* (Lawrence: University of Kansas Governmental Research Center, 1955).

62. Crumbine, *Frontier Doctor*, 115.

63. Crumbine, *Frontier Doctor*, 271.

64. Crumbine cartoon files, unprocessed, SJC.

65. S. J. Crumbine, "A Clean Kansas," *Club Member*, 1908, 5–6, KSHS. On the creation of race see David Roediger, *Wages of Whiteness: Race and the Making of the American Working Class* (New York: Verso, 1991); and Grace Elizabeth Hale, *Making Whiteness: The Culture of Segregation in the South, 1890–1940* (New York: Vintage, 1998).

66. Samuel Crumbine, "A Few Highlights in the History of Sanitation" [final draft, ca. 1953], 18–21, f. "Public Health Issues: Sanitation," series 1, subseries 2, SJC; and "A Few Highlights in the History of Sanitation" [draft, ca. 1953], 10, f. "Public Health Issues: Sanitation," series 1, subseries 2, SJC.

67. The Governor's Trophy, 1916, f. 14, series 14, SJC.

68. Crumbine, final draft, "Few Highlights," 21–22.

69. Kansas State Board of Health, *Fourth Biennial Report of the Kansas State Board of Health, 1907–1908* (Topeka: Kansas State Printing Office, 1909).

70. Robert E. McDonnell, "Sewage Disposal Plants in Kansas," Eighth Annual Meeting of the Kansas Gas, Water and Electric Association Proceedings, 1905, p. 16, KSHS.

71. Crumbine, draft, "Few Highlights," 7, 9; and final draft, "Few Highlights," 17.

72. Quote from *Star* article in "In Kansas City, Missouri," *Club Member* 17, no. 13 (1909): 4, KSHS. For examples of women's input on municipal services like gas and electric plants in Kansas City, Kansas, see Lacy Haynes, "A City's Aid in Housekeeping," *American City* 10, no. 3 (1914): 243–246.

73. Quotes from *Star* article in "In Kansas City, Missouri," *Club Member* 17, no.

13 (1909): 4, KSHS. Kathryn Kish Sklar, "Two Political Cultures in the Progressive Era: The National Consumers' League and the American Association for Labor Legislation," and Judith Walzer Leavitt, "Gendered Expectations: Women in Early Twentieth-Century Public Health," both in U.S. History as Women's History: New Feminist Essays, ed. Linda K. Kerber, Alice Kessler-Harris, and Kathryn Kish Sklar (Chapel Hill: University of North Carolina Press, 1995). Maureen A. Flanagan, "The City Profitable, the City Livable: Environmental Policy, Gender, and Power in Chicago in the 1910s," Journal of Urban History 22, no. 2 (1996): 163–190; and Suellen Hoy, "'Municipal Housekeeping': The Role of Women in Improving Urban Sanitation Practices, 1880–1917," in Pollution and Reform in American Cities, 1870–1930, ed. Martin V. Melosi (Austin: University of Texas Press, 1980), 173–198.

74. For examples, see Club Member publications at KSHS.

75. Crumbine, Frontier Doctor, 147–152; and Pfister, Kansas State Board of Health, 39.

76. "River Pollution and Water Purification," Engineering Record 60, no. 25 (1909): 676–677.

77. Ohio State Board of Health, "General Report," in Twenty-Third Annual Report of the Ohio State Board of Health, 1908 (Springfield, OH: Ohio State Board of Health, 1909), 12.

78. Ohio State Board of Health, Twenty-Seventh Annual Report of the Ohio State Board of Health, 1912 (Columbus, OH: Ohio State Board of Health, 1913), 600.

79. Ohio State Board of Health, "General Report," in Twenty-Seventh Annual Report, 10–11.

80. Crumbine, Frontier Doctor, 140.

81. Paul D. Haney, "The Missouri River—A Vital Resource" (paper presented at 58th Annual Conference of the Water Pollution Control Federation, Kansas City, MO, 1985), KCPL.

82. Crumbine, Frontier Doctor, 234–237; Crumbine memoir manuscripts, series 8, subseries 1, f. "Misc.," 290, 300; and "He Gave Us the Fly Swatter," Long Island Sunday Press, Jan. 23, 1949, f. "unprocessed," SJC.

83. This story from his memoir is bolstered by a 1918 newspaper story reporting that the federal government threatened to step in if the city would not control disease and prostitution. It would make sense if Crumbine had aligned himself with the federal level, although he does not explain having done so in his memoirs. See also Schirmer, City Divided, 21.

84. Sarah S. Elkind, Bay Cities and Water Politics: The Battle for Resources in Boston and Oakland (Lawrence: University Press of Kansas, 1998), 164.

85. "Big Bond Issue," Journal, Jan. 1, 1905, scrapbook vol. 26, p. 1, CCS. See Common Council agendas from 1925–1926 for discussion on sewers and regionalism, f. 75, ABP.

86. R. E. McDonnell, "Should Kansas City Supply Water for Jackson County?," *Citizens' League Bulletin*, Mar. 23, 1935, 461, KCPL.

87. Letter from Raytown Water Co., Feb. 3, 1925, f. 101, ABP.

88. This relationship with Kansas City, Missouri, began as early as 1913 and continued until 1923, shortly after the annexation of Rosedale into Kansas City, Kansas. National Board of Fire Underwriters Committee on Fire Prevention and Engineering Standards, *Report on the City of Kansas City, Mo.*, no. 52 (Feb. 1924), 6, KCPL.

89. "Water Bonds Needed," *Citizens' League Bulletin*, Mar. 21, 1931, 299, KCPL.

90. Melvin P. Hatcher, "How Kansas City, Mo., Sells Water to Its Suburbs," *American City* 67, no. 5 (1952): 99–101.

91. In the 1920s one service that had not become regional was firefighting, which was still city-specific because not all cities used the same couplings to connect to hydrants. Therefore, a fire truck from Independence could not hook up to a fire hydrant in Kansas City, Missouri. Kansas City, Kansas, however, could offer mutual aid in a fire because they were supplied with both couplings. Fire Underwriters, *Report on City of Kansas City, Mo.*, 23, KCPL.

92. *American Civic Association* 3, no. 1 (1925): 22.

93. S. Herbert Hare, "Regional Survey for Kansas City and Vicinity," *Citizens' League Bulletin*, Mar. 13, 1926, 1–2, KCPL; and Hare, "Regional Planning for Greater Kansas City," *Citizens' League Bulletin*, Mar. 16, 1929, 473–474, KCPL.

94. S. Herbert Hare, "Regional Survey for Kansas City and Vicinity," *Citizens' League Bulletin*, Mar. 13, 1926, 1–2, KCPL; and Hare, "Regional Planning for Greater Kansas City," *Citizens' League Bulletin*, Mar. 16, 1929, 473–474, KCPL.

95. "Memorandum on Conference with Kansas City, Kansas Officials Relative to Improvements of Kansas Side of C.I.D.," Aug. 20, 1954, f. 22, box 8, CIDA.

96. McDowell, *Building a City*, 36, 49, 66. When the Kansas side wanted to annex the Fairfax industrial district along the river, it took two decades because the industrial district fought to remain independent. The city finally annexed it in the 1960s, knowing that Fairfax waste content would be a challenge, especially owing to anticipated federal pollution regulations. Jersey Creek, running through Fairfax, was an open industrial sewer. In the postwar regulatory era, federal agencies concerned with water quality made numerous mentions of how difficult it was to control or regulate pollution in these industrial areas because they were not connected to city systems.

97. R. E. McDonnell, "Should Kansas City Supply Water for Jackson County?," *Citizens' League Bulletin*, Mar. 23, 1935, 461–462, KCPL.

98. Hatcher, "How Kansas City, Mo., Sells Water," 99–101.

99. The state of Kansas had required all new systems to treat sewage since the

282 | NOTES TO PAGES 131–134

1950s. Therefore, Kansas suburban areas built in the postwar period had sufficient sewerage and treatment works. These areas were attractive to annex because they were not liabilities. McDowell, *Building a City*, 53.

100. "MARC Studies Water Quality Management Plan" (June 1978), in *Water Quality Management 208 Final Plan, Kansas City Metropolitan Region*, Mid-America Regional Council, Dec. 1978, 221–222, WPC, DEQ, Missouri DNR, MSA. MARC history document from Barbara Hensley, personal communication with author, Jan. 2006.

CHAPTER SIX: SOCIAL INNARDS

Epigraph: W. E. B. Du Bois, *The Souls of Black Folk* (1903; Project Gutenberg, 2008), http://www.gutenberg.org/files/408/408-h/408-h.htm.

1. The sufferer probably referenced the 1908 flood, quoted in Darlene Isaacson and Elizabeth Wallace, *Kansas City in Vintage Postcards* (Charleston, SC: Arcadia, 2003), 27.

2. See Washington, *Packing Them In*, vii, 5, 10–11, 18.

3. Maury Klein and Harvey Kantor, *Prisoners of Progress: American Industrial Cities, 1850–1920* (New York: Macmillan, 1976), 343–344. See Brown and Dorsett, *K.C.*; and Reddig, *Tom's Town*.

4. For a demographic overview, see the second chapter in Schirmer, *City Divided*. Schirmer looks at population changes from 1900 to 1920 in river ward neighborhoods that had a sizable number of black residents. The neighborhoods in 1900 were usually quite mixed by race and class. The trend was for less diversity; white citizens and middle classes tended to move southward and people became more segregated by race and class within these older neighborhoods.

5. US Census data reported in "Kansas City Most American of Cities," *Star*, Jan. 30, 1908, scrapbook vol. 38, pp. 23–24, CCS; and editorial, *Kansas City Sun*, Apr. 3, 1915, SHSM; Klein and Kantor, *Prisoners of Progress*, 185; Serda, "Blow to the Spirit"; Loren L. Taylor, *The Consolidated Ethnic History of Wyandotte County* (Kansas City, KS: Kansas City, Kansas, Ethnic Council, 2000); and Schirmer, *City Divided*.

6. See Steven J. Diner, *A Very Different Age: Americans of the Progressive Era* (New York: Hill and Wang, 1998), 98–101.

7. "Kansas City Most American of Cities," Jan. 30, 1908, scrapbook vol. 38, pp. 23–24, CCS.

8. On immigration, see Diner, *Very Different Age*; Alan M. Kraut, *Silent Travelers: Germs, Genes, and the "Immigrant Menace"* (New York: Basic Books, 1994); Hoy, *Chasing Dirt*; Sklar, "Two Political Cultures," 36–62; and Flanagan, "City Profitable," 163–190.

9. Serda, "Blow to the Spirit."

10. Mildred Ray, "Kansas City Livestock Exchange Building," *Kansas City Times*, Oct. 19, 1968, SC58, Missouri Valley Room, KCPL.

11. "Clifford Naysmith," p. 14A, box 1, f. 4, A. Theodore Brown Collection, SHSM. Naysmith cites interview with Mrs. Fleming.

12. "They Say," *Kansas City Sun* [hereafter *Sun*], Oct. 17, 1914, SHSM; and editorial, *Sun*, Nov. 14, 1914, SHSM.

13. "The Patch under the Hammer," *Times*, Jan. 4, 1910; and "Patch Ordered Vacated," *Journal*, Apr. 9, 1910, KCPL.

14. A number of sources confirm this, including Board of Public Welfare maps, also found in Martin's *Our Negro Population: A Sociological Study of the Negroes of Kansas City, Missouri* (1913; repr., New York: Negro University Press, 1969). See also *Manual and Directory of the Public Schools of Kansas City, Mo., 1912–13*, SHSM. Sanborn Fire Insurance Company maps from 1909 reveal that black schools were located in all parts of the city (Special Collections, University of Missouri, Columbia, MO). If black citizens owned, operated, worshipped at, or resided in a business, church, or home, it was usually labeled "colored" on the maps.

15. Advertisement, *Sun*, 1916, SHSM.

16. Advertisement, *Rising Son*, 1907, SHSM.

17. Joe W. Trotter, "African Americans in the City: The Industrial Era, 1900–1950," *Journal of Urban History* 21, no. 4 (1995): 438–457; and Henry Louis Taylor Jr. and Walter Hill, eds., *Historical Roots of the Urban Crisis: African Americans in the Industrial City, 1900–1950* (New York: Garland, 2000), 8–9.

18. Martin, *Our Negro Population*, 34, 86–106.

19. See "Society for the Suppression of Vice," 1924, f. 271, ABP.

20. "'Vice District' is Finally Defined," Sept. 19, 1913, scrapbook vol. 1, p. 141, HJP; and Schirmer, *City Divided*, 18–21. See Kevin J. Mumford, *Interzones: Black/White Sex Districts in Chicago and New York in the Early Twentieth Century* (New York: Columbia University Press, 1997).

21. The term "sacrifice zone" comes from Robert D. Bullard, "Urban Infrastructure: Social, Environmental, and Health Risks to African Americans," in *Handbook of Black American Health: The Mosaic of Conditions, Issues, Policies, and Prospects*, ed. Ivor Lensworth Livingston (Westport, CT: Greenwood Press, 1994), 323; and Steve Lerner, *Sacrifice Zones: The Front Lines of Toxic Chemical Exposure in the United States* (Cambridge, MA: MIT Press, 2010), xiiv–xiv, 7–9.

22. Quotes from "'Vice District' is Finally Defined," and "Italians Protest Vice Location among Them," *Post*, Sept. 30, 1913, scrapbook vol. 4, p. 55, HJP.

23. See Kansas City Board of Public Welfare, *Third Annual Report of the Board of Public Welfare, 1911–1912*, 22–24.

24. Clippings, *Post*, Oct. 6 and 12, 1913, scrapbook vol. 4, pp. 21, 49, 58, HJP.

25. "'Vice District' is Finally Defined," Sept. 30, 1913, scrapbook vol. 4, p. 55, HJP.

26. "23 Women Protest against Franchise"; and "'Over Our Bodies,' is Priest's Retort," Feb. 19, 1913, scrapbook vol. 1, pp. 19, 23, HJP.

27. Problems of "uneven development" fed on themselves, as shown by Gotham, *Uneven Development*, 1–4; and James W. Button, *Blacks and Social Change: Impact of the Civil Rights Movement in Southern Communities* (Princeton, NJ: Princeton University Press, 1989); Schirmer, *City Divided*; and Kenneth Jackson, *Crabgrass Frontier: The Suburbanization of the United States* (New York: Oxford University Press, 1985). For examples in other cities see Craig E. Colten, "Basin Street Blues: Drainage and Environmental Equity in New Orleans, 1890–1930," *Journal of Historical Geography* 28, no. 2 (2002): 237–257; and Andrew Hurley, *Environmental Inequalities: Class, Race, and Industrial Pollution in Gary, Indiana, 1945–1980* (Chapel Hill: University of North Carolina, 1995).

28. Judith F. Laird, "Argentine, Kansas: The Evolution of a Mexican-American Community, 1905–1940" (PhD diss., University of Kansas, 1975), 44; and Valerie M. Mendoza, "The Creation of a Mexican Immigrant Community in Kansas City, 1890–1930" (PhD diss., University of California, Berkeley, 1997), 72.

29. Kansas City Board of Public Welfare Research Bureau, *Social Prospectus of Kansas City, Missouri* (Kansas City, MO: Kansas City Board of Public Welfare, 1912), 35, SHSM.

30. "Tuberculosis and Bad Housing" [graph], by Kansas City Tuberculosis Society and Kansas City Consumers' League, 1927, scrapbook vol. 1, f. 9, Guadalupe Collection, KCPL.

31. "Negroes Want Old Hospital," *Rising Son*, Nov. 9, 1907, SHSM.

32. Joan E. Lynaugh, *The Community Hospitals of Kansas City, Missouri, 1870–1915* (New York: Garland, 1989), 38, 46.

33. "Separate Hospital Question," *Rising Son*, Mar. 10, 1905, SHSM.

34. Lynaugh, *Community Hospitals of Kansas City*, 114, 127n32.

35. "History of the Urban League of Kansas City," 51st Annual Report, 1970, 2, Urban League of Kansas City Records, BAMA.

36. "The Negro City Hospital," *Sun*, Feb. 6, 1915, SHSM.

37. "New County Home," *Sun*, Feb. 20, 1915, SHSM.

38. Cleaver quoted in Sallie Han, "Partnership Turns 'War Zone' into Homes: Public and Private Funding Helped Rebuild an Eastside Complex," *Star*, Dec. 17, 1992, C2; and Gotham, *Uneven Development*, 91.

39. Race is a culturally constructed concept rooted in the body. See Barbara

Fields, "Slavery, Race, and Ideology in the United States of America," *New Left Review* 181, May–June 1990, 95–118.

40. Kansas City Board of Public Welfare Research Bureau, *Does Kansas City Have a Housing Problem?* (Kansas City, MO: 1912), 26, KCPL; and Sherry Lamb Schirmer, "Landscape of Denial: Space, Status and Gender in the Construction of Racial Perceptions among White Kansas Citians, 1900–1958" (PhD diss., University of Kansas, 1995), 47–48.

41. Kansas City Board of Public Welfare, *Does Kansas City Have a Housing Problem?*, 26.

42. Jacob A. Riis, *How the Other Half Lives: Studies among the Tenements of New York*, ed. Sam Bass Warner Jr. (Cambridge, MA: Belknap Press of Harvard University Press, 1970).

43. "Expose Bad Housing Here," Feb. 12, 1913, scrapbook vol. 53, p. 24, CCS.

44. Gotham, *Uneven Development*, 37; and Schirmer, *City Divided*, 71.

45. Kansas City Board of Public Welfare Research Bureau, *Social Prospectus of Kansas City*, 83.

46. Kansas City Board of Public Welfare Research Bureau, *Social Prospectus of Kansas City*, 35.

47. National Board of Fire Underwriters Committee on Fire Prevention and Engineering Standards, *Report on the City of Kansas City, Mo.*, no. 2 (Feb. 1924), 24–25, KCPL; and "Seek to Regulate Tenement Building," *Journal*, July 4, 1912, box 4, scrapbook vol. 1, p. 90, HJP.

48. Schirmer, *City Divided*, 73–76; and D. J. Haff, "Where May Kansas City Negroes Build Homes?" *Citizens' League Bulletin*, Apr. 8, 1922, 1, KCPL. Housing advertisements in local Kansas City newspapers and the Sanborn Fire insurance maps also demonstrate this shift in boundaries. "Civic Club Wants No Negro College," Jan. 24, 1913, scrapbook vol. 1, p. 11, HJP; and *Sun* editorials, Aug. 22 and 29, 1914, Sept. 5 and 26, 1914, Oct. 10, 1914, SHSM.

49. "Civic Club Wants No Negro College," Jan. 24, 1913, scrapbook vol. 1, p. 11, HJP.

50. Raymond A. Mohl and James F. Richardson, *The Urban Experience: Themes in American History* (Belmont, CA: Wadsworth, 1973), 214–215; Jon C. Teaford, *The Twentieth-Century American City*, 2nd ed. (Baltimore, MD: Johns Hopkins University Press, 1993), 68–69; and Brown and Dorsett, *K.C.*, 172–178.

51. Contested spaces included cemeteries, theaters, and railcars. See editorial, *Rising Son*, Mar. 2, 1907, SHSM.

52. Scott B. Child, "The Lowering of a City's Death Rate," *Citizens' League Bulletin*, Mar. 24, 1923, 2, KCPL.

53. "Environment," *Rising Son*, July 20, 1907, SHSM.

54. Editorial, *Sun*, Apr. 18, 1914, and Sept. 16, 1916, SHSM.

55. Article reprinted from the *State Journal* (Columbus, OH), in *Sun*, Jan. 31, 1914, SHSM.

56. See Char Miller, "Martin Luther King, Social Justice, and Streetscape Environmentalism," KCET, Apr. 2, 2014, accessed Jan. 2017, https://www.kcet.org/redefine/martin-luther-king-social-justice-and-streetscape-environmentalism; and Char Miller, "Streetscape Environmentalism: Floods, Social Justice, and Political Power in San Antonio, 1921–1974," *Southwestern Historical Quarterly* 118, no. 2 (Oct. 2014): 158–177.

57. Gotham, *Uneven Development*, 39, 42; and William S. Worley, *J. C. Nichols and the Shaping of Kansas City: Innovation in Planned Residential Communities* (Columbia: University of Missouri Press, 1990).

58. Gotham, *Uneven Development*, 38, 42–46, 56, 159n9, 162n6; and Thomas J. Sugrue, *The Origins of the Urban Crisis: Race and Inequality in Postwar Detroit* (Princeton, NJ: Princeton University Press, 1996).

59. Mohl and Richardson, *Urban Experience*, 214–215; Teaford, *Twentieth-Century American City*, 68–69; Brown and Dorsett, *K.C.*, 172–178; J. C. Nichols Company, *Country Club Plaza, Kansas City* (Kansas City, MO: [J. C. Nichols Co.], 1913), KCPL; and Ralph H. Ochsner, "The Firm of Hare & Hare and the Art of Landscape Architecture," *Kawsmouth* 3, no. 1 (Winter–Spring 2001): 46–47.

60. See, for example, communication between Nichols and Beach administration, Aug. 8, 1924, and Aug. 12, 1924, f. 246, ABP; for photos of infrastructure development see AIA, SHSM.

61. Advertisement, *Kansas City Spirit*, May 1909, 13, KCPL.

62. Schirmer, *City Divided*, 18.

63. "A Gasoline Filling Station in Fashionable Architectural Garb," *Star*, Apr. 9, 1916, 12A; and "Standard Oil Company" [photo], 1903, P8, box 1, f. 18, no. 88, KCPL.

64. "To Promote Friendly Relations between the Races," *Citizens' League Bulletin*, Apr. 8, 1922, 2, KCPL.

65. D. J. Haff, "Where May Kansas City Negroes Build Homes?" *Citizens' League Bulletin*, Apr. 8, 1922, 1, KCPL.

66. "Great Prosperity among the Negroes," *Rising Son*, June 29, 1907, SHSM; and "Buy Realty by the Week," *Star*, Mar. 6, 1927, in "Kansas City: Mecca of the New Negro," by Sonny Gibson (scrapbook, 1997), 123, Black Archive of Mid-America [hereafter BAMA].

67. Clipping, *Post* [1914], scrapbook vol. 4, p. 115, HJP.

68. Martin, *Our Negro Population*, 94, 152; and Beach correspondence, "Welfare Board," f. 291, ABP.

69. Letters from Hartman to Beach, Aug. 27, 1928, and Sept. 6 and 13, 1928; and letter from Beach to Cavaness, Sept. 4, 1928, "Health Director," f. 137, ABP.

70. Letter from Mrs. M. J. Loftus to Beach, June 17, 1927, f. 251, ABP.

71. "Golf Appeal to High Court," Star, Aug. 15, 1928, clipping in "Kansas City: Mecca of the New Negro," by Sonny Gibson (scrapbook, 1997), 120, BAMA.

72. Clipping, July 14, 1913, scrapbook vol. 1, p. 94, HJP.

73. Interview with Autumn Saxton-Ross by Shereen Marisol Meraji, "Outdoor Afro: Busting Stereotypes That Black People Don't Hike or Camp," on *Codeswitch*, July 12, 2015, National Public Radio, accessed Dec. 2016, http://www.npr.org/sections/codeswitch/2015/07/12/421533481/outdoor-afro-busting-stereotypes-that-blacks-dont-hike-or-camp.

74. Assessments done by Van Brunt, f. "The Paseo," Kansas City Board of Park Commissioners, Resolutions and Ordinances (roll PA-72), KCPR.

75. Letter from Mrs. M. J. Loftus to Beach, June 17, 1927, f. 251, ABP.

76. Schirmer, *City Divided*, 16–17; and K. Vaughan, A. Kaczynski, et al., "Exploring the Distribution of Park Availability, Features, and Quality across Kansas City, Missouri by Income and Race/Ethnicity: An Environmental Justice Investigation," *Annals of Behavioral Medicine* 45, no. 1 (2013): 28–38.

77. Mayor Albert Beach speech, "Kansas City Plan" [ca. 1930], f. 272, ABP.

78. Waring, *Report on the Social Statistics of Cities*, 2:557.

79. Letter from Beach to the magazine *Building Age*, Mar. 1, 1929, p. 1, f. 236, ABP.

80. "Civic Club Wants No Negro College," Jan. 24, 1913, scrapbook vol. 1, p. 11, HJP.

81. "Parkway of Negro Blocks?" Sept. 22, 1915, scrapbook vol. 2, n.p., HJP; and Schirmer, *City Divided*, 112–116.

82. "Bombing Threats," *Citizens' League Bulletin*, Dec. 4, 1926, 51, KCPL.

83. Emphasis mine. Letter from Beach to *Building Age*, Mar. 1, 1929, p. 2, f. 236, ABP.

84. Gotham, *Uneven Development*, 151–152; and Richard Rothstein, *The Color of Law: A Forgotten History of How Our Government Segregated America* (New York: Liveright, 2017). For interrelationships, see Mumford, *Interzones*.

85. Kansas City Chamber of Commerce Minutes, 1904, vol. 18, p. 164, CCM.

86. Letter from Beach to McQueeny, Oct. 13, 1925, f. 52, ABP.

87. For example, "Poverty Causes Illness," *Citizens' League Bulletin*, June 11, 1938, 520, KCPL.

88. Rick Halpern and Roger Horowitz, *Meatpackers: An Oral History of Black Packinghouse Workers and Their Struggle for Racial and Economic Equality* (New York: Twayne, 1996), 85–100; and Martin, *Our Negro Population*, 50–51.

89. "Kansas City Negroes," *Citizens' League Bulletin*, Nov. 12, 1927, 216, KCPL.

90. "Negroes Ask Consideration," *Citizens' League Bulletin*, Nov. 26, 1932, 48, KCPL.

91. "Negro Health Concerns All," *Citizens' League Bulletin*, Apr. 18, 1931, 315, KCPL.

92. Martin, *Our Negro Population*, 117, 56. See Jacqueline Jones, *Labor of Love, Labor of Sorrow: Black Women, Work, and the Family from Slavery to the Present* (New York: Vintage, 1985).

93. Kansas City Urban League, "Industrial Relations," *Pilot*, Dec. 1936, 3, Royal Fleming Collection, BAMA.

94. Martin, *Our Negro Population*, 51.

95. William Young and Nathan Young Jr., *Your Kansas City and Mine* (Kansas City, MO: published by the author, 1950), 112, 162, BAMA.

96. Letter from Adkins to Beach, Mar. 14, 1925, f. 218, ABP.

97. E. S. Lewis, "Negroes as Workers in the Industries," *Citizens' League Bulletin*, Apr. 26, 1930, 145–146, KCPL.

98. Werner Troesken, "Limits of Jim Crow: Race and the Provision of Water and Sewerage Services in American Cities, 1880–1925," *Journal of Economic History* 62, no. 3 (2002): 735.

99. "Health Problem Acute," *Citizens' League Bulletin*, Apr. 30, 1927, 135; "Unhealthful Conditions," *Citizens' League Bulletin*, June 4, 1927, 156, KCPL; and see Leavitt, *Healthiest City*. Kansas City public health reformers saw Milwaukee as exemplary.

100. J. E. Perry, "Local Disease is a Menace to All Parts of the City," *Citizens' League Bulletin*, May 11, 1935, 489, KCPL; "Urban League of Kansas City," in William Young and Nathan Young Jr., *Your Kansas City and Mine* (Kansas City, MO: 1950), 139, BAMA; and Martin, *Our Negro Population*, 121.

101. "Health Problem Acute," *Citizens' League Bulletin*, Apr. 30, 1927, 135, KCPL; and "Unhealthful Conditions," *Citizens' League Bulletin*, June 4, 1927, 156, KCPL. See Tomes, *Gospel of Germs*; and Hunter, *To 'Joy My Freedom*, esp. chap. 9, "Tuberculosis as the 'Negro Servants' Disease.'"

102. "The Negro City Hospital," *Sun*, Feb. 6, 1915, SHSM.

103. "The Negro City Hospital," *Rising Son*, Feb. 6, 1915, SHSM; Laird, "Argentine, Kansas," 120, 192, n66, and esp. chap. 4; and Sherry Schirmer, "Overview of the Mexican-American Community," 1976, 13, v.f. "Mexican Americans," KCPL. Due to employment and housing risks, Mexicans were prone to disease and their death rates were much higher. Laird, "Argentine, Kansas," 201.

104. On Gallagher and Guadalupe Center, see Schirmer, "Overview of the Mexican American Community," 13, v.f. "Mexican American"; and the Guadalupe Collection, KCPL.

105. "Mexicans to Get Garbage Pails," Apr. 27, 1926, scrapbook vol. 1, f. 9, p. 23A, Guadalupe Collection, KCPL.

106. Mendoza, "Mexican Immigrant Community," 129–130; and Schirmer, City Divided, 71–73.

107. David Nasaw quoted in Strasser, Waste and Want, 139.

108. Board of Director's Minutes, box 1, f. 2; box 1, f. 3, p. 129; box 1, f. 4, pp. 177–179; quote "hard times" in 1931 Annual Meeting, p. 4; Executive Board of Committees Minutes, box 3, f. 53, pp. 36, 68, 81; box 3, f. 54, p. 131; WCC Bulletin, 6, no. 4 (1923), box 16, f. 345; Bulletin, 10, no. 1 (1927), box 16, f. 349; Bulletin (1931), p. 7, box 16, f. 353; "regret" quote in Bulletin (1938), p. 18, box 16, f. 357; scrapbooks, 1935–1937, vol. 8; "major concern" quote in article "Women Enlisting," scrapbooks, 1937–1939, vol. 9, all in Woman's City Club Records, SHSM.

109. Kansas City Public Service Institute, Public Health in Kansas City: A Study of the Present Health Situation and Public Health Work (Kansas City, MO: Public Service Institute, 1922), 5; and Hospital and Health Committee of Kansas City Public Improvement Association, "Public Health Needs" (final report, June 15, 1925), f. 137, ABP.

110. The typhoid fever death rate for Kansas City was 11.7 per 100,000 in 1921, up from 10.9 in 1919. The national average for 1919 was a lower 9.2 per 100,000. Kansas City Public Service Institute, Public Health in Kansas City, 10.

111. Hospital and Health Committee of Kansas City Public Improvement Association, "Public Health Needs" (final report, June 15, 1925), f. 137, ABP.

112. "Public Health," in Bulletin 5, no. 5 (1923), scrapbook vol. 1, WCC.

113. "Public Health," in Bulletin 5, no. 5 (1923), scrapbook vol. 1, WCC.

114. Quoted in Sun, Jan. 23, 1915, SHSM.

115. "Health Pays Dividends," Citizens' League Bulletin, Nov. 12, 1932, 40, KCPL.

116. "A Bad Health Record," Citizens' League Bulletin, Feb. 21, 1931, 284, KCPL.

117. Edward P. Heller, "What Health Prospects in K.C.?," Citizens' League Bulletin, Apr. 13, 1940, 209, KCPL.

118. "Health Department Being Revised," Citizens' League Bulletin, June 1, 1940, 338, KCPL.

119. R. E. McDonnell, "How Good Health Can Be Purchased," Citizens' League Bulletin, Apr. 1, 1922, 4, KCPL; "Health Problem Acute," Citizens' League Bulletin, Apr. 30, 1927, 135, KCPL; and "Unhealthful Conditions," Citizens' League Bulletin, June 4, 1927, 156, KCPL.

120. Kate L. Cowick, The Story of Kansas City (booklet of articles published by the Kansas City Kansan, 1924–1926), 30, KSHS.

121. Vaughan, Kaczynski, et al., "Exploring the Distribution of Park Availability," 28–38.

122. Gottlieb, *Forcing the Spring*, 8.

123. Gottlieb, *Forcing the Spring*; Hurley, *Environmental Inequalities*; Robert Bullard, *Dumping in Dixie: Race, Class and Environmental Quality* (Boulder, CO: Westview Press, 1990); Giovanna Di Chiro, "Nature as Community: The Convergence of Environmental and Social Justice," in Cronon, *Uncommon Ground*, 298–320; Dianne D. Glave and Mark Stoll, eds., *"To Love the Wind and the Rain": African Americans and Environmental History* (Pittsburgh, PA: University of Pittsburgh Press, 2006); Washington, *Packing Them In*; Button, *Blacks and Social Change*; Troesken, "Limits of Jim Crow," 734–772; Sellers, *Hazards of the Job*; Gregg Mitman, "In Search of Health: Landscape and Disease in American Environmental History," *Environmental History* 10, no. 2 (2005): 184–209; and Colten, *Unnatural Metropolis*.

124. Bullard, "Urban Infrastructure," 327.

125. Hurley, *Environmental Inequalities*; and Hal K. Rothman, *Greening of a Nation? Environmentalism in the United States since 1945* (Fort Worth, TX: Harcourt Brace, 1998), 161–165.

126. Bullard, "Urban Infrastructure," 318.

Part III. Basin: Health and Wealth

First Epigraph: Gov. John Burke, in *Proceedings of a Conference of Governors in the White House, May 13–15, 1908* (Washington, DC: US Government Printing Office, 1909), online at Library of Congress, accessed Oct. 2006, http://memory.loc.gov /cgi-bin/query/r?ammem/consrv:@field(DOCID+@lit(amrvgvg16div73)).

Second Epigraph: Editorial, *Municipal Sanitation* 8, no. 4 (1937): 242.

Chapter Seven: A Broader Vision

Epigraph: Company letterheads from the first decade of twentieth century, Kansas City District, 1907–1930, box 17, f. 86 and box 5, f. 1585, USACE, NARA-CPR.

1. Champion S. Chase quoted in *Official Report of the Proceedings of the Missouri River Convention*, compiled by H. M. Kirkpatrick (Kansas City: Lawton and Hayens, 1885), 24.

2. Lawrence M. Jones, "The Importance of Missouri River Improvement" (address to Kansas City Business Men at Midland Hotel, Kansas City, Missouri, Oct. 18, 1906), 1–8, quotes 8, MHS.

3. Diane Oerly and Jane Brengarth from Boonville, Missouri, conversations with author, 2005 and 2006.

4. Susan Jezak Ford, "Biography of Herbert S. Hadley (1872-1927), Governor," 2003, Missouri Valley Special Collections, KCPL, accessed Apr. 5, 2018, http://

kchistory.org/content/biography-herbert-s-hadley-1872-1927-governor; and Parrish, Jones, and Christensen, *Missouri*, 249–253.

5. Thelen, *Paths of Resistance*; and see Samuel P. Hays, *Conservation and the Gospel of Efficiency: The Progressive Conservation Movement, 1890–1920* (Cambridge, MA: Harvard University Press, 1959); Robert Wiebe, *The Search for Order: 1877–1920* (New York: Hill and Wang, 1967); and Robert L. McCormick, "Public Life in Industrial America, 1877–1917," in *The New American History*, rev. ed., ed. Eric Foner (Philadelphia: Temple University Press, 1997), 107–132.

6. David Benac, "Whose Forest Is This?: Hill Folk, Industrialists, and Government in the Ozarks," *Missouri Historical Review* 101, no. 1 (2006): 17–35, esp. 21–29 on forestry.

7. On Chicago, see Washington, *Packing Them In*, esp. chap. 3.

8. Libby Hill, *The Chicago River: A Natural and Unnatural History* (Chicago: Lake Claremont, 2000), 119–135.

9. "Chicago Sewage," *Waterways Journal*, Dec. 9, 1899, 5, Herman T. Pott National Inland Waterways Library, Mercantile Library [hereafter MERC].

10. "Pollution of the Mississippi," *Waterways Journal*, Dec. 23, 1899, 7, MERC.

11. "Facts about the Drainage Canal," *Waterways Journal*, Sept. 2, 1899, 8, MERC.

12. "Pollution of the Mississippi," *Waterways Journal*, Dec. 23, 1899, 7, MERC.

13. "To Enjoin Canal's Operation," *Waterways Journal*, July 22, 1899, 5, MERC.

14. Harold L. Platt, "Chicago, the Great Lakes, and the Origins of Federal Urban Environmental Policy," *Journal of the Gilded Age and Progressive Era*, 1, no. 2 (2002): 122–153.

15. Thelen, *Paths of Resistance*, 227.

16. Craig Colten, "Environmental Justice in the American Bottom: The Legal Response to Pollution, 1900–1950," in Hurley, *Common Fields*, 167; and Justice Oliver Wendell Holmes, opinion of the United States Supreme Court, *State of Missouri v. State of Illinois and the Sanitary District of Chicago* 200 US 496 (1906), accessed July 2006, http://caselaw.lp.findlaw.com/cgi-bin/getcase.pl?friend=nytimes&court=us&vol=200&invol=496.

17. Ralph Chester Williams, *The United States Public Health Service, 1798–1950* (Washington, DC: Commissioned Officers Association of the USPHS, 1951), 316–317.

18. "The Effect of Chicago Sewage," *Waterways Journal*, Dec. 21, 1901, 8, MERC.

19. Hill, *Chicago River*, 97–152. A 1911 study, however, said that the Sanitary and Ship Canal made the area thirty miles downstream "putrescent." See Christopher Theriot and Kelly Tzoumis, "Bankside Chicago," in Kibel, *Rivertown*, 67–84.

20. "The Sanitary District of Chicago," correspondence 10, July 27, 1903, General Administration Files, Sioux City District Engineer Office, 1903–1906, box 16, f. 173, USACE.

21. See items on the Wixford process, 1904–1912, Olhausen Papers, MHS.

22. Graphs in Saint Louis health board reports, "Water Commission Report," in *The Mayor's Message with Accompanying Documents, to the Municipal Assembly of the City of St. Louis 1910–1911* (Saint Louis, MO: Municipal Assembly, 1911), 121, MHS; and John C. Crighton, *The History of Health Services in Missouri* (Omaha, NE: Barnhart Press, 1993).

23. US Supreme Court, *Missouri v. Illinois* (1906).

24. US Supreme Court, *Missouri v. Illinois* (1906).

25. US Supreme Court, *Missouri v. Illinois* (1906); and see Theodore Steinberg, *Nature Incorporated: Industrialization and the Waters of New England* (Cambridge: Cambridge University Press, 2004).

26. US Supreme Court, *Missouri v. Illinois* (1906).

27. The longer-term solution for Saint Louis would be a new waterworks on the Missouri River. The Howard Bend plant opened in 1929 and is still in operation today.

28. "St. Louis Water Supply," *Waterways Journal*, Feb. 3, 1900, 8, MERC.

29. Washington, *Packing Them In*, 78–80.

30. Missouri WPC, DEQ, *Water Quality Standards Reports: Water Quality of the Lower Missouri River, Gavins Point Dam to Mouth* (Jefferson City: Missouri Department of Natural Resources, 1982), 1, DNR, MSA. The worst months for typhoid fever tended to be from late fall to early spring, when water levels were lower and dilution was diminished. The public incorrectly believed that the cool temperatures of winter hampered bacterial growth, but because there was less oxidation in wintertime, bacteria survived the cold weather. Ice, necessary for icebox refrigeration in homes, businesses, and railroad cars, was cut from the river and other bodies of water with horse-drawn saws. Therefore, state boards of health ran "pure ice" campaigns to inform consumers that bacteria could survive freezing and that they should select suppliers (and therefore the geographic origin of the ice) carefully.

31. Letter from F. O. Marvin, Dean of Engineering, University of Kansas, to Gov. Herbert Hadley, Nov. 29, 1910, f. 186, Herbert S. Hadley Papers, SHSM [hereafter HHP].

32. Letter from W. C. Hoad to Gov. Hadley, Jan. 3, 1911, f. 195, p. 3, HHP.

33. Hoad to Hadley, Jan. 3, 1911, f. 195, p. 2, HHP.

34. R. E. McDonnell to Mayor Darius A. Brown, Dec. 29, 1910, f. 194, HHP.

35. Hoad to Gov. Hadley, Jan. 3, 1911, f. 195, 2–3, HHP.

36. Kansas State Board of Health, *Sixth Biennial Report*, 20–21.

37. Crumbine, *Frontier Doctor*, 139–140. See Tarr, *Search for the Ultimate Sink*, esp. chaps. 4 and 6.

38. Crumbine manuscript, f. "Miscellaneous Pages," 200, series 8, subseries 1, SJC.

39. Crumbine, in Kansas State Board of Health, *Sixth Biennial Report*, 20.

40. Tarr, *Search for the Ultimate Sink*, 169–173, 192.

41. George C. Whipple, "The Policy of Water Filtration," *Engineering Record* 60, no. 26 (1909): 718.

42. "Notes and Comments," *Engineering Record* 60, no. 7 (1909): 170.

43. "National and State Control over Sewage Disposal," *Engineering Record* 60, no. 26 (1909): 704; and Rudolph Hering, letter to the editor, "The Passaic Valley Sewerage Project," *Engineering Record* 60, no. 7 (1909): 196.

44. Tarr, *Search for the Ultimate Sink*, 167, 175; McLaughlin, *Sewage Pollution*, 48; and "History of ASTHO," *Association of State and Territorial Health Officials*, accessed Apr. 5, 2018, http://www.astho.org/About/History/.

45. Beginning in 1909, the United States and Canada began discussing an international treaty in the boundary waters of the Great Lakes that included concern for pollution and public health. Lee Botts and Paul Muldoon, *Evolution of the Great Lakes Water Quality Agreement* (East Lansing: Michigan State University Press, 2005).

46. Ohio State Board of Health, *Twenty-Fourth Annual Report of the Ohio State Board of Health, 1909* (Columbus, OH: Ohio State Board of Health, 1910), 13–14.

47. "Notes and Comments," *Engineering Record* 60, no. 7 (1909): 170.

48. For information about the 1908 conference, see Richard N. L. Andrews, *Managing the Environment, Managing Ourselves: A History of American Environmental Policy* (New Haven, CT: Yale University Press, 1999), 152; Tarr, *Search for the Ultimate Sink*, 165; *Proceedings of a Conference of Governors in the White House, May 13–15, 1908* (Washington, DC: US Government Printing Office, 1909), available online in The Evolution of the Conservation Movement, 1850–1920, American Memory Collection, Library of Congress, accessed Apr. 5, 2018, http://memory.loc.gov/cgi-bin/query/r?ammem/consrv:@field(DOCID+@lit(amrvgvg16)):@@@REF; and *Report of the National Conservation Commission, February 1909* (Washington, DC: US Government Printing Office, 1909), available online in The Evolution of the Conservation Movement, 1850–1920, American Memory Collection, Library of Congress, accessed Apr. 5, 2018, http://memory.loc.gov/cgi-bin/query/r?ammem/consrv:@field(DOCID+@lit(amrvgvg38)):@@@REF. For a Mississippi River comparison, see John O. Anfinson, *The River We Have Wrought: A History of the Upper Mississippi* (Minneapolis: University of Minnesota Press, 2003).

49. Gov. Joseph W. Folk, "Address by Joseph W. Folk, Governor of Missouri," in *Proceedings of a Conference of Governors*, 159–161, accessed Oct. 2006, http://memory.loc.gov/cgi-bin/query/r?ammem/consrv:@field(DOCID+@lit(amrvgvg16div64)).

50. Schneiders, *Unruly River*, 91–94; and Branyan, *Taming the Mighty Missouri*, 9.

51. Missouri Waterway Commission, *First Biennial Report of the Missouri Waterway Commission*, 1911, found in appendix, *Forty-Sixth Missouri Legislature General Assembly*, part 2 (Jefferson City, MO: Forty-Sixth Missouri General Assembly, 1911), 8–17, quote 16, SHSM.

52. In comparison, sanitary reform was brought about through the leadership of the Cincinnati Commercial Club. Edward J. Cleary, *The ORSANCO Story: Water Quality Management in the Ohio Valley under an Interstate Compact* (Baltimore, MD: Resources for the Future/Johns Hopkins University Press, 1967).

53. Melosi, *Sanitary City: Environmental Services*, 111; Duffy, *Sanitarians*; Sellars, *Hazards of the Job*; and Hays, *Gospel of Efficiency*.

54. Hospital and Health Board, *Fourth Annual Report of the Hospital and Health Board of Kansas City, Missouri, 1911–1912*, 71–72; and see also "Keep the Missouri Clean," Nov. 18, 1908, scrapbook, vol. 40, p. 94, CCS.

55. See Kansas State Board of Health, *Sixth Biennial Report*, 73–75, 95. Quote to Crumbine, *Frontier Doctor*, 74.

56. *Seventeenth Biennial Report of the Iowa State Board of Health* (Des Moines, IA: State Printer, 1915), 46; Crumbine, in Kansas State Board of Health, *Sixth Biennial Report*, 20; Williams, *United States Public Health Service*, 421–450; and Botts and Muldoon, *Evolution of the Great Lakes Water Quality Agreement*.

57. Tarr, *Search for the Ultimate Sink*, 364–365.

58. *Seventeenth Biennial Report of the Iowa State Board of Health*, 46.

59. *Seventeenth Biennial Report of the Iowa State Board of Health*, 46–48; and Kansas State Board of Health, *Sixth Biennial Report*, 15; and also WPC, DEQ, *Water Quality of the Lower Missouri River*, 1.

60. McLaughlin, *Sewage Pollution*.

61. McLaughlin, *Sewage Pollution*, 45.

62. For example, see Paul D. Haney, "The Missouri River—A Vital Resource" (paper presented at 58th Annual Conference of the Water Pollution Control Federation, Kansas City, MO, 1985), KCPL.

63. Kansas State Board of Health, *Sixth Biennial Report*, 15; and WPC, DEQ, *Water Quality of the Lower Missouri River*.

64. McLaughlin, *Sewage Pollution*, 48–49.

65. *Seventeenth Biennial Report of the Iowa State Board of Health*, 46.

66. A cautionary note: a 1948 article recounting Crumbine's work called the MRSC "diplomacy" and mistakenly said that Crumbine "single-handedly wheedled the states of Iowa, Nebraska, Missouri and Kansas into joint treaties governing sanitary control of the Missouri River." This was his unaccomplished wish. The article mistakenly goes on to say that in celebration of this achievement for

improved sanitation in the lower basin, other sanitarians "honored" the Crumbine family with a dinner held in the Water and Sewage Laboratory—the aforementioned one that used laboratory instruments for cooking and serving. The dinner really did happen, but it was not celebrating the victory of pollution control on the Missouri. In his memoirs, Crumbine says they were celebrating a similar victory on the Ohio River. Records of what was happening on the Ohio at that time show that his memory was not accurate about the cause for celebration. Research shows no agreement was reached on the Missouri or Ohio in the 1910s; rather, the Ohio health board had one of their empowering water quality laws upheld in court. They celebrated that a like-minded state health board retained power to affect water quality. "Doctor Recalls Early Health Crusades," *Single Service News* 5, no. 9 (1948): 1, 3, f. unprocessed, SJC. See also Crumbine's memoir, *Frontier Doctor*.

67. Editorial, Municipal Sanitation 8, no. 4 (1937): 242.

68. *Seventeenth Biennial Report of the Iowa State Board of Health*, 46.

69. *Seventeenth Biennial Report of the Iowa State Board of Health*, 57–58.

70. *Seventeenth Biennial Report of the Iowa State Board of Health*, 60–61.

71. *Seventeenth Biennial Report of the Iowa State Board of Health*, 63.

72. *Seventeenth Biennial Report of the Iowa State Board of Health*, 62.

73. Paquin, a physician, hygienist, and lecturer, died in 1916 while he was director of health. Katherine Baxter, "Notable Kansas Citians, 1915–1918" (1925), 143–144, Native Sons Collection, KCPL.

74. "Farmers Can Help Keep Cities Well," Sept. 29, 1915, scrapbook vol. 2, HJP.

75. McLaughlin, *Sewage Pollution*, 14.

76. R. E. McDonnell, "Sewers and Water Works and Their Relation to Public Health" (conference proceeding reprint of address, Missouri Valley Public Health Association, Sept. 28, 1915), 1, B&M.

77. Philip V. Scarpino, *Great River: An Environmental History of the Upper Mississippi, 1890–1950* (Columbia: University of Missouri Press, 1985), 101–102.

78. Duffy, *Sanitarians*, 176–178.

79. Elkind, *Bay Cities and Water Politics*, 5.

80. See Dale H. Porter, *The Thames Embankment: Environment, Technology and Society in Victorian London* (Akron, OH: University of Akron Press, 1998).

81. Paul D. Haney, "The Missouri River—A Vital Resource" (paper presented at 58th Annual Conference of the Water Pollution Control Federation, Kansas City, MO, 1985), 18–22, 28.

82. The first graph showed the high rates of disease in Kansas City, Kansas, until filtration began in 1915. The second showed that Kansas City, Missouri, had the highest per capita typhoid fever rates in comparison to the thirty largest American cities. See f. 25 [ca. 1922], B&V.

83. James L. Barron, ed., *Proceedings of Seventh Annual Water Works School* (Kansas Water Works Association, 1929), 1:110.

84. Kansas State Board of Health, *Sixth Biennial Report*, 50.

85. WJ McGee, "Water Resources," in *Report of the National Conservation Commission to the Senate, February 1909* (Washington, DC: Government Printing Office, 1909), 46.

86. Tarr, *Search for the Ultimate Sink*, 293–308.

87. Ways and Means Committee of the Commercial Club, Feb. 13, 1912, scrapbook vol. 30, p. 89, CCM.

88. Tarr, *Search for the Ultimate Sink*, 293–308.

89. R. E. McDonnell, "Progress in the Disposal of Human Wastes, *Canadian Engineer* (1931): 15–16, 51–52.

90. Williams, *United States Public Health Service*, 318.

91. Kansas State Board of Health, *Twelfth Biennial Report of Kansas State Board of Health, 1923–1924* (Topeka: Kansas State Printing Office, 1924).

92. Earle G. Brown, letter to the editor, "State Board of Health Removing Pollution from Rivers Thruout Kansas," *Topeka State Journal*, Jan. 11, 1930, Kansas State Board of Health clippings, vol. 1, pp. 208–209, KSHS.

93. Paul D. Haney, "The Missouri River—A Vital Resource" (paper presented at 58th Annual Conference of the Water Pollution Control Federation, Kansas City, MO, 1985), KCPL, 28.

94. Kansas State Board of Health, *Fourth Biennial Report*, 13.

95. Kansas State Board of Health, *Fourth Biennial Report*, 12–13; and for Mississippi comparison see Scarpino, *Great River*, esp. chap. 3, "Shells, Sewage and Silt."

96. Earnest Boyce, "30 Years' Progress in Sewerage Practice," *Kansas Government Journal* 37 (Sept. 1951): 542–543.

97. Earle G. Brown, letter to the editor, "State Board of Health Removing Pollution from Rivers Thruout Kansas," *Topeka State Journal*, Jan. 11, 1930, Kansas State Board of Health clippings, vol. 1, pp. 208–209, KSHS.

98. M. M. Ellis of the Federal Bureau of Fisheries focused on the health of fish and wildlife on the Missouri River and produced reports in 1937 and 1944 that investigated pollution effects on aquatic life and made recommendations for water quality. Williams, *United States Public Health Service*, 332, 538; and Haney, "Missouri River," 31.

99. Correspondence from Carter to Garrison, and from H.D. to Carter, May 19 to July 16, 1915, Kansas City District, 1907–1930, box 17, f. 96, USACE.

100. "Sewage disposal," Nov.–Dec. 1935, General Administration, Kansas City District, box 41, f. 2340, USACE.

101. Quote from Missouri State Board of Health, *Annual Report of the Division*

of Public Health, Engineering, and Sanitation (Jefferson City, 1934), 3, 6, box 60, f. 14, MSA. The majority of facilities built in Missouri and Kansas in the 1930s received federal grants and loans. On Missouri, see Abel Wolman, "What Progress in Stream Sanitation," *Municipal Sanitation* 8, no. 1 (1937): 40–41. On Kansas, see Pfister, *Kansas State Board of Health*, 14; and Boyce, "30 Years' Progress," 542–543. On financing and better facilities, see McDonnell, "R.E. McDonnell," *Municipal Sanitation* 8, no. 1 (1937): 57. On increased drinking of public water, see "Status of Municipal Ownership of Water Works in the United States: Abstract of the Report by Burns & McDonnell, of Kansas City, Mo." *American City* 53, no. 7 (1938): 53–54.

102. McDonnell, "Progress in the Disposal of Human Wastes," 15–16, 51–52.

103. On early sanitary control, see "State Control of Public Water Supplies," *Engineering Record* 60, no. 7 (1909): 393–394. The initial Ohio River sanitary meetings occurred at the same time as the MRSC. See Cleary, *ORSANCO Story*, 8, 20–23, and chaps. 11–14; and Tarr, *Search for the Ultimate Sink*, 363.

104. Cleary, *ORSANCO Story*, 36.

105. "Polluting Rivers," *Citizens' League Bulletin*, Nov. 28, 1926, 4, KCPL; and "Rentals for Sewer Service," *Citizens' League Bulletin*, Apr. 17, 1937, 314, KCPL.

106. Series of letters on sanitation between Corps of Engineers and attorneys dated 1933 to 1935, "Kansas City District General Administration," box 41, f. 2340, USACE.

107. Series of letters on sanitation between Corps of Engineers and attorneys dated 1933–1935, "Kansas City District General Administration," box 41, f. 2340, USACE.

108. Griffith, *Missouri River*, 84.

109. Griffith, *Missouri River*, 85.

110. Branyan, *Taming the Mighty Missouri*; Schneiders, *Unruly River*, esp. chap. 5; and Thorson, *River of Promise, River of Peril*.

111. Schneiders, *Unruly River*, 88.

112. Schneiders, *Unruly River*, esp. chap. 6, "The River Rediscovered."

113. Martin Reuss and Paul K. Walker, "Financing Water Resources Development: A Brief History," US Army Corps of Engineers, July 1983, 27–31, accessed May 2018, https://permanent.access.gpo.gov/lps106084/entire.pdf; and Rosemary Feurer, "River Dreams: St. Louis Labor and the Fight for a Missouri Valley Authority," in Hurley, *Common Fields*.

114. See Schneiders, *Big Sky Rivers*.

115. Ivan Doig, *Bucking the Sun: A Novel* (New York: Simon & Schuster, 1996), 64.

116. Schneiders, *Unruly River*; Ferrell, *Big Dam Era*; Hart, *Dark Missouri*; Thorson, *River of Promise, River of Peril*; Feurer, "River Dreams"; Branyan, *Taming the Mighty Missouri*; and Martin Reuss, "The Pick-Sloan Plan," in *Builders and Fighters: U.S.*

Army Engineers in World War II, ed. Barry Fowle (Fort Belvoir, VA: US Army Corps of Engineers, 1992), 233–244.

117. McDonnell, "R.E. McDonnell," 57.

118. Lewis A. Pick, "The Missouri River Development Program," *American Water Works Association Journal* 38, no. 7 (1946), 859–867; and WPC, DEQ, *Water Quality of the Lower Missouri River*, 1.

119. Pick, "Missouri River Development Program," 862.

120. Lawson, *Dammed Indians*, 55.

121. Tribal land flooded by the Oahe Dam is the same land where, decades later, the Sioux and their pan-Indian supporters would contest the siting of the Dakota Access Pipeline. Standing Rock, the origin of the #NoDAPL protests and encampment, was one of the communities that had opposed the Pick-Sloan vision of the river. See Mark Sundeen, "What's Happening in Standing Rock?" *Outside*, Sept. 2, 2016, accessed Aug. 2017, https://www.outsideonline.com/2111206/whats-happening-standing-rock.

122. "Flood Wall Work Begun on Missouri," *New York Times*, Mar. 22, 1946, 28.

CHAPTER EIGHT: A VIEW FROM THE BLUFFS, 1951

1. *Topeka State Journal*, July 13, 1951.

2. Dwight F. Metzler, *Kansas Public Water Supplies—A Century of Progress*, Kansas Water Environment Association, accessed Mar. 31, 2018, http://www.kwea.net/images/About/documents/ks-public-water-supplies.pdf. Numbers differ among sources: Metzler cites one million; the Missouri Basin Survey Commission, *Missouri: Land and Water* (Washington, DC: US Government Printing Office, 1953), says 510,000, which is supported elsewhere.

3. The general narrative of the flood is drawn from *Kansas City Times* and *Kansas City Star*, July 7–18, 1951, SHSM.

4. William Blair, "Course of Ruin," *Star*, July 13, 1951, 2A.

5. Photograph, "View of high water in Kansas City, Kansas, during the flood of 1951," accession no. 2008-52, f. Missouri-Kansas Flood, 1951, Harry S. Truman Library & Museum, National Archives and Records Administration.

6. *Times* and *Star*, July 13, 1951.

7. A few buildings survived, like the church, but most homes and buildings in this area were built after 1951. Kansas City, Kansas, established a public housing authority in 1957 and located many of the complexes in this recently flooded area. Today, this flood-prone area has a large Hispanic population.

8. "City Escapes from under Threat of Imperiled Water Supply," *Kansas Citian* (Aug. 1951), 13–14, box 3, f. 10, CIDA.

9. Quote to *Star*, July 15, 1951, 2A; and generally, see *Times* and *Star*, July 13–18, 1951.

10. "City Escapes from under Threat of Imperiled Water Supply," *Kansas Citian*, Aug. 1951, 13–14, box 3, f. 10, CIDA.

11. *Star*, July 15, 1951, 1; Branyan, *Taming the Mighty Missouri*, 88.

12. Otis Dudley Duncan, W. Richard Scott, et al., eds., *Metropolis and Region* (Baltimore, MD: Johns Hopkins University Press, 1960), 352–355.

13. John Thornberry, "Rivers in a Rage," *Swing*, Aug. 1951, 317, KCPL.

14. *Star*, July 16, 1951.

15. *Star*, July 15, 1951, 9A.

16. Federal Writers' Project, *Kansas: A Guide to the Sunflower State* (New York: Viking, 1939), 211.

17. "City Escapes from under Threat of Imperiled Water Supply," *Kansas Citian*, Aug. 1951, 13–14, box 3, f. 10, CIDA.

18. *Times*, July 19, 1951.

19. Dwight F. Metzler, *Kansas Public Water Supplies—A Century of Progress*, Kansas Water Environment Association, accessed Mar. 31, 2018, http://www.kwea.net/images/About/documents/ks-public-water-supplies.pdf.

20. Pfister, *Kansas State Board of Health*, 50, 96.

21. Samuel J. Crumbine, "Frontier Doctor," draft, 10; final draft, 18, SJC.

22. Pfister, *Kansas State Board of Health*, 50, 96.

23. "Flood-Routed Rats Menace in Kansas City," *Moberly Monitor*, July 20, 2015, 10.

24. The hog-waste system included prisoners at Kansas City's Municipal Farm along the Blue River. "Kansas City, Mo., Develops Garbage-Cooking Program to Feed Its Hogs," *American City* 64, no. 7 (1949): 98–100.

25. *Times*, July 14, 1951, 2.

26. Thornberry, "Rivers in a Rage," 318.

27. *Times*, July 19, 1951, 1.

28. *Times*, July 17, 1951.

29. *Times*, July 14, 1951, 4.

30. Thornberry, "Rivers in a Rage," 317–318. Thornberry's comments are from a broadcast on WHB radio on July 16, 1951, and reprinted in *Swing*, a publication for a male audience.

31. See Carolyn Merchant, *The Death of Nature: Women, Ecology, and the Scientific Revolution* (New York: HarperCollins, 1980); and *Ecological Revolutions: Nature, Gender, and Science in New England* (Chapel Hill: University of North Carolina Press, 1989).

32. *Times*, July 18, 1951.

33. "Pick Uses Floods to Back Dam Plea," *New York Times*, July 21, 1951.

34. Quote to Hart, *Dark Missouri*, 148.

35. "Never Again!" *Star*, July 15, 1951, 1. The Tuttle Creek Reservoir was built upstream from Manhattan.

36. "The U.S. Masters the Big Muddy," *Life*, Aug. 22, 1955, 21–27. The magazine cover reads: "Wild Missouri Tamed at Last"; the table of contents declares "Big Muddy Tamed at Last."

37. "Flood in K.C.," *Journal-Post*, Jan. 26, 1937, box 4, f. 1648-97, Kansas City District, USACE, NARA-CPR.

38. Branyan, *Taming the Mighty Missouri*, 88.

39. Schneiders, *Unruly River*, shows how adamant the corps was that the river could be controlled. On philosophy of control and management, see Paul W. Hirt, *A Conspiracy of Optimism: Management of the National Forests since World War Two* (Lincoln: University of Nebraska, 1994).

40. Steven L. Driever and Danny M. Vaughn, "Flood Hazard in Kansas City since 1880," *Geographical Review* 78, no. 1 (1988): 1–19.

41. Commentary found in *Swing*, Oct. 1951, KCPL.

42. Brian Burnes, "Flood of 1951: K.C. Changed as Tales of Misery, Valor Live On," *Star*, July 13, 2001.

43. Thomas Hart Benton, letter to Congress, Oct. 13, 1951, available at "1951 Kaw River Flood," *Missouri State Parks*, accessed July 7, 2015, https://mostateparks.com/sites/mostateparks/files/Benton%20letter1_0.pdf.

44. *Star*, July 15, 1951, 8A.

45. Maultsby Company, Housing (pamphlet, ca. 1950s), box 39, f. 21, AIA.

46. Mark Fiege, *The Republic of Nature: An Environmental History of the United States* (Seattle: University of Washington Press, 2012), esp. chap. 8.

47. A series of *Star* articles on the decline of downtowns provide good data on Kansas City's urban decline. See William D. Tammeus, "Downtown 'Democracy' in Trouble Here," *Star*, Aug. 13, 1972, 1A, 14A; and "Downtown Decay Suburbia's Concern," *Star*, Aug. 15, 1972, both found in box 39, f. 9, AIA. Gotham, *Uneven Development*, outlines the larger demographic changes in class and race in Kansas City; and Richard Rothstein, "The Making of Ferguson: The Public Policies at the Root of its Troubles," Economic Policy Institute, Oct. 15, 2014, accessed Jan. 2016, http://www.epi.org/publication/making-ferguson/.

48. Crumbine, "Frontier Doctor," draft, 10; and final draft, 18, SJC.

49. Branyan, *Taming the Mighty Missouri*, 88.

50. Harry Truman, "Special Message to the Congress . . . Rehabilitation of the Flood Stricken Areas of the Midwest," Aug. 20, 1951, *Public Papers of the Presidents of the United States: Harry S. Truman, 1951* (Washington, DC: US Government Printing Office, 1965), 7:471–474.

51. "Flood Protection for the Kansas Cities," May 1947, box 3, f. 14, CIDA. For flood comparisons between 1951 and 1993, including costs, see Robert Cox, Ernest Kary, et al., "The 1951 Kansas-Missouri Floods... Have We Forgotten?" *National Oceanic Atmospheric Association*, accessed Nov. 2006, http://www.noaanews .noaa.gov/stories/images/kansasflood1951.pdf. The total cost of 1993 flooding was estimated at $16 billion.

52. Schneiders, *Unruly River*, esp. chap. 10, "The Mighty Missouri and the Final Quest for Control."

53. Bill Graham, "Riverfront Flood Plaques Dedicated," *Star*, July 26, 2003.

54. Ted Steinberg, *Acts of God: The Unnatural History of Natural Disaster in America*, 2nd ed. (New York: Oxford University Press, 2006); and Christine A. Klein and Sandra B. Zellmer, *Mississippi River Tragedies: A Century of Unnatural Disaster* (New York: New York University Press, 2014).

55. For comparison to another city, see Andrew Hurley, "Floods, Rats, and Toxic Waste: Allocating Environmental Hazards since World War II," in Hurley, *Common Fields*.

CHAPTER NINE: DOWNSTREAMERS AND POSTWAR POLLUTION IN A FEDERAL ERA

Epigraph: Rachel Carson, *Silent Spring* (1962; repr., Boston: Houghton Mifflin, 2002), 44.

1. Andrews, *Managing the Environment, Managing Ourselves*, 383–384; and Melosi, *Sanitary City: Urban Infrastructure*, 315–319.

2. WPC, DEQ, *Water Quality of the Lower Missouri River*, 2.

3. Glen Hopkins, "Conference at St. Joseph, Missouri on Interstate Pollution in the Missouri River," June 11, 1957, 5, St. Joseph Area, Missouri River Basin, River Basin Files, Predecessor Water Pollution Control Organizations, EPA, NARA-CPR; Chairman Stein, "Pollution of Interstate Waters: Missouri River, Kansas City Metropolitan Area," Dec. 3, 1957, 4, Missouri River Basin, River Basin Files, Predecessor Water Pollution Control Organizations, EPA, NARA-CPR; Williams, *United States Public Health Service*, 793; and WPC, DEQ, *Water Quality of the Lower Missouri River*, 2.

4. Williams, *United States Public Health Service*, 797.

5. Melvin P. Hatcher, "Effects of Missouri River Basin Control on Water Quality: Introduction," *American Water Works Association Journal* 50, no. 9 (1958): 1186–1187; and WPC, DEQ, *Water Quality of the Lower Missouri River*, 3.

6. Kansas State Board of Health, "Statement for the Hearing," 1959, 2, Lower Missouri Pollution Data, St. Joseph Hearing Witness File, St. Joseph Area,

Missouri River Basin, River Basin Files, Predecessor Water Pollution Control Organizations, EPA. On bacterial counts doubling between 1951–1965 at Saint Louis County waterworks, see table I in Herbert O. Hartung, "Comprehensive Planning for Protection and Improvement of Public Water Supply," *Water Pollution Control Federation Journal* 39, no. 12 (1967): 1999.

7. J. K. Neel studied this for the USPHS in 1963 and is cited in WPC, DEQ, *Water Quality of the Lower Missouri River*.

8. WPC, DEQ, "Water and Waste Digest" (1969–), DNR, MSA.

9. Kansas State Board of Health statement, "Lower Missouri Pollution Data . . . Witness file," 1959, 1, St. Joseph Area, Missouri River Basin, River Basin Files, EPA.

10. Glen Hopkins, "Conference at St. Joseph, Missouri on Interstate Pollution in the Missouri River, June 11, 1957," 5, St. Joseph Area, Missouri River Basin, River Basin Files, Predecessor Water Pollution Control Organizations, EPA.

11. These reports are discussed in retrospect in WPC, DEQ, *Water Quality of the Lower Missouri River*, 1–2.

12. On FWPCA see Andrews, *Managing the Environment, Managing Ourselves*; and Melosi, *Sanitary City: Urban Infrastructure*.

13. C. M. Walter in EPA, *Everyone Can't Live Upstream*, 2.

14. Murray Stein, "The Value and Use of Water Quality Criteria in the Federal Enforcement Program" (presented 1962), in *Biological Problems in Water Pollution: Third Seminar, 1962*, Public Health Service Publication no. 99-WP-25 (Cincinnati, OH: US Department of Health, 1965), 5–7, available at http://nepis.epa.gov/Adobe/PDF/20014YFO.PDF.

15. Stein, "Conference at St. Joseph Missouri on Interstate Pollution in the Missouri River, June 11, 1957," 2–3, St. Joseph Area, Missouri River Basin, River Basin Files, Predecessor Water Pollution Control Organizations, EPA.

16. Testimony of Stein and Hopkins, "Conference at St. Joseph, Missouri on Interstate Pollution in the Missouri River," June 11, 1957, 8, St. Joseph Area, Missouri River Basin, River Basin Files, Predecessor Water Pollution Control Organizations, EPA.

17. Hopkins, "Conference at St. Joseph, Missouri on Interstate Pollution in the Missouri River," June 11, 1957, 5, St. Joseph Area, Missouri River Basin, River Basin Files, Predecessor Water Pollution Control Organizations, EPA.

18. "Metzler, Health Advocate, Dies at 85," *Topeka Capital Journal*, Nov. 1, 2001, accessed July 2015, cjonline.com/stories/110101/kan_metzler.shtml#.VZ2_SqZNIgQ; and Stephen J. Randtke, "Fifty Years of Progress and Challenges for the Next Century" (presentation at 50th Annual Environmental Engineering Conference, University of Kansas, Lawrence, KS, 2000), accessed July 2015, http://

kuscholarworks.ku.edu/bitstream/handle/1808/1037/Randtke%20History%20Rev1.pdf.

19. "Legal War on Water Pollution," *Business Week*, July 16, 1960, 132, 134.

20. Formed under the FWPCA, this agency was a precursor to the Missouri Water Pollution Board.

21. Clifford Summers, "Conference at St. Joseph, Missouri on Interstate Pollution in the Missouri River, June 11, 1957," 9, St. Joseph Area, Missouri River Basin, River Basin Files, Predecessor Water Pollution Control Organizations, EPA.

22. Mayor Bartle, "Pollution of Interstate Waters, Missouri River, Kansas City Metropolitan Area," Dec. 3, 1957, 8, Kansas City, Mo. Area, Missouri River, River Basin Files, Predecessor Water Pollution Control Organizations, EPA.

23. EPA, *Everyone Can't Live Upstream*, 72–73; WPC, DEQ, *Water Quality of the Lower Missouri River*, 2.

24. "Council Proceedings," *Kansas City Times*, July 11, 1953, 6.

25. Bartle, "Pollution of Interstate Waters," 9.

26. Bartle, "Pollution of Interstate Waters," 10–14.

27. Kansas State Board of Health testimony, "Lower Missouri River Pollution Data, St. Joseph Hearing Witness File," 1959, 1, River Basin Files, Predecessor Water Pollution Control Organizations, EPA. Melosi, *Sanitary City: Urban Infrastructure*, 314–315, singles out Kansas, saying that it was "moderately active" in passing laws to abate water pollution and "quite active" in litigating for state enforcement of those laws.

28. "Otherwise," as a historian wrote of the Kansas health board's philosophy, "if one city acts to treat its sewage before discharging it in the stream at one point, this action will be in vain if a city or cities upstream do not carry out similar measures." Pfister, *Kansas State Board of Health*, 82–84.

29. KSBH testimony, "Lower Missouri River Pollution Data, St. Joseph Hearing Witness File," 1959, 1, River Basin Files, Predecessor Water Pollution Control Organizations, EPA; and Dwight Metzler, in Chester S. Wilson, chair, "Hearing at Kansas City, Missouri Concerning Pollution of the Interstate Waters of Missouri River—Turkey Creek Sewer, Kansas Cities Metropolitan Area" (proceedings of Committee Investigating Pollution of Interstate Waters for Department of Health, Education and Welfare, Public Health Service and Water Pollution Control, Kansas City, MO, June 13, 1960), 324.

30. Metzler, in Wilson, "Hearing at Kansas City," 322–323. Primary treatment removed solids and the majority of bacteria through a short one- or two-day process of settling and aeration. Secondary treatment employed a more thorough purification process. Treatment largely relied on the natural process of microbes breaking down the sewage.

31. Pfister, *Kansas State Board of Health*, 82–84.

32. EPA, *Everyone Can't Live Upstream*, 12.

33. Benjamin Powers, in Wilson, "Hearing at Kansas City," 275, 258.

34. Claude Relf, in Wilson, "Hearing at Kansas City," 314.

35. Ora J. Wheeler and George Frazier, in Wilson, "Hearing at Kansas City," 168–173.

36. KSBH testimony, "Lower Missouri River Pollution Data, St. Joseph Hearing Witness File," 1959, 5, River Basin Files, Predecessor Water Pollution Control Organizations, EPA; and "Progress Evaluation Meeting in the Matter of Pollution of the Interstate Waters of the Missouri River and Tributary Waters, Kansas City Metropolitan Area, June 30, 1960," 6, Kansas City, Missouri Area, Missouri River Basin, River Basin Files, Predecessor Water Pollution Control Organizations, EPA.

37. Joe Becker and George Cord, "Lower Missouri River Pollution Data, St. Joseph Hearing Witness File," River Basin Files, Predecessor Water Pollution Control Organizations, EPA.

38. Division of Water Pollution Control, USPHS, "Preliminary Summary, A Study of Pollution in Interstate Waters of the Missouri River between St. Joseph, Missouri and Kansas City, Kansas" (June 1959), 7a, St. Joseph area (1957–60), Missouri River Basin, River Basin Files, Predecessor Water Pollution Control Organizations, EPA.

39. Mayor Bartle, "Pollution of Interstate Waters, Missouri River, Kansas City Metropolitan Area," Dec. 3, 1957, 13, Kansas City, Missouri Area, Missouri River, River Basin Files, Predecessor Water Pollution Control Organizations, EPA.

40. The Fairfax industrial district opposed being annexed by Kansas City, Kansas, in the 1950s and 1960s largely because it did not want the imposition of city waste regulations. One source discussing this is former mayor Joseph H. McDowell, *Building a City*, 18–20. The list of polluters in the Fairfax industrial district included Sinclair Refining, General Motors, and Procter & Gamble.

41. Hartung, "Comprehensive Planning," 2007, 1997, 1999; and see also WPC, DEQ, *Water Quality of the Lower Missouri River*, 2.

42. "Pollution of Interstate Waters, Missouri River and Connecting or Tributary Waters in or Adjacent to the Kansas City Metropolitan Area, June 12–17, 1960," Kansas City, Missouri Area, Missouri River Basin, River Basin Files, Predecessor Water Pollution Control Organizations, EPA.

43. WPC, DEQ, *Water Quality of the Lower Missouri River*, 3–4, 7.

44. Hopkins, "Conference at St. Joseph, Missouri on Interstate Pollution in the Missouri River, June 11, 1957," 7, St. Joseph Area, Missouri River Basin, River Basin Files, Predecessor Water Pollution Control Organizations, EPA.

45. Hatcher, "Effects of Missouri River Basin Control," 1185–1187.

46. Hatcher, "Effects of Missouri River Basin Control," 1186.

47. Joseph F. Erdei, "Effects of Missouri River Basin Control on Water Quality: Water Quality at Omaha, Neb.," *American Water Works Association Journal* 50, no. 9 (1958): 1196–1198; Herbert O. Hartung, "Water Quality at St. Louis," *American Water Works Association Journal* 50, no. 9 (1958): 1198–1200; and Hatcher, "Effects of Missouri River Basin Control," 1185–1187.

48. Conway Briscoe testimony, in Wilson, "Hearing at Kansas City," 268.

49. Wilson, "Hearing at Kansas City," 267–271. For further evidence that sediment had purifying properties, see Paul D. Haney, "The Missouri River—A Vital Resource" (paper presented at 58th Annual Conference of the Water Pollution Control Federation, Kansas City, MO, 1985), 30, KCPL.

50. Hartung, "Comprehensive Planning."

51. Workers of the Writers' Program of the Works Projects Administration, *The W.P.A. Guide to 1930s Missouri* (Lawrence: University Press of Kansas, 1986), 4.

52. Numbers differed by source but six hundred tons per day was the smallest estimate: Letter from E. E. Harper to I. N. Watson, May 7, 1924, p. 1, f. 105, ABP; and "Kansas City Annual Meeting," 990.

53. Ferrell, *Soundings*, 137.

54. Hartung, "Comprehensive Planning," 2002; Hatcher, "Effects of Missouri River Basin Control"; and Haney, "Missouri River."

55. R. W. Love, "The Missouri River of Today," *Water Pollution Control Federation Journal* 39, no. 12 (1967): 1986.

56. Love, "Missouri River of Today," 1990–1991.

57. Melosi, *Sanitary City: Urban Infrastructure*, chap. 15.

58. Quoted in Jennifer Howe, "Mighty Missouri," *Star*, Sept. 16, 1990, L1, 4.

59. See tables 9 and 10 in "Pollution of Interstate Waters, Missouri River, Kansas City Metropolitan Area," Dec. 3, 1957, 13, Kansas City Area, Missouri River, River Basin Files, Predecessor Water Pollution Control Organizations, EPA; and WPC, DEQ, *Water Quality of the Lower Missouri River*, 4–6, 7.

60. Martin V. Melosi, *Effluent America: Cities, Industry, Energy, and the Environment* (Pittsburgh, PA: University of Pittsburgh Press, 2001).

61. I. H. Reed, in Wilson, "Hearing at Kansas City," 271–273. Chlorination costs increased for waterworks downstream from Kansas City. Crighton, *History of Health Services in Missouri*, 247. On average, the Jefferson City plant used three times more chlorine. See "Legal War on Pollution," *Business Week*, July 16, 1960, 132, 134.

62. Bishop, in Wilson, "Hearing at Kansas City," 280–281.

63. Briscoe, in Wilson, "Hearing at Kansas City," 270.

64. Williams, *United States Public Health Service*, 796.

65. R. W. Love, Lewis A. Young, and Herbert O. Hartung, "Water Quality in the

Missouri River: Progress and Prospects," *Water Pollution Control Federation Journal* 39, no. 12 (1967): 2005, 2006, 1999.

66. Haney, "Missouri River," 30. Drinking water standards in the late twentieth century came to measure pollutants in parts per million. More recently, the scale of study has been reduced to the *nano*, or one-billionth; moreover, in nanotechnology atoms and molecules are altered. See Hope Shand and Kathy Jo Wetter, "Shrinking Science: An Introduction to Nanotechnology," in *State of the World 2006* (New York: W. W. Norton, 2006), 78–95.

67. Hartung, "Comprehensive Planning," 2003, 2005.

68. Haney, "Missouri River," 37. The MRPWSA continues to represent water suppliers along the river today.

69. Linda Lear, *Rachel Carson: Witness for Nature* (New York: Henry Holt, 1997); and Andrews, *Managing the Environment, Managing Ourselves*, chap. 11, "The Rise of Modern Environmentalism," and 199.

70. Hartung, "Comprehensive Planning," 1996–1997.

71. "Legal War on Water Pollution," 132, 134.

72. Metzler, in Wilson, "Hearing at Kansas City," 323; Hopkins, in Wilson, "Hearing at Kansas City," 42–43.

73. KSBH, "Lower Missouri River Pollution Data, St. Joseph Hearing Witness File," 1959, 5, River Basin Files, Predecessor Water Pollution Control Organizations, EPA; Metzler, in Wilson, "Hearing at Kansas City," 323.

74. Metzler, in Wilson, "Hearing at Kansas City," 325; and McDowell, *Building a City*, 18–20.

75. Hopkins, in Wilson, "Hearing at Kansas City," 43.

76. Crighton, *History of Health Services in Missouri*, 247–249.

77. Hopkins, in Wilson, "Hearing at Kansas City," 43.

78. "Legal War on Water Pollution," 132, 134; Andrews, *Managing the Environment, Managing Ourselves*, 204–207; and US Army Corps of Engineers, Northwest Division, *Missouri River Mainstem Reservoir System Master Water Control Manual: Missouri River Basin*, 2006, accessed Jan. 2007, http://www.nwd-mr.usace.army.mil/rcc/reports/mmanual/MasterManual.pdf.

79. Thelen, *Paths of Resistance*.

80. "Legal War on Pollution," *Business Week*, July 16, 1960, 132, 134.

81. Gerald Gorman, "Ilus Davis: Exemplar of the 'Greatest Generation'" (Charles N. Kimball Lecture, Apr. 24, 2000), SHSM, accessed June 2006, http://shs.umsystem.edu/kansascity/kimball/Gorman-04-24-2000.pdf.

82. USPHS, Division of Water Supply and Pollution Control, "Sewage Treatment Plant Construction Cost Index: Construction Cost Trends, Municipal Waste

Treatment Works," no. 1069 (Washington, DC: US Government Printing Office, 1963).

83. EPA, *Everyone Can't Live Upstream*, 7–8.

84. President Richard M. Nixon, "38 - Special Message to the Congress on Environmental Quality," Feb. 10, 1970, online at American Presidency Project, accessed Apr. 5, 2018, http://www.presidency.ucsb.edu/ws/?pid=2757; "26 - Annual Message to the Congress on the State of the Union," Jan. 22, 1971, online at American Presidency Project, accessed Apr. 5, 2018, http://www.presidency.ucsb.edu/ws/?pid=3110.

85. "Keeping the Water Clean No Problem for Kansas City," March 1978, in Mid-America Regional Council, "Water Quality Management 208 Final Plan, Kansas City Metropolitan Region," Dec. 1978, 218, WPC, DEQ, DNR, MSA.

86. Crighton, *History of Health Services in Missouri*, 249. Mississippi River cities upstream from Saint Louis built treatment plants earlier because pollution backed up behind the lock, dam, and reservoir system. Scarpino, *Great River*, 107.

87. EPA, *Everyone Can't Live Upstream*, 3–10; and McDowell, *Building a City*, 53.

88. Murray Stein, US Department of Health, Education and Welfare, "Progress Evaluation Meeting in the Matter of Pollution of the Interstate Waters of the Missouri River and Tributary Waters, Kansas City's Metropolitan Area," Apr. 21, 1965, 4, EPA.

89. Mitchell Wendell, "Legal Aspects of Water Pollution Control," *Water Pollution Control Federation Journal* 39, no. 12 (1967): 1945–1950.

90. Paul D. Haney, "The Missouri River—A Vital Resource" (paper presented at 58th Annual Conference of the Water Pollution Control Federation, Kansas City, MO, 1985), 35, KCPL; and Wendell, "Legal Aspects," 1946.

91. The phrase "live, work, and play" is from Gottlieb, *Forcing the Spring*, 8. Other historians who examine the origins of environmentalism are Karl Brooks, *Before Earth Day: The Origins of American Environmental Law, 1945–1970* (Lawrence: University Press of Kansas, 2009); Paul C. Milazzo, *Unlikely Environmentalists: Congress and Clean Water, 1945–1972* (Lawrence: University Press of Kansas, 2006); Samuel P. Hays, *Beauty, Health and Permanence: Environmental Politics in the United States, 1955–1985* (New York: Cambridge University Press, 1987); and Andrews, *Managing the Environment, Managing Ourselves*.

92. WPC, DEQ, *Water Quality of the Lower Missouri River*, 6, 26–28.

93. WPC, DEQ, "Water Quality Management Basin Plan for Lower Missouri River Basin," June 1976, 71, DNR, MSA; and Andrews, *Managing the Environment, Managing Ourselves*, 236–237.

94. EPA, *Everyone Can't Live Upstream*, 11.

95. Rothman, *Greening of a Nation*, 111.

96. See Donald Worster, *Nature's Economy: A History of Ecological Ideas*, 2nd ed. (Cambridge: University of Cambridge Press, 1985); Hirt, *Conspiracy of Optimism*; Susan Flader, *Thinking Like a Mountain: Aldo Leopold and the Evolution of an Ecological Attitude toward Deer, Wolves and Forests* (Madison: University of Wisconsin Press, 1994); and Andrews, *Managing the Environment, Managing Ourselves*.

97. Lambrecht, *Big Muddy Blues*, xxvi.

98. Hartung, "Comprehensive Planning," 2007.

99. Rothman, *Greening of a Nation*, 110.

100. Stephen J. Randtke, "Fifty Years of Progress and Challenges for the Next Century" (presentation at 50th Annual Environmental Engineering Conference, University of Kansas, Lawrence, KS, 2000), 20, accessed Oct. 2006, https://ku scholarworks.ku.edu/dspace/bitstream/1808/1037/3/Randtke+History+Rev1.pdf; Andrews, *Managing the Environment, Managing Ourselves*, 206; and Melosi, *Sanitary City: Urban Infrastructure*, 315–319.

CHAPTER TEN: CONCLUDING WITH A VIEW FROM THE RIVER

Epigraph: Donella Meadows, "Lines in the Mind, Not in the World," *Donella Meadows Institute*, Dec. 24, 1987, accessed Aug. 2016, http://donellameadows.org /archives/lines-in-the-mind-not-in-the-world/.

1. Advertisement for Ashe Lockhart Inc. in "Alumni News," *Iowa State University Veterinarian*, Winter 1946, 175, accessed May 28, 2018, https://lib.dr.iastate.edu /cgi/viewcontent.cgi?article=1453&context=iowastate_veterinarian.

2. Freeman, "Trailing History Down the Big Muddy," 73–120.

3. Bruce Katz, "Kansas City: Region on the Rise" (presentation to Mid-America Regional Council, Brookings Institution Center on Urban and Metropolitan Policy, June 4, 2004), accessed Apr. 3, 2018, https://www.brookings.edu/wp-content /uploads/2016/06/20040604_KansasCity.pdf.

4. Heat-Moon, *River-Horse*, 226.

5. Jennifer Howe, quoting Bob Gardner and Richard Lynn, "Mighty Missouri," *Kansas City Star*, Sept. 16, 1990, L1, 4.

6. Lambrecht, *Big Muddy Blues*, 34.

7. Peter Carrels (lecture, University of Missouri-Columbia, 2002).

8. Gerald Mestl and Eugene Zuerlein, "Missouri River Navigation" (paper presented at Eighth Annual Missouri River Natural Resources Conference, May 2004, Columbia, MO); C. Phillip Baumel and Jerry Van Der Kamp, *Past and Future Grain Traffic on the Missouri River* (Minneapolis, MN: Institute for Agriculture and Trade Policy, 2003), accessed May 23, 2013, http://www.iatp.org/files

/Past_and_Future_Grain_Traffic_on_the_Missouri_.pdf; and Environmental Defense Fund, "Missouri River Barges Provide Little Benefit to Farmers," Aug. 24, 1998, accessed May 2013, http://www.edf.org/news/edf-report-missouri-river-barges-provide-little-benefit-farmers. Also see "Executive Summary" in National Research Council, *The Missouri River Ecosystem: Exploring the Prospects for Recovery* (Washington, DC: National Academy Press, 2002), 5.

9. Environmental Defense Fund, "Missouri River Barges Provide Little Benefit to Farmers"; and "Army Corps' Options Dwindle along the Mississippi River," *Weekend Edition*, National Public Radio, Jan. 13, 2013.

10. Kansas City, Missouri Parks Department and City Planning Department, "Proposed Major Parks, Boulevards, Parkways and Greenways," 1965, 42, box 32, f. 23, AIA.

11. Branyan, *Taming the Mighty Missouri*, 93–94.

12. John L. Funk and John W. Robinson, *Changes in the Channel of the Lower Missouri River and Effects on Fish and Wildlife* (Jefferson City: Missouri Department of Conservation, Nov. 1974).

13. For images of the 1977 flood see Leland Hauth et al., "Floods in Kansas City, Missouri and Kansas, September 12–13, 1977," USGS Paper 1169 (1981), 20–24, accessed May 2018, https://pubs.usgs.gov/pp/1169/report.pdf.

14. Missouri Water Pollution Board, Missouri Water Pollution Board Biennial Report, 1969–1970, Missouri Department of Health, MSA.

15. Mid-America Regional Council, "Water Quality Management, 208 Final Plan, Kansas City Metropolitan Region," Dec. 1978, WPC, DEQ, Department of Natural Resources, MSA.

16. Lambrecht, *Big Muddy Blues*. On the master manual and decision-making, including court cases, see US Army Corps of Engineers, Northwestern Division, Missouri River Master Manual, accessed Nov. 2006, http://www.nwd-mr.usace.army.mil/mmanual/mast-man.htm.

17. Michael Mansur, "Riverfront Planners Unearth Old Hazard," *Star*, May 28, 2001, A1; and Lynn Horsley, "Debris Cleanup Gets Under Way," *Star*, May 28, 2001, 8.

18. Alice Thorson, "Road Art," *Star*, July 6, 2002; "Enlarging Green Infrastructure," *Star*, Jan. 21, 2001; and "Riverfront Redevelopment," *Star*, July 18, 2000.

19. Lambrecht, *Big Muddy Blues*, xxiv.

20. Conversation with author, ca. 2005.

21. Michael Grunwald, "Washed Away," *New Republic*, Oct. 27, 2003, 16.

22. National Research Council, *Missouri River Ecosystem*.

23. National Research Council, "Executive Summary," *Missouri River Ecosystem*, 3.

24. National Research Council, "Executive Summary," *Missouri River Ecosystem*, 2–3.

25. Letter to Interested Missouri River Parties from Gregg Martin, US Army Brigadier General, July 14, 2006, US Army Corps of Engineers, Northwestern Division, Missouri River Basin, accessed Dec. 2006, http://www.nwd-mr.usace.army.mil/rcc/mrric.html.

26. Steinberg, *Acts of God*; and Klein and Zellmer, *Mississippi River Tragedies*.

27. "Missouri Tops Risk List," Star, Apr. 2, 2002; and American Rivers, *America's Most Endangered Rivers of 2001*, 3–4, accessed June 15, 2006, http://www.americanrivers.org/site/DocServer/mer2001WEB_new.pdf?docID=2141.

28. American Rivers, *America's Most Endangered Rivers of 2002*, 14, accessed June 15, 2006, http://www.americanrivers.org/site/DocServer/mer02final.pdf?docID=671.

29. Editorial, "Protect Both River, People," *Omaha World-Herald*, quoted in American Rivers, *America's Most Endangered Rivers of 2002*, 10.

30. Joe Engeln, Missouri River Institute presentation, Feb. 13, 2002, University of Missouri, Columbia.

31. Mike Hendricks, "Help Give Big Muddy a Makeover," Star, May 23, 2003; Paul Hansen, "Restore the Missouri," Star, Sept. 24, 2003; and Bill Graham, "Riverfront Flood Plaques Dedicated," Star, July 26, 2003.

32. "Officials Break Ground on Riverfront Park in KCK," Star, Sept. 27, 2003; and "KCK Working to Get Riverfront Park," Star, Apr. 9, 2003.

33. Dan Sturdevant, interview by Daniel Becton, Ubuntu Project, 2013, accessed June 2013, http://kcdv.tv/river-heroes/dan-sturdevant/riverbend-chapter.html.

34. Robert Kelley Schneiders, "Everyone's River: Democratizing the Lower Missouri," Eco in the Know, June 18, 2011, accessed Mar. 2013, http://ecointheknow.com/our-rivers/everyones-river-democratizing-the-lower-missouri-after-the-great-flood-of-2011/.

35. Chad Pegracke, "Saving America's Rivers" (lecture, University of Missouri, Columbia, Apr. 19, 2001).

36. Conversation with author, 2013 and 2018.

37. River Relief, "September 13, 2003: Looking to the River in Kansas City," accessed June 26, 2006, www.riverrelief.org/previous030913.html; and "Many Brave Rain to Clean Up River," Star, Sept. 14, 2003, B3.

38. Photos of crowds (estimated at ten to fifteen thousand people) can be seen in newspaper articles—see Times, June 14, 1935, f. 1648, box 4, Kansas City District, General Administration, USACE.

39. Bill Graham, "Volunteers Gather for River Cleanup," Star, Sept. 13, 2003, B1.

40. Scott Mansker, interview by Daniel Becton, Project Ubuntu, 2013, accessed June 2013, http://kcdv.tv/river-heroes/scott-mansker/river-miles.html.

41. Steven Schnarr, interview by Daniel Becton, Project Ubuntu, 2013, accessed June 2013, http://kcdv.tv/river-heroes/steven-schnarr/river-relief.html.

42. Karen O'Neill, "Fluvial Confluences" (roundtable, American Society for Environmental History annual conference, Toronto, 2013).

43. Bruce Babbitt, "Whenever Land Divides Us, Water Unites Us" (remarks at National Association of Counties, Baltimore, MD, July 14, 1997), online at William J. Clinton Digital Library, 2, accessed May 2013, http://www.clintonlibrary.gov/assets/DigitalLibrary/AdminHistories/Box%20011–020/Box%20012/1225030-interior-reclamation-2.pdf. Quotation corrected for grammar.

44. EPA "Kansas City, Mo., to Spend $2.5 Billion to Cut Sewer Overflows" (news release), May 18, 2010, accessed May 22, 2013, http://yosemite.epa.gov/opa/admpress.nsf/names/r07_2010-5-18_kansas_city_mo_sewer_overflows; Burns & McDonnell, "Controlling the Flow: Cities Find Innovative Ways to Implement Overflow Control Programs," *BenchMark* no. 2, 2012, 9–12.

BIBLIOGRAPHY

Archives and Collections

BAMA	Black Archives of Mid-America, Kansas City, Missouri
	Royal Fleming Collection
B&M	Burns & McDonnell Library, Kansas City, Missouri
KCPL	Kansas City Public Library, Kansas City, Missouri
	Missouri Valley Room Special Collections
	Guadalupe Collection
KCPR	Kansas City, Missouri, Parks and Recreation Archives
KSHS	Kansas State Historical Society, Topeka
	WJB W. J. Bailey Papers
	WSP William R. Stubbs Papers
LHL	Linda Hall Library of Science, Engineering and Technology, Kansas City, Missouri
MERC	Mercantile Library, University of Missouri, St. Louis
	Herman T. Pott National Inland Waterways Library
MHS	Missouri Historical Society, St. Louis
MSA	Missouri State Archives, Jefferson City
	DH Department of Health and Senior Services, Communicable Disease and Environmental Public Health
	DNR Department of Natural Resources
NARA-CPR	National Archives and Records Administration, Central Plains Region, Kansas City, Missouri
	EPA Environmental Protection Agency
	USACE United States Army Corps of Engineers
SHSM	State Historical Society of Missouri, Columbia (includes the former Western Historical Manuscript Collection from both Columbia and Kansas City)
	ABP Albert I. Beach Papers
	AIA Kansas City Chapter, American Institute of Architects Records
	ATB A. Theodore Brown Collection
	B&V Black & Veatch Engineers/Architects Records
	CCM Kansas City Chamber of Commerce Minutes

	CCS	Commercial Club of Kansas City Scrapbooks
	CIDA	Central Industrial District Association Records
	DAR	Daughters of the American Revolution, Benton Chapter Records
	HHP	Herbert S. Hadley Papers
	HJP	Henry Jost Papers and Scrapbooks
	JCMS	Jackson County Medical Society Records
	JRS	James A. Reed Scrapbooks
	SJS	St. Joseph Stockyards Company Records
	TYP	Kansas City's Ten-Year Plan Records
	WCC	Kansas City Women's Chamber of Commerce Records
	WCCR	Woman's City Club Records
	WKP	Wynkoop Kiersted Papers
SJC		Samuel J. Crumbine Collection, Clendening History of Medicine Library, University of Kansas Medical Center, Kansas City, Kansas
UMSC		University of Missouri Special Collections, Columbia, Missouri
		Sanborn Fire Insurance Maps
WCHS		Wyandotte County Historical Society, Kansas City, Kansas

Serials

Newspapers
KSHS
   University Daily Kansan, Lawrence, Kansas
SHSM
   Journal, Kansas City, Missouri
   Journal-Post, Kansas City, Missouri
   Liberator, Kansas City, Missouri
   Kansas City Post, Kansas City, Missouri
   Kansas City Star, Kansas City, Missouri
   Kansas City Sun, Kansas City, Missouri
   Reform, Kansas City, Missouri
   Rising Son, Kansas City, Missouri

Periodicals
KCPL
   Citizens' League, *Citizens' League Bulletin*
   Kansas City Chamber of Commerce, *Kansas Citian*

KSHS
   Kansas Federation of Women's Clubs, *The Club Member*

Reports
   Annual Report of the Missouri State Board of Health
   Biennial Report of the State Board of Health of the State of Iowa
   Kansas City Board of Park Commissioners Report
   Kansas City Board of Public Welfare Annual Report
   Kansas City Hospital and Health Board Annual Report
   Kansas City, Missouri Water Services Department, Annual Report, 1998–1999

ADDITIONAL PRIMARY SOURCES

"Alumni News." *Iowa State University Veterinarian*, Winter 1946. Accessed May 28, 2018. https://lib.dr.iastate.edu/cgi/viewcontent.cgi?article=1453&context=iowastate_veterinarian.

Babbitt, Bruce. "Whenever Land Divides Us, Water Unites Us." Remarks at National Association of Counties, Baltimore, MD, July 14, 1997. Accessible at William J. Clinton Digital Library, 34. Accessed May 23, 2013. http://www.clintonlibrary.gov/assets/DigitalLibrary/AdminHistories/Box%20011-020/Box%20012/1225030-interior-reclamation-2.pdf.

Barron, James L., ed. *Proceedings of Seventh Annual Water Works School*. Vol. 1. 1929.

Bartling, Howard G. *Kansas City in Caricature*. Kansas City, MO: 1912.

Baumel, C. Phillip, and Jerry Van Der Kamp. "Past and Future Grain Traffic on the Missouri River." Institute for Agriculture and Trade Policy. July 2003. Accessed May 23, 2013. http://www.iatp.org/files/Past_and_Future_Grain_Traffic_on_the_Missouri_.pdf.

Bone, D. M., ed. *Annual Review of Greater Kansas City*. Kansas City, MO: Bishop Press, 1908.

Boyce, Earnest. "30 Years' Progress in Sewerage Practice." *Kansas Government Journal* 37 (1951): 542–543.

Burns & McDonnell. "Controlling the Flow: Cities Find Innovative Ways to Implement Overflow Control Programs." *BenchMark* no. 2 (2012): 9–12.

Chanute, Octave. "The Sewerage of Kansas City." *Kansas City Review of Science and Industry* 7, no. 9 (1884): 519–527.

Engeln, Joe. Missouri River Institute. Presentation, University of Missouri, Columbia. Feb. 13, 2002.

Environmental Defense Fund. "Missouri River Barges Provide Little Benefit to

Farmers." Aug. 24, 1998. Accessed May 2013. http://www.edf.org/news/edf-report-missouri-river-barges-provide-little-benefit-farmers.

Environmental Protection Agency. "Kansas City, Mo., to Spend $2.5 Billion to Cut Sewer Overflows." News release. May 18, 2010. Accessed May 22, 2013. http://yosemite.epa.gov/opa/admpress.nsf/names/r07_2010-5-18_kansas_city_mo_sewer_overflows.

Federal Writers' Project. *Kansas: A Guide to the Sunflower State.* New York: Viking, 1939.

———. "Water Quality at St. Louis." *American Water Works Association Journal* 50, no. 9 (1958): 1198–1200.

Hatcher, Melvin P. "Effects of Missouri River Basin Control on Water Quality: Introduction." *American Water Works Association Journal* 50, no. 9 (1958): 1185–1187.

———. "How Kansas City, Mo., Sells Water to Its Suburbs." *American City* 67, no. 5 (1952): 99–101.

"Kansas City Annual Meeting." *American Journal of Public Health* 28, no. 8 (1938): 989–995.

Kansas City Board of Public Welfare Research Bureau. *Does Kansas City Have a Housing Problem?* 1912.

———. *Social Prospectus of Kansas City.* Kansas City, MO: Kansas City Board of Public Welfare, 1912.

Kansas City Chamber of Commerce. *Kansas City Health and Hospital Survey.* Kansas City, MO: Lechtman Printing, 1931.

Kansas City Public Service Institute. *Public Health in Kansas City: A Study of the Present Health Situation and Public Health Work.* Kansas City, MO: Kansas City Public Service Institute, 1922.

Kansas State Board of Health. *Fourth Biennial Report of the Kansas State Board of Health, 1907–1908.* Topeka: Kansas State Printing Office, 1909.

———. *Seventh Biennial Report of the Kansas State Board of Health, 1912–1914.* Topeka: Kansas State Printing Office, 1914.

———. *Sixth Biennial Report of the Kansas State Board of Health, 1911–1912.* Topeka: Kansas State Printing Office, 1912.

———. *Third Biennial Report of the Kansas State Board of Health, 1905–1906.* Topeka: Kansas State Printing Office, 1907.

———. *Twelfth Biennial Report of Kansas State Board of Health, 1923–1924.* Topeka: Kansas State Printing Office, 1924.

Katz, Bruce. "Kansas City: Region on the Rise." Presentation to Mid-America Regional Council, Brookings Institution Center on Urban and Metropolitan Policy. June 4, 2004.

Kirkpatrick, H. M., ed. *Official Report of the Proceedings of the Missouri River Convention.* Kansas City: Lawton and Hayens, 1885.

"Legal War on Water Pollution." *Business Week*, July 16, 1960, 132, 134.
Love, R. W., Lewis A. Young, and Herbert O. Hartung. "Water Quality in the Missouri River: Progress and Prospects." *Water Pollution Control Federation Journal* 39, no. 12 (1967): 1986–2007.
Ludlow, Alfred D. "Turkey Creek Sewer One of the Largest Yet Built." *Engineering News-Record* 87, no. 20 (1921): 817.
Mann, A. H., and W. C. Hoad. "Gagings of Sewage Flow at Lawrence, Kan." *Transactions of the Kansas Academy of Science* 20, no. 2 (1906): 281–283.
McDonnell, Robert E. "Are Water Rates Adequate for the Service Now Demanded?" *Public Service Management* 44, no. 6 (1928): 213.
———. "The Bane of Politics in the Water Department and the Remedies." *Water Works Engineering* 84, no. 10 (1931): 659–660.
———. "The Engineer Looks at Management." *Journal of the American Waterworks Association* 32, no. 6 (1940): 923–932.
———. "How Water Works Can Give Real Service." *Water Works Engineering* 82, no. 24 (1929): 1665–1666.
———. "Money Used for Water Works Brings Satisfactory Rewards." *Water Works Engineering* 81, no. 17 (1928): 1195.
———. "Progress in the Disposal of Human Wastes." *Canadian Engineer* (1931): 15–16, 51–52.
———. "R.E. McDonnell." *Municipal Sanitation* 8, no. 1 (1937): 57.
———. "Sewage Disposal Plants in Kansas." In Eighth Annual Meeting of the Kansas Gas, Water and Electric Association Proceedings. 1905, 15–20.
McDowell, Joseph H. *Building a City: A Detailed History of Kansas City, Kansas*. Reprinted articles from Kansas City Kansan, [n.d., ca. 1970].
McLaughlin, Allan J. *Sewage Pollution of Interstate and International Waters with Special Reference to the Spread of Typhoid Fever: The Missouri River from Sioux City to its Mouth*. Washington, DC: US Government Printing Office, 1913.
"Meeting of the Kansas Water and Sewage Works Association." *Water Works and Sewerage* 83 no. 6 (1936): 221.
Metzler, Dwight F. "Emergency Sanitation Lessons from the 1951 Flood in Kansas." *American Journal of Public Health* 42, no. 4 (1952): 364–372.
"Metzler, Health Advocate, Dies at 85." *Topeka Capital Journal*, Nov. 1, 2001.
Missouri Basin Survey Commission. *Missouri: Land and Water*. Washington, DC: US Government Printing Office, 1953.
Missouri Department of Environmental Quality–Water Pollution Control. *Water Quality Standards Reports: Water Quality of the Lower Missouri River, Gavins Point Dam to Mouth*. Jefferson City: Missouri Department of Natural Resources, 1982.

Missouri Water Pollution Board. *Missouri Water Pollution Board Biennial Report, 1969–1970*. Missouri Department of Health and Senior Services, 1970.

Missouri Waterway Commission. *First Biennial Report of the Missouri Waterway Commission*. In *Forty-Sixth Missouri Legislature General Assembly*, part 2, appendix. Jefferson City: Forty-Sixth Missouri General Assembly, 1911.

National Board of Fire Underwriters Committee on Fire Prevention and Engineering Standards. *Report on the City of Kansas City, Mo.*, no. 52 (Feb. 1924).

National Research Council. *The Missouri River Ecosystem: Exploring the Prospects for Recovery*. Washington, DC: National Academy Press, 2002.

Ohio State Board of Health. *Twenty-Fourth Annual Report of the Ohio State Board of Health, 1909*. Columbus, OH: Ohio State Board of Health, 1910.

———. *Twenty-Seventh Annual Report of the Ohio State Board of Health, 1912*. Columbus, OH: Ohio State Board of Health, 1913.

———. *Twenty-Third Annual Report of the Ohio State Board of Health, 1908*. Springfield, OH: Ohio State Board of Health, 1909.

O'Neill, Karen. "Fluvial Confluences." Roundtable, Annual Conference of the American Society for Environmental History, Toronto, 2013.

Pegracke, Chad. "Saving America's Rivers." Lecture, University of Missouri, Columbia, Apr. 19, 2001.

Pick, Lewis A. "The Missouri River Development Program." *American Water Works Association Journal* 38, no. 7 (1946): 859–867.

Proceedings of a Conference of Governors in the White House, May 13–15, 1908. Washington, DC: US Government Printing Office, 1909.

Samuel, T. D., Jr. "The Water Supply System of Kansas City, Missouri." *American Water Works Association Journal* 22, no. 9 (Sept. 1930): 1236–1246.

State of Missouri v. State of Illinois and the Sanitary District of Chicago. 200 US 496: United States Supreme Court, 1906.

Steinbeck, John. *Travels with Charley: In Search of America*. New York: Viking Press, 1962.

Thornberry, John. "Rivers in a Rage." *Swing*, Aug. 1951, 314–318.

Truman, Harry. "Special Message to the Congress . . . Rehabilitation of the Flood Stricken Areas of the Midwest," Aug. 20 1951. *Public Papers of the Presidents of the United States: Harry S. Truman, 1951*. Vol. 7. Washington, DC: US Government Printing Office, 1965.

United States Environmental Protection Agency, Office of Water Quality, Region VII. *Everyone Can't Live Upstream: A Contemporary History of Water Quality Problems on the Missouri River, Sioux City, Iowa to Hermann, Missouri*. Kansas City, MO: Environmental Protection Agency, 1971.

"The U.S. Masters the Big Muddy." *Life*, Aug. 22, 1955, 21–27.

Waring, George E., Jr. *Report on the Social Statistics of Cities*. Vol. 2. Washington, DC: US Government Printing Office, 1887.
Whipple, George C. "The Policy of Water Filtration." *Engineering Record* 60, no. 26 (1909): 718–719.
Wilson, Chester S., chair. "Hearing at Kansas City, Missouri Concerning Pollution of the Interstate Waters of Missouri River-Turkey Creek Sewer, Kansas Cities Metropolitan Area." Proceedings of Committee Investigating Pollution of Interstate Waters for Department of Health, Education and Welfare, Public Health Service and Water Pollution Control, Kansas City, MO, June 13, 1960.

Secondary Sources

Ambrose, Stephen. *Undaunted Courage: Meriwether Lewis, Thomas Jefferson, and the Opening of the American West*. New York: Simon & Schuster, 1996.
Andrews, Richard N. L. *Managing the Environment, Managing Ourselves: A History of American Environmental Policy*. New Haven, CT: Yale University Press, 1999.
Anfinson, John O. *The River We Have Wrought: A History of the Upper Mississippi*. Minneapolis: University of Minnesota Press, 2003.
Baker, Paula. "The Domestication of Politics: Women and American Political Society, 1790–1920." *American Historical Review* 89, no. 3 (1984): 620–647.
Baldwin, Peter. *Domesticating the Street: The Reform of Public Space in Hartford, 1850–1930*. Columbus: Ohio State University Press, 1999.
Baumhoff, Richard G. *The Dammed Missouri Valley: One Sixth of Our Nation*. New York City: Knopf, 1951.
Benac, David. "Whose Forest Is This?: Hill Folk, Industrialists, and Government in the Ozarks." *Missouri Historical Review* 101, no. 1 (2006): 17–35.
Blake, Nelson Manfred. *Water for the Cities: A History of the Urban Water Supply Problem in the United States*. Syracuse, NY: Syracuse University Press, 1956.
Botts, Lee, and Paul Muldoon. *Evolution of the Great Lakes Water Quality Agreement*. East Lansing: Michigan State University Press, 2005.
Branyan, Robert L. *Taming the Mighty Missouri: A History of the Kansas City District Corps of Engineers, 1907–1971*. Kansas City, MO: US Army Corps of Engineers, 1974.
Brooks, Karl Boyd. *Before Earth Day: The Origins of American Environmental Law, 1945–1970*. Lawrence: University Press of Kansas, 2009.
Brown, Andrew Theodore, and Lyle W. Dorsett. *K.C.: A History of Kansas City, Missouri*. Boulder, CO: Pruett, 1978.
Brown, Andrew Theodore, and Richard Wohl. "The Usable Past: A Study of Historical Traditions in Kansas City." In *The Pursuit of Local History: Readings on Theory and Practice*, edited by Carol Kammen, 145–163. Walnut Creek, CA: Altamira Press, 1966.

Bruce, Janet. *The Kansas City Monarchs: Champions of Black Baseball*. Lawrence: University Press of Kansas, 1985.

Bullard, Robert D. *Dumping in Dixie: Race, Class and Environmental Quality*. Boulder, CO: Westview Press, 1990.

———. "Urban Infrastructure: Social, Environmental, and Health Risks to African Americans." In *Handbook of Black American Health: The Mosaic of Conditions, Issues, Policies, and Prospects*, edited by Ivor Lensworth Livingston, 315–330. Westport, CT: Greenwood Press, 1994.

Button, James W. *Blacks and Social Change: Impact of the Civil Rights Movement in Southern Communities*. Princeton, NJ: Princeton University Press, 1989.

Carson, Rachel. *Silent Spring*. 1962. Reprint, Boston: Houghton Mifflin, 2002.

Cashill, Jack. *A Century of Excellence: Burns & McDonnell*. Kansas City, MO: Burns & McDonnell, 1998.

———. "Kansas City—America's Engineering Mecca." *Ingram's*, Oct. 2002, 28–33.

Chafe, William H. "Women's History and Political History: Some Thoughts on Progressivism and the New Deal." In *Visible Women: New Essays on American Activism*, edited by Nancy Hewitt and Suzanne Lebsock, 101–118. Urbana: University of Illinois Press, 1993.

Cioc, Marc. *The Rhine: An Eco-Biography, 1815–2000*. Seattle: University of Washington Press, 2002.

Ciucci, Georgio, et al. *The American City: From the Civil War to the New Deal*. Cambridge, MA: MIT Press, 1979.

Cleary, Edward J. *The ORSANCO Story: Water Quality Management in the Ohio River Valley under an Interstate Compact*. Baltimore, MD: Resources for the Future/Johns Hopkins University Press, 1967.

Colten, Craig E. "Basin Street Blues: Drainage and Environmental Equity in New Orleans, 1890–1930." *Journal of Historical Geography* 28, no. 2 (2002): 237–257.

———. "Environmental Justice in the American Bottom: The Legal Response to Pollution, 1900–1950." In *Common Fields: An Environmental History of St. Louis*, edited by Andrew Hurley, 163–175. Saint Louis: Missouri Historical Society, 1997.

———, ed. *Transforming New Orleans and Its Environs: Centuries of Change*. Pittsburgh, PA: University of Pittsburgh Press, 2000.

———. *An Unnatural Metropolis: Wresting New Orleans from Nature*. Baton Rouge: Louisiana State University Press, 2005.

Corbett, Katharine. "Draining the Metropolis: The Policy of Sewers in Nineteenth Century St. Louis." In *Common Fields: An Environmental History of St. Louis*, edited by Andrew Hurley, 107–125. Saint Louis: Missouri Historical Society Press, 1997.

Cox, Robert, Ernest Kary, et al., "The 1951 Kansas-Missouri Floods . . . Have we

Forgotten?" National Oceanic Atmospheric Association, Nov. 2006. http://www.noaanews.noaa.gov/stories/images/kansasflood1951.pdf.
Crighton, John C. *The History of Health Services in Missouri*. Omaha, NE: Barnhart Press, 1993.
Cronon, William. *Nature's Metropolis: Chicago and the Great West*. New York: W. W. Norton, 1991.
———, ed., *Uncommon Ground: Rethinking the Human Place in Nature*. New York: W. W. Norton, 1996.
Crumbine, Samuel J. *Frontier Doctor: The Autobiography of a Pioneer on the Frontier of Public Health*. New York: Dorrance, 1948.
Davis, Mike. "Slum Ecology: Poverty's Niche in the Ecology of the City." *Orion*, Mar./Apr. 2006, 16–23.
Di Chiro, Giovanna. "Nature as Community: The Convergence of Environmental and Social Justice." In *Uncommon Ground: Rethinking the Human Place in Nature*, edited by William Cronon, 298–320. New York: W. W. Norton, 1996.
Diner, Steven J. *A Very Different Age: Americans of the Progressive Era*. New York: Hill and Wang, 1998.
Doig, Ivan. *Bucking the Sun: A Novel*. New York: Simon & Schuster, 1996.
Dorsett, Lyle W. *The Pendergast Machine*. New York: Oxford University Press, 1968.
Dowden, Priscilla A. "'Over This Point We Are Determined to Fight': The Urban League of St. Louis in Historical Perspective." *Gateway Heritage* 13, no. 4 (1993): 32–47.
Driever, Steven L., and Danny M. Vaughn. "Flood Hazard in Kansas City since 1880." *Geographical Review* 78, no. 1 (1988): 1–19.
Driggs, Frank, and Chuck Haddix. *Kansas City Jazz: From Ragtime to Bebop—A History*. New York: Oxford University Press, 2005.
Du Bois, W. E. B. *The Souls of Black Folk*. 1903. Reprint, Project Gutenberg, 2008. http://www.gutenberg.org/files/408/408-h/408-h.htm.
Duffy, John. *The Sanitarians: A History of American Public Health*. Urbana: University of Illinois Press, 1990.
Duncan, Otis Dudley, W. Richard Scott, et al., eds. *Metropolis and Region*. Baltimore, MD: Johns Hopkins University Press, 1960.
Ehrlich, George. *Kansas City, Missouri: An Architectural History, 1826–1990*. Revised ed. Columbia: University of Missouri Press, 1992.
Elkind, Sarah S. *Bay Cities and Water Politics: The Battle for Resources in Boston and Oakland*. Lawrence: University Press of Kansas, 1998.
Ellis, Roy. *A Civic History of Kansas City, Missouri*. Springfield, MO: Elkins-Swyers, 1930.
Erdei, Joseph F. "Effects of Missouri River Basin Control on Water Quality: Water

Quality at Omaha, Neb." *American Water Works Association Journal* 50, no. 9 (1958): 1196–1198.

Felton, Jean S., and Alfred Katz, eds. *Health and the Community: Readings in the Philosophy and Sciences of Public Health*. New York: Free Press, 1965.

Ferrell, John R. *Big Dam Era: A Legislative and Institutional History of the Pick-Sloan Missouri Basin Program*. Omaha, NE: Missouri River Division, United States Army Corps of Engineers, 1993.

———. *Opposites*. Omaha, NE: Feather Works Books, 2000.

———. *Soundings: One Hundred Years of the Missouri River Navigation Project*. [Kansas City, MO]: United States Army Corps of Engineers, 1996.

Feurer, Rosemary. "River Dreams: St. Louis Labor and the Fight for a Missouri Valley Authority." In *Common Fields: An Environmental History of St. Louis*, edited by Andrew Hurley, 221–241. Saint Louis: Missouri Historical Society Press, 1997.

Fiege, Mark. *The Republic of Nature: An Environmental History of the United States*. Seattle: University of Washington Press, 2012.

Fields, Barbara Jeanne. "Slavery, Race, and Ideology in the United States of America." *New Left Review* 181, May–June 1990, 95–118.

Flader, Susan L. *Thinking Like a Mountain: Aldo Leopold and the Evolution of an Ecological Attitude toward Deer, Wolves and Forests*. Madison: University of Wisconsin Press, 1994.

Flanagan, Maureen. "The City Profitable, the City Livable: Environmental Policy, Gender, and Power in Chicago in the 1910s." *Journal of Urban History*, 22, no. 2 (1996): 163–190.

Flores, Dan. *Horizontal Yellow: Nature and History in the Near Southwest*. Albuquerque: University of New Mexico Press, 1999.

———. "Place: An Argument for Bioregional History." *Environmental History Review* 18, no. 4 (1994–1995): 1–18.

Freeman, Lewis R. "Trailing History Down the Big Muddy." *National Geographic Magazine*, July 1928, 73–120.

Funk, John L., and John W. Robinson. *Changes in the Channel of the Lower Missouri River and Effects on Fish and Wildlife*. Jefferson City: Missouri Department of Conservation, 1974.

Glaab, Charles N. *The American City: A Documentary History*. Homewood, IL: Dorsey Press, 1963.

———. "The Historian and the American City: A Bibliographic Survey." In *American Urban History: An Interpretive Reader with Commentaries*, edited by Alexander B. Callow, 654–673. New York: Oxford University Press, 1969.

———. *Kansas City and the Railroads: Community Policy in the Growth of a Regional Metropolis*. Madison: State Historical Society of Wisconsin, 1962.

Glave, Dianne, and Mark Stoll, eds. "To Love the Wind and the Rain": African Americans and Environmental History. Pittsburgh, PA: University of Pittsburgh Press, 2006.

Goings, Kenneth W., and Raymond A. Mohl. "Toward a New African American Urban History." Journal of Urban History 21, no. 3 (1995): 283–295.

Gorman, Gerald. "Ilus Davis: Exemplar of the 'Greatest Generation.'" Charles N. Kimball Lecture, Apr. 24, 2000. Accessed July 2015. http://shs.umsystem.edu/kansascity/kimball/Gorman-04-24-2000.pdf.

Gotham, Kevin Fox. Race, Real Estate, and Uneven Development: The Kansas City Experience, 1900–2000. Albany: State University of New York Press, 2002.

Gottlieb, Robert. Forcing the Spring: The Transformation of the American Environmental Movement. Washington, DC: Island Press, 1993.

Griffith, Cecil. The Missouri River: A River Rat's Guide to Missouri River History and Folklore, edited by K. R. Canfield and R. L. Sutton. Leawood, KS: n.p., 1974.

Grunwald, Michael. "Washed Away." New Republic, Oct. 27, 2003, 16–18.

Hale, Grace Elizabeth. Making Whiteness: The Culture of Segregation in the South, 1890–1940. New York: Vintage, 1998.

Halpern, Rick, and Roger Horowitz. Meatpackers: An Oral History of Black Packinghouse Workers and Their Struggle for Racial and Economic Equality. New York: Twayne, 1996.

Haney, Paul. "The Missouri River—A Vital Resource." Paper presented at 58th Annual Conference of the Water Pollution Control Federation, Kansas City, MO, 1985. KCPL.

Hanson, Joseph Mills. Conquest of the Missouri: Being the Story of the Life and Exploits of Captain Grant Marsh. 1909. Reprint, New York: Murray Hill Books, 1946.

Harlan, James, and James Denny. Atlas of Lewis and Clark in Missouri. Columbia: University of Missouri Press, 2003.

Hart, Henry C. The Dark Missouri. Madison: University of Wisconsin Press, 1957.

Haskell, Henry C., Jr., and Richard B. Fowler. City of the Future: A Narrative History of Kansas City, 1850–1950. Kansas City, MO: Frank Glenn, 1950.

Hays, Samuel P. Beauty, Health and Permanence: Environmental Politics in the United States, 1955–1985. New York: Cambridge University Press, 1987.

———. Conservation and the Gospel of Efficiency: The Progressive Conservation Movement, 1890–1920. Cambridge, MA: Harvard University Press, 1959.

Heat-Moon, William Least. River-Horse: The Logbook of a Boat across America. Boston: Houghton Mifflin, 1999.

Heiser, Elizabeth Isabel. "Public Health Administration in Nebraska." MA thesis, University of Nebraska, 1920.

Herron, John P. Science and the Social Good: Nature, Culture, and Community, 1865–1965. New York: Oxford University Press, 2010.

Hill, Libby. *The Chicago River: A Natural and Unnatural History*. Chicago: Lake Claremont, 2000.

Hirt, Paul W. *A Conspiracy of Optimism: Management of the National Forests since World War Two*. Lincoln: University of Nebraska Press, 1994.

Hohl, Paul. "City beneath a City." *Kansas City Magazine*, Sept. 1979, 68–69, 73, 78, 82.

Hoy, Suellen. *Chasing Dirt: The American Pursuit of Cleanliness*. New York: Oxford University Press, 1995.

———. "'Municipal Housekeeping': The Role of Women in Improving Urban Sanitation Practices, 1880–1917." In *Pollution and Reform in American Cities, 1870–1930*, edited by Martin V. Melosi, 173–198. Austin: University of Texas Press, 1980.

Hunter, Tera W. *To 'Joy My Freedom: Southern Black Women's Lives and Labors after the Civil War*. Cambridge, MA: Harvard University Press, 1997.

Hurley, Andrew, ed. *Common Fields: An Environmental History of St. Louis*. Saint Louis: Missouri Historical Society Press, 1997.

———. *Environmental Inequalities: Class, Race, and Industrial Pollution in Gary, Indiana, 1945–1980*. Chapel Hill: University of North Carolina Press, 1995.

———. "The Social Biases of Environmental Change in Gary, Indiana, 1945–1980." *Environmental Review* 12, no. 4 (1998): 1–19.

Isaacson, Darlene, and Elizabeth Wallace. *Kansas City in Vintage Postcards*. Charleston, SC: Arcadia, 2003.

Jackson, Kenneth. *Crabgrass Frontier: The Suburbanization of the United States*. New York: Oxford University Press, 1985.

Jacoby, Karl. *Crimes against Nature: Squatters, Poachers, Thieves, and the Hidden History of American Conservation*. Berkeley: University of California Press, 2001.

Jones, Jacqueline. *Labor of Love, Labor of Sorrow: Black Women, Work, and the Family from Slavery to the Present*. New York: Vintage, 1985.

Kansas City Chamber of Commerce. *Where These Rocky Bluffs Meet: Including the Story of the Kansas City Ten-Year Plan*. Kansas City, MO: Chamber of Commerce, 1938.

Kelman, Ari. *A River and Its City: The Nature of Landscape in New Orleans*. Berkeley: University of California Press, 2003.

Kerber, Linda K., Alice Kessler-Harris, and Kathryn Kish Sklar, eds. *U.S. History as Women's History: New Feminist Essays*. Chapel Hill: University of North Carolina Press, 1995.

Kibel, Paul Stanton, ed. *Rivertown: Rethinking Urban Rivers*. Cambridge, MA: MIT Press, 2007.

Klein, Christine A., and Sandra B. Zellmer. *Mississippi River Tragedies: A Century of Unnatural Disaster*. New York: New York University Press, 2014.

Klein, Maury, and Harvey A. Kantor. *Prisoners of Progress: American Industrial Cities, 1850–1920.* New York: Macmillan, 1976.

Klingle, Matthew. *Emerald City: An Environmental History of Seattle.* New Haven, CT: Yale University Press, 2007.

Kraut, Alan M. *Silent Travelers: Germs, Genes, and the "Immigrant Menace."* New York: Basic Books, 1994.

Kuhn, Thomas S. *The Structure of Scientific Revolutions.* 2nd ed. Chicago: University of Chicago Press, 1970.

Laird, Judith Fincher. "Argentine, Kansas: The Evolution of a Mexican-American Community, 1905–1940." PhD diss., University of Kansas, 1975.

Lambrecht, Bill. *Big Muddy Blues: True Tales and Twisted Politics along Lewis and Clark's Missouri River.* New York: St. Martin's Press, 2005.

Landis, Margaret. "The Winding Valley and the Craggy Hillside: A History of the City of Rosedale, Kansas." 1976. Kansas Collection. Kansas City, Kansas, Public Library. Accessed Apr. 4, 2018. https://www.kckpl.org/kansas/documents/winding-valley-rosedale-history.pdf.

Lang, William, and Robert Carriker, eds. *Great River of the West: Essays on the Columbia River.* Seattle: University of Washington Press, 1999.

Larsen, Lawrence H., and Nancy J. Hulston. *Pendergast!* Columbia: University of Missouri Press, 1997.

Lawson, Michael. *Dammed Indians: The Pick-Sloan Plan and the Missouri River Sioux, 1944–1980.* Norman: University of Oklahoma Press, 1982.

Lear, Linda. *Rachel Carson: Witness for Nature.* New York: Henry Holt, 1997.

Leavitt, Judith Walzer. "Gendered Expectations: Women in Early Twentieth-Century Public Health." In *U.S. History as Women's History: New Feminist Essays,* edited by Linda K. Kerber, Alice Kessler-Harris, and Kathryn Kish Sklar. Chapel Hill: University of North Carolina Press, 1995.

———. *The Healthiest City: Milwaukee and the Politics of Health Reform.* Princeton, NJ: Princeton University Press, 1982.

Lerner, Steve. *Sacrifice Zones: The Front Lines of Toxic Chemical Exposure in the United States.* Cambridge, MA: MIT Press, 2010.

Logan, Michael F. *Desert Cities: The Environmental History of Phoenix and Tucson.* Pittsburgh, PA: University of Pittsburgh Press, 2006.

Lynaugh, Joan E. *The Community Hospitals of Kansas City, Missouri, 1870–1915.* New York: Garland, 1989.

Maher, Neil. "A New Deal Body Politic: Landscape, Labor, and the Civilian Conservation Corps." *Environmental History* 7, no. 3 (July 2002): 435–461.

Mallea, Amahia. "Progressive Black Kansas City." Paper presented at Missouri Conference on History, Kansas City, MO, 2002.

———. "Progressive Kansas City and the Missouri River." MA thesis, University of Missouri, Columbia, 2001.

Martin, Asa E. *Our Negro Population: A Sociological Study of the Negroes of Kansas City, Missouri*. 1913. Reprint, New York: Negro University Press, 1969.

McCormick, Robert L. "Public Life in Industrial America, 1877–1917." In *The New American History*, rev. ed., edited by Eric Foner, 107–132. Philadelphia: Temple University Press, 1997.

McKelvey, Blake. *The Urbanization of America, 1860–1915*. New Brunswick, NJ: Rutgers University Press, 1963.

Melosi, Martin V. *Effluent America: Cities, Industry, Energy, and the Environment*. Pittsburgh, PA: University of Pittsburgh Press, 2001.

———. *Garbage in the Cities: Refuse, Reform, and the Environment, 1880–1980*. College Station: Texas A&M University Press, 1981.

———. "The Place of the City in Environmental History." *Environmental History Review* 17, no. 1 (1993): 1–23.

———. *Pollution and Reform in American Cities, 1870–1930*. Austin: University of Texas Press, 1980.

———. *The Sanitary City: Environmental Services in Urban America from Colonial Times to the Present*, abridged ed. Pittsburgh: University Press of Pittsburgh, 2008.

———. *The Sanitary City: Urban Infrastructure in America from Colonial Times to the Present*. Baltimore, MD: Johns Hopkins University Press, 2000.

Melosi, Martin V., and Joseph A. Pratt. *Energy Metropolis: An Environmental History of Houston and the Gulf Coast*. Pittsburgh, PA: University of Pittsburgh Press, 2007.

Mendoza, Valerie M. "The Creation of a Mexican Immigrant Community in Kansas City, 1890–1930." PhD diss., University of California, Berkeley, 1997.

Meraji, Shereen Marisol. "Outdoor Afro: Busting Stereotypes That Black People Don't Hike or Camp." *Codeswitch*, July 12, 2015. National Public Radio. Accessed July 2015. http://www.npr.org/sections/codeswitch/2015/07/12/421533481/outdoor-afro-busting-stereotypes-that-blacks-dont-hike-or-camp.

Merchant, Carolyn. *The Death of Nature: Women, Ecology, and the Scientific Revolution*. New York: HarperCollins, 1980.

———. *Ecological Revolutions: Nature, Gender, and Science in New England*. Chapel Hill: University of North Carolina Press, 1989.

Mestl, Gerald, and Eugene Zuerlein. "Missouri River Navigation." Eighth Annual Missouri River Natural Resources Conference, Columbia, MO, May 2004.

Metzler, Dwight F. *Kansas Public Water Supplies—A Century of Progress*. Kansas Water Environment Association. Accessed Mar. 31, 2018. http://www.kwea.net/images/About/documents/ks-public-water-supplies.pdf.

Milazzo, Paul Charles. *Unlikely Environmentalists: Congress and Clean Water, 1945–1972*. Lawrence: University Press of Kansas, 2006.

Miller, Char. "Martin Luther King, Social Justice, and Streetscape Environmentalism." Apr. 2, 2014. KCET. Accessed Jan. 2017. https://www.kcet.org/redefine/martin-luther-king-social-justice-and-streetscape-environmentalism.

———, ed. *On the Border: An Environmental History of San Antonio*. Pittsburgh, PA: University of Pittsburgh Press, 2001.

———. "Streetscape Environmentalism: Floods, Social Justice, and Political Power in San Antonio, 1921–1974." *Southwestern Historical Quarterly* 118, no. 2 (Oct. 2014): 158–177.

Mitman, Gregg. "In Search of Health: Landscape and Disease in American Environmental History." *Environmental History* 10, no. 2 (2005): 184–210.

Mohl, Raymond A., and James F. Richardson. *The Urban Experience: Themes in American History*. Belmont, CA: Wadsworth, 1973.

Moulton, Gary E., ed. *The Journals of the Lewis and Clark Expedition*. Vol. 2. Lincoln: University of Nebraska Press, 2001.

Muir, John. *My First Summer in the Sierra*. Boston: Houghton Mifflin, 1944.

Mumford, Kevin J. *Interzones: Black/White Sex Districts in Chicago and New York in the Early Twentieth Century*. New York: Columbia University Press, 1997.

Mumford, Lewis. *The City in History: Its Origins, Its Transformations, and Its Prospects*. New York: Harcourt, 1961.

Nash, Linda. "The Changing Experience of Nature: Historical Encounters with a Northwest River." *Journal of American History* 86, no. 4 (2000): 1600–1629.

———. "Finishing Nature: Harmonizing Bodies and Environments in Late-Nineteenth-Century California." *Environmental History* 8, no. 1 (2003): 25–52.

Neihardt, John G. *The River and I*. New York: Macmillan, 1927.

Novak, William J. *The People's Welfare: Law and Regulation in Nineteenth-Century America*. Chapel Hill: University of North Carolina Press, 1996.

———. "Private Wealth and Public Health: A Critique of Richard Epstein's Defense of the 'Old' Public Health." *Perspectives in Biology and Medicine* 46, no. 3 (2003): 176–198.

Ochsner, Ralph H. "The Firm of Hare & Hare and the Art of Landscape Architecture." *Kawsmouth* 3, no. 1 (2001): 43–56.

Parrish, William E., Charles T. Jones Jr., and Lawrence O. Christensen. *Missouri: The Heart of the Nation*. 2nd ed. Wheeling, IL: Harlan Davidson, 1992.

Peiss, Kathy. *Cheap Amusements: Working Women and Leisure in Turn-of-the Century New York*. Philadelphia: Temple University Press, 1986.

Pfister, Harriet S. *Kansas State Board of Health*. Lawrence: University of Kansas Governmental Research Center, 1955.

Pisani, Donald J. *Plumbing the Truckee: Water, Diversion and the Creation of Community along the Truckee River, Nevada*. ProQuest, 2007.

———. "The Polluted Truckee: A Study in Interstate Water Quality, 1870–1934." *Nevada Historical Society Quarterly* 20, no. 3 (1977): 151–166.

Platt, Harold L. "Chicago, the Great Lakes, and the Origins of Federal Urban Environmental Policy." *Journal of the Gilded Age and Progressive Era*, 1, no. 2 (2002): 122–153.

Polanyi, Karl. *The Great Transformation: The Political and Economic Origins of Our Time*. 2nd ed. Boston: Beacon Press, 2001.

Porter, Dale H. *The Thames Embankment: Environment, Technology, and Society in Victorian London*. Akron, OH: University of Akron Press, 1998.

Potter, David. *People of Plenty: Economic Abundance and the American Character*. Chicago: University of Chicago Press, 1954.

Price, Jennifer. "Remaking American Environmentalism: On the Banks of the L.A. River." *Environmental History* 13, no. 3 (July 2008): 536–555.

———. "Thirteen Ways of Seeing Nature in L.A." *Believer Magazine* 4 (Apr.–May 2006). Accessed May 2018. https://www.believermag.com/issues/200604/?read=article_price.

Randtke, Stephen J. "Fifty Years of Progress and Challenges for the Next Century." Presentation at 50th Annual Environmental Engineering Conference, University of Kansas, Lawrence, KS, 2000. Accessed Oct. 2006. https://kuscholarworks.ku.edu/dspace/bitstream/1808/1037/3/Randtke+History+Rev1.pdf.

Reddig, William M. *Tom's Town: Kansas City and the Pendergast Legend*. Philadelphia: Lippincott, 1947. Reprint, Columbia: University of Missouri Press, 1986.

Reid, Robert L., ed. *Always a River: The Ohio River and the American Experience*. Bloomington: Indiana University Press, 1991.

Reuss, Martin, and Paul K. Walker. "Financing Water Resources Development: A Brief History." US Army Corps of Engineers. July 1983, 27–31. Accessed May 2018. https://permanent.access.gpo.gov/lps106084/entire.pdf.

———. "The Pick-Sloan Plan." In *Builders and Fighters: U.S. Army Engineers in World War II*, edited by Barry W. Fowle, 233–244. Fort Belvoir, VA: US Army Corps of Engineers, 1992.

Rhodes, Richard. *A Hole in the World: An American Boyhood*. New York: Simon & Schuster, 1990.

Riis, Jacob A. *How the Other Half Lives: Studies among the Tenements of New York*. Edited by Sam Bass Warner Jr. Cambridge, MA: Belknap Press of Harvard University Press, 1970.

Roediger, David. *Wages of Whiteness: Race and the Making of the American Working Class*. New York: Verso, 1991.

Rose, Mark, and J. G. Clark. "Light, Heat, and Power: Energy Choices in Kansas City, Wichita, and Denver, 1900–1935." *Journal of Urban History* 5, no. 3 (1979): 340–364.

Rosen, Christine Meisner. *The Limits of Power: Great Fires and the Process of City Growth in America.* Cambridge: Cambridge University Press, 1986.

Rothman, Hal K. *The Greening of a Nation? Environmentalism in the United States since 1945.* Fort Worth, TX: Harcourt Brace, 1998.

Rothstein, Richard. *The Color of Law: A Forgotten History of How Our Government Segregated America.* New York: Liveright, 2017.

———. "The Making of Ferguson: The Public Policies at the Root of its Troubles." Economic Policy Institute. Oct. 15, 2014. Accessed Jan. 2016. http://www.epi.org/publication/making-ferguson/.

Scarpino, Philip V. *Great River: An Environmental History of the Upper Mississippi, 1890–1950.* Columbia: University of Missouri Press, 1985.

Schirmer, Sherry Lamb. *A City Divided: The Racial Landscape of Kansas City, 1900–1960.* Columbia: University of Missouri Press, 2002.

———. "Landscape of Denial: Space, Status and Gender in the Construction of Racial Perceptions among White Kansas Citians, 1900–1958." PhD diss., University of Kansas, 1995.

———. "Overview of the Mexican-American Community." Typescript held by Kansas City, KS, Public Library. 1976.

Schneiders, Robert Kelley. *Big Sky Rivers: The Yellowstone and Upper Missouri.* Lawrence: University Press of Kansas, 2003.

———. "Everyone's River: Democratizing the Lower Missouri." Eco in the Know. June 18, 2011. Accessed Mar. 15, 2013. http://ecointheknow.com/our-rivers/everyones-river-democratizing-the-lower-missouri-after-the-great-flood-of-2011/.

———. *Unruly River: Two Centuries of Change along the Missouri.* Lawrence: University Press of Kansas, 1999.

Sellers, Christopher C. *Hazards of the Job: From Industrial Disease to Environmental Health Science.* Chapel Hill: University of North Carolina Press, 1997.

———. "Thoreau's Body: Towards an Embodied Environmental History." *Environmental History* 4, no. 4 (1999): 486–514.

Serda, Daniel. "A Blow to the Spirit: The Kaw River Flood of 1951 in Perspective." Paper presented at Midcontinent Perspectives lecture series, Midwest Research Institute, Kansas City, MO, Oct. 28, 1993. Accessed June 30, 2006. html://www.umkc.edu/whmckc/PUBLICATIONS/MCP/MCPPDF/serda-10-28-93.pdf.

Shand, Hope, and Kathy Jo Wetter. "Shrinking Science: An Introduction to Nanotechnology." In *State of the World 2006*, edited by Worldwatch Institute, 78–95. New York: W. W. Norton, 2006.

Sheriff, Carol. *The Artificial River: The Erie Canal and the Paradox of Progress, 1817–1862.* New York: Hill and Wang, 1996.

Shortridge, James. *Kansas City and How It Grew, 1822–2011.* Lawrence: University Press of Kansas, 2012.

Sklar, Kathryn Kish. *Florence Kelley and the Nation's Work: The Rise of Women's Political Culture, 1830–1900.* New Haven, CT: Yale University Press, 1995.

———. "Two Political Cultures in the Progressive Era: The National Consumers' League and the American Association for Labor Legislation." In *U.S. History as Women's History: New Feminist Essays*, edited by Linda K. Kerber, Alice Kessler-Harris, and Kathryn Kish Sklar, 36–62. Chapel Hill: University of North Carolina Press, 1995.

Smith-Howard, Kendra. *Pure and Modern Milk: An Environmental History since 1900.* New York: Oxford University Press, 2014.

Spain, Daphne. *How Women Saved the City.* Minneapolis: University of Minnesota Press, 2001.

Spirn, Anne Whiston. *The Granite Garden: Urban Nature and Human Design.* New York: Basic Books, 1984.

Steinberg, Ted. *Acts of God: The Unnatural History of Natural Disaster in America.* 2nd ed. New York: Oxford University Press, 2006.

Steinberg, Theodore. *Nature Incorporated: Industrialization and the Waters of New England.* Cambridge: Cambridge University Press, 2004.

Still, Bayrd. *Urban America: A History with Documents.* Boston: Little, Brown and Company, 1974.

Strasser, Susan. *Waste and Want: A Social History of Trash.* New York: Metropolitan Books, 1999.

Sugrue, Thomas J. *The Origins of the Urban Crisis: Race and Inequality in Postwar Detroit.* Princeton, NJ: Princeton University Press, 1996.

Tarr, Joel A., ed. *Devastation and Renewal: An Environmental History of Pittsburgh and Its Region.* Pittsburgh, PA: University of Pittsburgh Press, 2003.

———. *The Search for the Ultimate Sink: Urban Pollution in Historical Perspective.* Akron, OH: University of Akron Press, 1996.

Taylor, Henry Louis, Jr., and Walter Hill, eds. *Historical Roots of the Urban Crisis: African Americans in the Industrial City, 1900–1950.* New York: Garland, 2000.

Taylor, Loren L. *The Consolidated Ethnic History of Wyandotte County.* Kansas City, KS: Kansas City, KS, Ethnic Council, 2000. Held by Wyandotte County Historical Society, Kansas City, KS.

Teaford, Jon C. *The Twentieth-Century American City: Problem, Promise, and Reality.* 2nd ed. Baltimore, MD: Johns Hopkins University Press, 1993.

Thelen, David. *Paths of Resistance: Tradition and Democracy in Industrializing Missouri*. Columbia: University of Missouri Press, 1991.

Theriot, Christopher, and Kelly Tzoumis. "Bankside Chicago." In *Rivertown: Rethinking Urban Rivers*, edited by Paul Stanton Kibel, 67–84. Cambridge, MA: MIT Press, 2007.

Thorson, John E. *River of Promise, River of Peril: The Politics of Managing the Missouri River*. Lawrence: University Press of Kansas, 1994.

Tomes, Nancy. *The Gospel of Germs: Men, Women, and the Microbe in American Life*. Cambridge, MA: Harvard University Press, 1998.

Troesken, Werner. "The Limits of Jim Crow: Race and the Provision of Water and Sewerage Services in American Cities, 1880–1925." *Journal of Economic History* 62, no. 3 (2002): 734–772.

———. "Race, Disease, and the Provision of Water in American Cities, 1889–1921." *Journal of Economic History* 61, no. 3 (2001): 750–776.

Trotter, Joe W. "African Americans in the City: The Industrial Era, 1900–1950." *Journal of Urban History* 21, no. 4 (1995): 438–457.

Twain, Mark. *Life on the Mississippi*. New York: Harper and Brothers: 1896. Reprint, 1981.

Vaughan, K., A. Kaczynski, et al. "Exploring the Distribution of Park Availability, Features, and Quality across Kansas City, Missouri by Income and Race/Ethnicity: An Environmental Justice Investigation." *Annals of Behavioral Medicine* 45, no. 1 (2013): 28–38.

Vestal, Stanley. *The Missouri*. Lincoln: University of Nebraska Press, 1945. Reprint, Lincoln: Bison Books, 1964.

Walter, John. "Kansas City Engineering Firm Marks 100 Years of Service." *Jackson County Historical Society Journal* 38, no. 2 (1998): 6–7.

Washington, Sylvia Hood. *Packing Them In: An Archaeology of Environmental Racism in Chicago, 1865–1954*. Lanham, MD: Lexington, 2005.

Wendell, Mitchell. "Legal Aspects of Water Pollution Control." *Water Pollution Control Federation Journal* 39, no. 12 (1967): 1945–1950.

Wennersten, John R. *Anacostia: The Death and Life of an American River*. Baltimore, MD: Chesapeake Book Co., 2008.

White, Richard. *The Organic Machine: The Remaking of the Columbia River*. New York: Hill and Wang, 1995.

Wiebe, Robert. *The Search for Order: 1877–1920*. New York: Hill and Wang, 1967.

Williams, Ralph Chester. *The United States Public Health Service, 1798–1950*. Washington, DC: Commissioned Officers Association of the USPHS, 1951.

Wilson, William H. *The City Beautiful Movement*. Baltimore, MD: Johns Hopkins University Press, 1989.

———. *The City Beautiful Movement in Kansas City*. Columbia: University of Missouri Press, 1964.

Works Project Administration, Workers of the Writers' Program. *The W.P.A. Guide to 1930s Missouri*. Foreword by Charles van Ravenswaay. Introduction by Howard Wright Marshall and Walter A. Schroeder. Lawrence: University Press of Kansas, 1986.

Worley, William S. *J. C. Nichols and the Shaping of Kansas City: Innovation in Planned Residential Communities*. Columbia: University of Missouri Press, 1990.

Worster, Donald. *Dust Bowl: The Southern Plains in the 1930s*. New York: Oxford University Press, 1979.

———. *Nature's Economy: A History of Ecological Ideas*. 2nd ed. Cambridge: Cambridge University Press, 1985.

Zimmerman, Michael, and J. Baird Callicott, eds. *Environmental Philosophy: From Animal Rights to Radical Ecology*. Upper Saddle River, NJ: Prentice Hall, 1993.

INDEX

abatement, 77, 113, 124, 188, 189, 213, 214, 216, 225, 229, 232; pursuing, 215; resistance to, 226, 231
activism, 47, 161, 170
Addams, Jane, 46, 225
Admiral Hay, 19
African Americans, 133, 135, 138, 146, 162; community and, 109–110; death rate among, 157, 159; employment of, 155, 156; environment and, 145; flooding and, 24; healthcare and, 140, 141–142; movement of, 137; population of, 136, 142; restrictions on, 41
agriculture, 167, 209, 221, 250; water quality and, 196
algae, 213, 249, 250
American Creosoting, 85
American Medical Association (AMA), 60
American Public Health Association, 160
American Radiator, 85
American Rivers, 243–244
American Water Works Association, 43, 220
Andrews, Richard, 225
animal life, sewage and, 189
aquatic life, 188–189, 216, 243; sewage and, 189
Argentine, 26, 107, 113, 135, 209; flood in, 27, 199; problems for, 27
Armour, 18, 35, 136
Armourdale, 18, 135, 208; cleanup in, 25 (photo)
Ash, Louis R., 65, 66, 112, 117
Asian carp, 243
Atherton, John, 206; painting by, 207 (fig.)
Atlas Oats, 19

Babbitt, Bruce, 247
bacteria, 175, 176, 182, 184, 188, 196, 218, 223, 224, 249, 275n136, 292n30, 303n30; counts, 189 (table), 218, 302n6; data on, 192; fecal, 53, 55; typhoid-fever-causing, 39, 47, 55, 192
bacteriology, 52, 54, 55, 182
Bailey, Willis, 25–26
Baltimore Hotel, 176
barges, 4, 5, 196, 203, 210, 238, 246
Bartle, H. Roe, 215, 216, 227
baths, 2, 48, 49; photo of, 48
Beach, Albert, 64, 65, 85, 86, 89, 149, 152, 153, 261n60, 266n136; African American appointments by, 156; election of, 62–63; Hartman and, 150; Nichols and, 286n60; supervision and, 66–67; Turkey Creek project and, 117–118
Beardsley, Henry M., 78–79
Belt Line Railroad, 112
Belvidere Hollow, 136
Bense Act, Ohio State Board of Health and, 126
Benton, Thomas Hart, 96, 206, 210; painting by, 208 (fig.)
Berkeley Waterfront Park, 240–241
"Big Blue Bend," 96–97
Big Four Hundred, 148
Big Muddy National Fish and Wildlife Refuge, 241
Bissell Point waterworks, 172
Black & Veatch (firm), 65, 118, 224
Blake, Nelson Manfred, 34
Blue River, 26, 30, 80, 100–102, 127; aquatic life in, 216; cleanup of, 89, 235; debate over, 81; described, 83, 101; engineering on, 89; flood control for, 96–97; park system and, 82; pollution in, 101, 231; as recreation spot, 86–87, 87 (fig.), 88; sewage in, 22, 73, 87, 88, 92, 216; as sewer park, 81–89; sewer system and, 81–82, 100–101, 228; trails along, 251; treatment plant on, 13, 81, 228, 275n136

Blue River Valley, 82, 85; floods in, 96; industrial workforce in, 109
Blue Valley Flood Protection and Parkway, 96
Blue Valley Parkway, 83
Blue Valley Sewer, 96
boardinghouses, 47, 135, 136
Board of Park Commissioners, 77
Board of Public Utilities, 68
Board of Public Welfare, 48, 86, 143, 270n54, 283n14
Board of Public Works, 78, 118
boiling orders, 39, 60, 200, 201, 250
bonds, 37, 63, 79, 93, 116, 272n83; general obligation, 227; hospital, 141; issuing, 82, 216; sewerage, 228; waterworks, 59, 60, 61
Boonville, Missouri, 167, 218, 223
boosters, 9, 43, 45, 46, 108, 129, 136, 157, 168, 232, 245; civic, 41; commercial, 181, 197; economic, 42, 194, 205; lobbying by, 197; navigation, 187; Pick-Sloan, 195, 222
border war, 169–170, 172–175
bottoms, 8, 11, 17–18, 75, 109, 110, 137, 236, 251; disease in, 56; evolution of, 234–235; flooding of, 20, 26, 29, 107; heavy industry and, 107; inhospitable, 210; living in, 86, 134; oil- and gas-drenched, 203; photo of, 31; residents/workers of, 29, 135; samples from, 101; soils of, 5
Bowery, The, 136
Boyce, Earnest, 189
bridges, 20, 26, 40
Briscoe, Conway, 221, 223
Brokaw, Charles, 108–109
Brooks, Karl, 249
Brown, Darius, 40
Brown, Earle, 188, 189
Brown, Linda, 198
Brown v. Board of Education (1954), 208
Brush Creek, 80, 90, 96, 205, 251; bacteria in, 249; flooding of, 97–98, 239; park system and, 81, 98, 99; paving, 74, 97; photo of, 98; sewage in, 21, 71, 73, 101; as sewer park, 81–89; sewer system and, 98–99

Bryant, Hughes, 59, 64, 266n136
Buchholz, William, 64, 266n136
Bucking the Sun (Doig), 195
Bullard, Robert D., 162, 283n21
Burke, John, 165
Burns, Clinton, 33, 34–35, 133
Burns & McDonnell (firm), 37, 65, 67, 68, 118, 246, 249; survey by, 128; water plants by, 33
Business Week, 225, 226

Cahokia, 3
Call, 146
Carson, Rachel, 211, 224–225
Cavaness, Ernest, 149, 150
Central Park, 83
Centropolis, 3, 87
cesspools, 71, 91, 218
Chanute, Octave, 71, 72
charity, 26, 27, 46, 94
Charles River, urban planning and, 84
Chase, Champion S., 167, 168, 175, 194, 222
chemicals, 211, 223, 224; agricultural/industrial, 101, 232, 249; treatment, 53, 192
Chicago & Alton Railway, 18
Chicago Cubs, 198
Chicago River, 174
Child, Scott, 145
Chittenden, Hiram, 276n21
chlorination, 193, 194, 223, 267n153
chlorine, using, 53, 54, 57, 263n95
Cincinnati Chamber of Commerce, 191
Cincinnati Commercial Club, 294n52
Citizens' Historical Association, 43
Citizens' League, 43, 57, 59, 60, 61, 65, 66, 81, 129, 148, 155, 160, 161, 192
Citizens' League Bulletin, 66, 81, 155, 156, 157
"city beautiful" movement, 43
city building, 74–75, 94
city charters, 59, 136
City Ice Company, 97
city services, 6, 33, 86, 90, 209; planned regionalism and, 127–131
"Civics by Radio" (show), 43
civil rights movement, 91, 245
Civil Works Administration, 95

INDEX | 335

Clark, William, 35
class, 6, 90, 91, 148, 156; bias, 150; connections and, 133; inequalities, 5; infrastructure and, 163; power relations of, 90, 132; prejudices of, 153; race and, 162
cleanup, 14, 23, 24, 29, 30, 234, 235; photo of, 25; unions and, 203
Clean Water Act, 4, 212, 230, 248; loophole in, 250; oversight of, 228
Cleaver, Emanuel, 142
coagulation, 38, 39, 53, 174, 187, 225, 267n153
codes: building, 89, 143; housing, 144
Coleridge, Samuel Taylor, 28
Colgan, William, 206
Combined Sewer Overflows (CSOs), 248
Commercial Club. *See* Kansas City Chamber of Commerce
community: African American, 142, 145, 151, 156; economic diversity of, 110; environment and, 225
Conference of Governors, 105, 165, 180
Connell, Ralph W., 177
Consolidated Water Company, 258n16
contamination, 21, 52, 53, 55, 56, 85, 86, 126, 143, 176, 177, 192, 201, 213, 223, 240; risk of, 200; tolerance for, 215
Convention Hall, 49; photo of, 23
Cookingham, L. P., 200
cooperation, 4, 39, 40, 41, 105, 118, 132, 179, 180, 181, 182, 213, 217, 219, 226, 239, 244; interstate, 231; regional, 125, 127–128
Corps of Discovery, 10, 244
corruption, 9, 55, 117, 170–171
Country Club district, 89, 90, 146, 147, 153
Country Club Plaza, 41; photo of, 98
Cowgill, James, 59
Cowick, Kate, on Kansas City, 15
Craddock, Mayor, flood relief and, 22
Crittenden, Thomas T., 79, 81, 82
Cromwell, Frank, 61, 62, 65, 117
Cross, Walter, 39
Crumbine, Samuel J., 13–14, 114–115, 121–127, 168, 176, 177–178, 179, 180, 185, 188, 189, 217, 232, 233; campaigns by, 119, 121; cartoons of, 121; cross-border thinking and, 106; flood and, 28, 202; grassroots strength of, 124; Missouri River and, 125, 126; MRSC and, 178, 182; photo of, 120; pollution and, 184; progressivism of, 119, 121–122; sanitation and, 122, 123; Turkey Creek and, 125; vision of, 183; work of, 294n66
culture, 108, 109, 142; political, 105, 131; popular, 157; river, 245

Dakota Access Pipeline, 4, 298n121
dams, 222, 233; building, 195; drinking water and, 213
Davidson Construction, payments to, 118
Davidson Construction v. City of Rosedale, et al., 278n50
Davis, Ilus "Ike," 228
DDT, 203, 223
Dehoney, Carl, 108, 109, 119
Delano, D. J., 80
demographics, 30, 109, 138, 300n47
Department of Health, Education and Welfare, 213–214, 228
Department of the Interior, 228
Des Moines Water Works, 250
Des Plaines River, 172, 174
Dickey, Walter S., 74, 80, 92, 194, 266n135
Dickey Clay Manufacturing Company, 74
dilution, 73, 173, 175, 176, 179, 189, 213, 225
discrimination, 90, 140, 155, 162
disease, 141, 288n103; breeding, 56; combating, 125; controlling, 280n83; dump sites and, 56; heart, 159; outbreaks of, 202; poverty and, 55, 155; rates, 288n103; risk of, 210; source, 53; venereal, 142, 157; waterborne, 60; zymotic, 71–72
Division of Sanitation, 189
Division of Water and Sewage, 114
Doig, Ivan, on dam construction, 195
drainage, 72, 75, 83, 88, 97, 111, 112, 250; problem, 110; urban, 82
drinking water, 20, 35, 45, 51, 105, 127, 130, 162, 176, 177, 187, 196, 212, 224, 226; chlorinated, 194; connections

drinking water (continued)
and, 163; dams and, 213; data on, 220; delivering, 54; floods and, 21, 201; health risks with, 64; history of, 39; importance of, 43; infrastructure for, 67; lead levels in, 249–250; lines, 7; monitoring, 52; protecting, 188; public, 69; purifying, 251; quality of, 68, 173; as regional issue, 40–41; river water for, 4, 5; safe, 41, 55; systems, 33, 249–250; treatment of, 53
droughts, 32, 190, 195, 196, 240
Du Bois, W. E. B., 132
Duffy, John, 42
Dunn, W. H., 82, 84

Earth Day, 230
East Bottoms, 30, 67, 85, 135, 137, 150; flooding of, 256n15; pumping station for, 62, 200; survey of, 86
ecology, 6, 8, 14, 216, 238, 239, 248
economic conditions, 8, 9, 47, 49, 70, 85, 113, 129, 190
economic development, 11, 42, 50, 151, 160, 169, 215, 233
economic issues, 7–8, 94, 153, 157
Economic River, 168, 238, 239, 243–244; challenge to, 186, 242, 245; creation of, 232; environmental injustice and, 196–197; Healthy River and, 181, 182, 221; investing in, 208; public cost for, 243; responsibility for, 197; triumph of, 194; vision of, 10–11, 136, 143, 169, 191–192, 194, 195
economic units, 105; social units and, 109
economic wealth, 46, 168, 233; public health and, 27
ecosystems, 11, 232, 242
Edwards, George H., 59, 62, 64, 266n136
Eighteenth and Vine Streets, 136, 145, 146, 152, 209
electricity, 7, 15, 20, 21, 30, 33, 37, 195, 200, 220
Elkind, Sarah, 127
Ellis, M. M., 296n98
Ellis, Roy, 75, 84
Emerson, Ralph Waldo, 42

employment, 191, 288n103; African Americans and, 161
Endangered Species Act, 5, 242
endocrine-disrupting compounds, 232
Engineering Record, 125, 180
environment, 3, 55; African Americans and, 145; city/river and, 133; community and, 225; complexity of, 215; health and, 157, 161, 230; public health and, 169, 196; shaping, 144–145
environmental conditions, 190, 219, 225, 239
environmental damage, 219, 242
Environmental Engineering Conference, 215
environmental health, 8, 11, 176, 230, 241, 243; determining, 4; human health and, 14, 225; public health and, 217
environmental history, 10, 249, 253n6, 253–54n9, 255n14
environmentalism, 11, 105, 225, 239, 248, 307n91
environmental issues, 28–29, 102, 131, 153, 157, 162, 222, 230, 241, 251
environmental justice, 5, 162, 196–197
environmental movement, 162, 187, 230, 245; Endangered Species Act and, 242
Environmental Protection Agency (EPA), 4, 101, 229, 230, 231, 242, 248, 249
environmental reality, 178; regionalism and, 186
EPA. *See* Environmental Protection Agency
epidemics, 35, 39, 53, 60, 145; flooding and, 28
Erdei, Joseph F., 221
Ermine Case Jr. Park, 251
Etzanoa, 3
Everyone Can't Live Upstream (EPA), 231

Fairfax, 107, 108, 200, 208, 281n96, 304n40
Faxon, Frank, 46, 108
Federal Bureau of Fisheries, 296n98
federal jurisdiction, flirting with, 189–194
Federal Water Pollution Control Act (FWPCA; 1948), 212, 217, 225, 226, 230, 303n20; amendment for, 213–214

fertilizers, 219, 221, 223, 250
filtration, 53, 54, 59, 60, 174, 184, 187, 267n153; concentrating on, 178; rapid sand, 39
filtration plants, 38, 59, 62
Finn, Thomas M., 142
fish: health of, 296n98; invasive, 243
fish kills, 219
Fleming, Colonel, 50
Flint, Michigan, lead levels in, 249–250
flood control, 4, 88, 100, 111, 113, 117, 195, 196, 205, 233, 240, 244
Flood Control Act (1928), 195
Flood Control Act (1936), 195
Flood Disaster (Benton), 206, 208 (fig.)
flooding, 5, 6, 11, 19, 25, 99, 132, 190, 199, 201, 239, 240; bottoms and, 20, 26, 29, 107; challenges of, 203–204; comparing, 301n51; drinking water and, 21; exacerbating, 116; policy shifts and, 241; political vigilance and, 32; preventing, 205, 210, 245, 249; risk of, 17, 241; safeguards against, 113; severe, 238; stopping, 98, 195
flood of 1903, 41, 111, 194, 202, 205, 206; cleanup following, 24, 29, 30; as cultural benchmark, 13; deaths during, 30; epidemics following, 28; impact of, 17, 18, 20, 26, 30, 31–32; photo of, 19; social/environmental crisis of, 28–29; social status and, 23, 28
flood of 1951, 198–196, 208–210, 227, 236; impact of, 206, 210; photo, 202; pollution from, 203; waterworks and, 68
floodwaters, 24, 132, 199; contaminants in, 21, 22, 203; photo of, 31
flushers, 78, 79
Folk, Joseph, 170, 180
Food and Drug Act, 123
Fort Leavenworth, 20
Fort Peck Dam, 195
Fort Riley, 190
Fortune, 99
fountains, 2; construction of, 1; legacy of, 49
Fourteenth Ward Civic and Improvement Club, 144

Frankenfeld Sand and Fuel Company, 167
Franklin, Benjamin, 42
Freeman, Lewis, 236
Friends of Big Muddy, 241
Friends of the Kaw, 241
Fuller, George W., 62, 65, 66
Fuller & McClintock (firm), 62
FWPCA. *See* Federal Water Pollution Control Act

Gallagher, Dorothy, 158
garbage, 37, 78, 79, 80, 83, 122, 162, 177, 187, 219; dangers of, 185–186; disposal of, 202–203; as gateway issue, 246; hogs and, 203, 216
gas, 20, 101, 108, 279n72; services, 37; sewer, 72, 78
Gavins Point Dam, 220
General Motors, 200, 304n40
germs, 52, 55, 57, 157
"Get It Done" campaign, 93
Gillespie, G. L., 173
Gilpin, William, 3
Goodyear tire plant, 198
Gooseneck Sewer, 96
Gordon, Troy, 241
Gottlieb, Robert, 161
Governor's Trophy, 122, 123 (fig.)
Great Depression, 91, 94, 155, 159, 191
Great Lakes, 125, 172, 182, 183, 250, 293n45
Green, Charles W., Turkey Creek project and, 113, 114, 116
Green, T. W., 277n31
Greystone Heights, 112, 113, 116, 277n23
Griffith, Cecil, 193
"Grove, The," 77
Guadalupe Center, 158
Guinotte, James E., 87, 88
Gulf of Mexico, 10, 172, 222, 250

Hadley, Herbert S., 108, 119, 170, 174, 176, 180, 181; photo of, 171
Haff, D. J., 148
Halbert, Leroy A., 144
Hallmark Company, 206
Haney, Paul, 224, 230

Hannibal Bridge, 29, 71, 203
Hare, S. Herbert, 129
Harlem, 30
Hartman, J. T., 149, 150
Hartung, Herbert O., 220–221, 222; on drinking water, 232; on environment, 225; on pollution, 224; on water utility operators, 224
Hatcher, Melvin P., 14, 200, 211, 215, 220, 232, 233; pollution and, 212
health, 4, 6, 37, 45, 52, 221; campaigns, 56; economics of, 155–161; environment and, 14, 157, 161, 225, 230; improving, 42, 54, 89; influences on, 55; issues, 160, 161; municipal, 56; notions of, 34; protecting, 8; risks, 64, 157, 162, 194; river, 212–219; threats to, 92; urban, 90; wealth and, 11, 42, 43, 169, 178–182
healthcare, 4, 6; African Americans and, 109, 140, 142, 161–162; housing and, 140; private/public, 109, 141, 157, 159; segregated, 159; value of, 160
health professionals, segregation and, 142
Healthy River, 192, 232, 234, 237, 249; challenge from, 242; Economic River and, 181–182, 197, 221; iteration of, 245; responsibility of, 233; vision of, 11, 125, 169, 186; voice for, 212
Healthy Rivers Partnership, 23, 235, 238, 244
heart disease, deaths from, 159
Heart of America, 129, 134
Heat-Moon, William Least, 1, 69, 237
heavy metals, 101, 219, 223
Heim Brewery, 28
Hell's Half Acre, 134, 136
Hick's Hollow, 136
Higgs, Roy, 192, 193
Highway Act (1956), 209
Hill, Curtis, 79, 90, 117
Hoad, W. C., 114, 124, 176
Hoch, Edward W., 105
hogs: drugs for, 201; garbage-fed, 203, 216
Hohl, Paul, 70
Holly Street Reservoir, 37, 50
Holmes, Oliver Wendell, 174, 175

"Homecoming—Kaw Valley 1951," 206
homelessness, 24, 29, 199
Hoover, Herbert, 33, 121
Hopkins, Glen, 214, 220
hospitals, African American, 141–142
hospital staff, African American, 141 (photo)
housing, 42, 108; affordable, 90; African American, 133, 137, 148, 162; boxcar, 140; conditions, 157; costs, 135; good, 144; healthcare and, 140; improving, 44, 145; national market, 146; parks and, 151; public, 208, 209 (photo), 210; regulations, 48, 89; shortage of, 89; unhealthy, 149
Howard Bend, 221, 292n27
"How Our Town Saved the River" (exhibition), 239
Hoxie, George, 57
Hoxie, Mrs., 57
Hull House, 46
hydrants, 37, 74, 79, 138
hydroelectricity, 195
hypochlorite, 39, 57, 184

identity, 6, 27, 88, 109, 129, 134, 236; community, 30; preservation of, 140; regional, 110
Illinois River, 172
immigrants, 109, 134, 136, 140, 158, 175; diversity of, 27; social workers and, 138–139
incineration, 188, 216, 275n136
Independence, Missouri, 28n91; Blue River and, 88; pollution in, 212
industrial districts, 18, 85, 108, 133; photo of, 107
industrial economy, 4, 27, 39, 167, 175, 206, 225
industrialists, 41, 49, 187
industrialization, 35, 42, 46, 56, 139, 161, 169, 175, 185, 209, 210; costs of, 8; ecological impacts of, 248; urban, 124
industrial workers, 24, 27, 109
infrastructure, 7, 21, 79, 95, 99, 244; building, 74, 89, 91, 127, 152, 190; class and, 150, 163; disintegration of, 90,

133, 249; drinking water, 67; funding, 93; green, 240, 249; investment in, 147, 161, 249; public, 99–100; race and, 163; regional, 105; river ward, 102, 136; sewerage and, 80; topography and, 75; waste, 13
interconnections, 55, 101, 135, 156, 157, 225, 247
Iowa State Board of Health, 184
irrigation, 10, 195, 196, 221, 240, 241, 243

Jackson County, Missouri, 45, 100, 129, 170; sewer district, 92; water system for, 128, 130
Jefferson City, Missouri, 96, 215, 218, 223
Jim Crow, 144
Johnson County, Kansas, 110, 129; suburbs of, 237; wealth in, 8
Joint Regional Survey Committee, 128–129
Jones, Lawrence M., 167, 168, 194
Jost, Henry, 40, 79, 114, 116, 144
Journal of the American Water Works Association, 196
Jungle, The (Sinclair), 143

Kansas Cities: as economic/social/environmental unit, 7–8; historical understanding of, 7–8; map of, 16 (fig.), 104 (fig.); population of, 177; as regional hub, 3–4; water quality and, 213
Kansas City, Missouri: as "All-American" city, 26; growth of, 8, 9; illustration of, 12 (fig.); Kansas City, Kansas, and, 2–3; renewal of, 1–2; report by, 56; testing by, 181; wealth in, 8
Kansas City Athenaeum, 47
Kansas City Board of Fire and Water Commissioners, 37, 63, 64, 66, 59, 78, 267n150
Kansas City Board of Public Welfare, 47, 138, 139, 142
Kansas City Business Men, 167
Kansas City Chamber of Commerce, 46, 61–62, 91, 93, 97, 106, 108, 153, 199; Economic River and, 191–192. See also Kansas City Commercial Club

Kansas City Commercial Club, 9, 22–23, 45–46, 49, 50, 74, 77, 78, 85, 93, 108, 143, 144, 153, 168, 170, 180, 181, 188, 191–192, 194, 203; gift from, 26; meeting by, 19. See also Kansas City Chamber of Commerce
Kansas City Common Council, 64
Kansas City Consumers' League, 58
Kansas City Council of Clubs, 28, 125
Kansas City Department of Buildings, 143
Kansas City District Corps of Engineers, 192
Kansas City Gazette, on Commercial Club, 26
Kansas City Hospital and Health Board, 47–48, 56, 141, 142, 143, 161, 263n95
Kansas City in Caricature, 51
Kansas City Inter-Racial Committee, 148
Kansas City Livestock Exchange, 35, 135
Kansas City Magazine, 70
Kansas City Monarchs, 146
Kansas City Nut and Bolt, 86
Kansas City Park Board, 151, 154
Kansas City Parks Department, 238
Kansas City Port Authority, 244
Kansas City Post, 77
Kansas City Public Library, 256n15
Kansas City Public Service Institute, 92, 159
"Kansas City spirit," 20, 29, 108, 205
Kansas City Spirit, The (Rockwell and Atherton), 206; reproduction of, 207 (fig.)
Kansas City Star, 17, 43, 63, 64, 237, 240, 246, 256n15, 300n47; Crumbine and, 119; editorial cartoon from, 61; on flood, 25, 30–31, 205; on refugees, 24; on state line, 26
Kansas City Urban League, 156, 157
Kansas City Water Department, 65, 101
Kansas City Woman's City Club, 46
Kansas City Women's Chamber of Commerce, 159–160
Kansas River. *See* Kaw River
Kansas State Board of Health, 114, 115, 116, 176, 182, 215, 226; Crumbine and, 119, 121; survey by, 188; water plants and, 39; wildlife health and, 188–189

Kansas Supreme Court, 40
Kansas Women's Federation of Clubs, 125
Katy Trail State Park, 247
Kaw Point Park, 39, 234, 251; photo of, 235
Kaw Point Station, 35, 38, 39
Kaw River: aquatic life in, 216; engineering on, 98; flood relief along, 25–26; meatpacking along, 18, 106; Missouri confluence with, 1, 17; samples from, 183; sewage in, 80, 87–88, 114, 117, 129
Kaw Valley district, 107
Kaw Valley Drainage Board, 115
Kemp, William E., 199, 204
Kersey Coates Terrace North, 154
Kessler, George E., 82, 83, 129, 147; park and boulevard system and, 43–44
Kiersted, Wynkoop, 53, 57, 65, 72, 73

labor, 143; African American, 155, 156–157; diversity in, 8; gang, 156; shortages, 89, 211; strife, 27; transient, 109; wartime, 211
Lake Michigan, 169, 172
Lambrecht, Bill, 237, 241
Land and Water Conservation Fund, 238
landscapes, 3, 4, 10, 11, 129, 133, 151, 154, 195; aesthetic attributes of, 75; environmental, 152; industrial, 209; socioeconomic, 158; urban, 12, 41, 153
lead levels, 249–250
Leavenworth, Kansas, 108, 218
Leeds hospital unit, 159
levees, 5, 10, 68, 111–112, 130, 194, 195, 199, 200, 204, 205, 210, 237, 238, 240
Lewis, E. S., 156
Lewis and Clark, 10, 244
Lewis and Clark Trail Heritage Foundation, 244
Lexington, Missouri, 192, 218
Liberator, 30
Life Magazine, 205
lime, using, 39, 184
"Lines in the Mind, Not in the World" (Meadows), 234
Linwood Improvement Association, 152
Living Lands & Waters, 246

Loftus, M. J., 86
Loftus, Mrs. M. J., 151–152
"longer pipes" theory, 127, 130
Loose-Wiles Biscuit Company, 153
Love, R. W., 222
Ludlow, Alfred D., 117

Mahoning River, 125, 179
Main Street sewer, 71, 74, 75, 90; photo of, 76; smell of, 77
Maitland, Alexander, 59, 60, 62, 65, 66
Mandan villages, 3
manholes, 7, 21, 78
Mann, Conrad, 266n135
Mansker, Scott, 247
MARC, 131, 239
Martin, Asa, 138, 155
Marvin, Frank O., 124
Massman Construction Company, 192
Matscheck, Walter, 159
"Mayor's Christmas Tree, The," 61
McClure Flats, 49
McDonnell, Georgia Howlett, 42, 45, 50, 60; described, 47; drinking water and, 41; on soft water plan, 61
McDonnell, Robert E., 34–35, 37, 42, 45, 50, 62, 65, 80, 92, 119, 124, 130, 133, 148, 168, 176, 186, 189, 191, 192, 232; agriculture/water quality and, 196; caricature of, 51 (fig.); drinking water and, 41, 54; on good health, 160; letter from, 53; monumental achievement and, 67; municipal housekeeping and, 225; photo of, 34; progressivism and, 33; reform era and, 43; sewage treatment and, 72, 179; sunshine policy and, 58; on surface water, 188; typhoid fever and, 57; waste infrastructure and, 13, 58; on water purification, 60; water service and, 58; waterworks and, 59, 68
McElroy, Henry F., 67, 93
McGee, W. J., 180, 187–188
McLaughlin, Allan, 182, 183, 186
McQueeney, James, 153
Meadows, Donella, 234
meatpacking, 18, 25, 96, 106, 107, 169, 200, 256n15; African Americans in,

155; environment and, 50; waste from, 177
Meierhoffer Company, 167
Melosi, Martin, 7, 37, 222, 223
Meramec, 175
Metropolitan Railways Company, 21
Metropolitan Street Railway Company, 20
Metzler, Dwight, 215, 217
Mexican Americans, 91, 144, 157, 158, 199; disease/death rates for, 288n103
miasma theory, 52, 72
microbes, 8, 303n30
microbiology/microchemistry, 224
Mid-America Regional Council (MARC), 131, 239
Miller, L. M., 77
Miller, Mary, 60
Mississippi River, 3, 10, 107, 173; treatment plants on, 307n86; waste in, 172
Missouri Compromise, 244
Missouri Department of Conservation, 238, 241
Missouri Department of Health, 191
"Missouri Idea," 170
Missouri River: bacteria counts for, 189 (table); dependence on, 71; engineering on, 98; histories of, 11; illustration of, 12 (fig.); Kaw confluence with, 1, 17; as liquid freeway, 3–4; power of, 30–31; public perception of, 243–244; relationship with, 6, 132; sewage in, 81, 101, 129, 139, 168, 225; visions of, 10–11
Missouri River 340 Race, 247
Missouri River Basin, 185, 215; map of, 146
Missouri River Basin Association, 239
Missouri River Commission, 111
Missouri River Convention, 167
Missouri River Improvement Association, 194
Missouri River Public Water Suppliers Association (MRPWSA), 224, 232, 306n68
Missouri River Recovery Implementation Committee, 243

Missouri River Relief, 241, 247, 251; cleanup by, 234; organization of, 246; photo of, 235
Missouri River Sanitary Conference (MRSC), 125, 126, 168, 176–178, 184, 185, 186, 193–194, 231, 232, 240; diplomacy and, 294n66; health/wealth and, 178–182; USPHS and, 183
Missouri River System: Exploring the Prospects for Recovery, The (National Research Council), 242
Missouri State Board of Health, 92, 172; drinking water supplies and, 52; Turkey Creek and, 116; water samples and, 192
Missouri state constitution (1875), 227
Missouri Supreme Court, 61, 117
Missouri Valley Authority, 195
Missouri Valley Public Health Association (MVPHA), 185, 186
Missouri Valley River Improvement Association, 180
Missouri Valley Tunnel, 67
Missouri Water Pollution Board (MWPB), 226, 239, 303n20
Missouri Waterway Commission, 180, 181
MRPWSA, 224, 232, 306n68
MRSC. See Missouri River Sanitary Conference
Muir, John, 103
Mumford, Lewis, 7
municipal housekeeping, 46, 68, 124, 225
Municipal Sanitation, 43, 165, 183
municipal services, 91, 93, 133, 279n72
Murray, Matthew, 95
Muse of the Missouri (fountain), 1, 251; photo of, 3
mussels, 186, 249
MVPHA, 185, 186
MWPB, 226, 239, 303n20

nasal relief, calls for, 77–81
National Environmental Policy Act, 230
National Fire Underwriters Board, 49–50, 58
National Geographic, 236
National Guard, 199, 203

National Research Council (NRC), 241–242, 243
National Water Works Company, 36–37; advertisement by, 36 (fig.)
Naturalization Committee, 109
navigation, 136, 180, 187, 195, 205, 233, 240; barge, 4, 5, 196, 210, 238; pollution and, 190
Naysmith, Clifford, 135, 283n11
Neff, George, 78
Negro Health Week, 157
Negro Hyde Park, 136, 148
Negro Leagues, 146
neighborhoods, 91, 100; African American, 134, 137 (photo), 152–153; segregated, 146; wealthy, 135
Neihardt, John, on Missouri River, 31
Nelson, William Rockhill, 43, 44, 45, 128
Nelson-Atkins Museum of Art, 44
New Deal, 9, 188, 190, 191, 195, 206; economic/social reform of, 96; impact of, 99; projects of, 96, 100, 195; Ten-Year Plan and, 91–100, 130
newspapers: African American, 30, 109; variety of, 27, 30
Nichols, J. C., 41, 42, 61, 80, 90, 146, 159, 237, 239, 274n125; Beach and, 286n60; developments of, 147; Plaza and, 99
Nixon, Richard, 228
North End, 47, 48, 49, 133, 136, 139, 150, 151; population of, 138
North Kansas City, Kansas, 30, 200; filtration plant in, 59; tunnel in, 40; waterworks in, 62, 251
North Kansas City Waterworks, 64; plans for, 63 (map)
NRC, 241–242, 243

Oahe Dam, 298n121
odors, 72, 77, 213, 222
Ohio River, 125, 179, 183, 191, 295n66
Ohio River Valley Water Sanitation Commission (ORSANCO), 191
Ohio State Board of Health, 125, 126, 191
Ohio Supreme Court, Bense Act and, 126
oil refinery tanks, 18
O.K. Creek, 76, 80; drainage problem and, 110; sewage in, 21, 73, 79, 110, 112; sewer dispute and, 106, 113, 116; Turkey Creek and, 112–113, 115, 117, 118
Oklahoma! (Rodgers and Hammerstein), 9
O.K.-Turkey Creek Sewer, 112
Old City Hospital, 142
Olmsted, Frederick Law, 44
Omaha, Nebraska, 185, 210, 219, 229, 235; typhoid fever in, 177; waste from, 177
Omaha World-Herald, 244
100 Reasons Why 100 Cities Approve Municipal Ownership of the Public Utilities (McDonnell), 37
One Hundred Yards War, 110–119
O'Neill, Karen, 247
"One–Kansas City Idea," 109
"On the House Roof—A Memory of the Kaw River" (Parkhurst), first stanza of, 23
Osage Paint and Varnish Company, 199
Our Negro Population (Martin), 138

packing district, 18, 106
packinghouses, 75, 200, 201, 229, 262n72
Paquin, Paul, 185, 186
park and boulevard system, 43–44
Parkhurst, Ellen G., 23, 256n17
parks, 43–44, 100, 129, 162; access to, 150; housing development and, 151; planning, 238; property values and, 84
Paseo, The, 151, 152
Patch, The, 29, 134, 136
Pearse, Herman, 85, 89
Pegracke, Chad, 246
Pendergast, Jim, 9, 45, 117
Pendergast, Tom, 9, 45, 66, 92, 93; concrete and, 74; power of, 96; vice and, 127
Pendergast machine, 9, 45, 57, 58–59, 62, 74, 92, 94–95, 99, 210; fall of, 161; river wards and, 133–134
Penn Valley Park, 84
People's Ice, 19
Perry, J. E., 157
Perry, J. W., 266n135
pesticides, 223, 224
petroleum, 203, 223

pharmaceuticals, 232, 275n136
Phillips Refinery, 200, 203
Pick, Lewis A., 195, 196, 197, 204, 220; flood of 1951 and, 205
Pick-Sloan Plan, 10, 204, 236, 237, 238, 244, 298n121; adoption of, 197, 205, 210; aftermath of, 239; boosters and, 195, 222; commitment to, 215; described, 195–196; impact of, 196, 221; Missouri Basin of, 166 (map); public health and, 219–222
Pinchot, Gifford, 180
pipes, 2, 39, 133; sewer, 7, 71, 74
Plaza, The, 99, 146, 239, 251
plumbing, 79, 89, 90, 149
pneumonia, deaths from, 159
polio, African Americans and, 159
political boundaries, 5, 24, 26, 105, 110, 276n17; breaking down, 106; regionalism and, 131; social geographies and, 8
political machines. See Pendergast machine
political turnover, 91–92
politics, 57–58, 91–92, 225; national, 170; progressive, 121–122; urban, 7; waterworks and, 66
pollution, 11, 22, 35, 56, 84, 101, 136, 175, 187, 190, 192, 203; air, 225; chemical, 211; controlling, 126, 169, 214–215, 217, 229, 281n96; environmental, 222; impact of, 248; industrial, 111, 210; knowledge of, 232; minimum requirements for, 184; navigation and, 190; new, 222–225; nonpoint source of, 223, 231, 232; point source of, 231; poverty and, 162; preventing, 212, 231; public health and, 185, 189; railroads and, 111; rural, 186; sewage, 53, 100, 185, 187; smoke, 140; total load, 190; urban, 105, 186; water, 56, 72–73, 124, 179, 184, 213, 223, 224, 303n27
poverty, 11, 22, 27, 48, 56, 154; demographics of, 24; disease and, 55, 155; pollution and, 162; racism and, 162
power: consolidating, 58; economic, 71, 236; federal, 214; political, 71;

relations, 132, 133; social, 71, 90–91, 236; socioeconomic, 162
Powers, Benjamin, 218
Prevailing Theories and Practices Relating to Sewage Disposal (Kiersted), 72
Prier Brass Manufacturing, 85–86
privies, 71, 90, 91; photo of, 137
Procter & Gamble, 107, 199, 304n40
Progressive Era, 4, 37, 48, 124, 134, 137, 161, 178, 180, 181, 183; fountains from, 1; growth in, 56; water pollution and, 72–73
Progressive Party, 41
Progressives, 145–146, 148
progressivism, 41, 43, 68, 78, 85, 88, 108, 170; public health and, 157
property taxes, 40, 79
prostitutes, 138, 139, 204; dismembered bodies of, 169; regulating, 127, 280n83
public health, 6, 11, 13, 14, 28, 37, 42–43, 50, 54, 58, 90, 105, 117, 119, 126, 150, 159, 180, 181, 182, 187, 210, 220, 230, 233; campaigns, 34, 47, 185; economic wealth and, 27; environment and, 169, 196, 217; importance of, 5, 113; improving, 49, 113, 160, 188, 217; investing in, 161; issues, 114, 176, 201–202; Pick-Sloan and, 219–222; planning, 125; pollution and, 185, 189; progressivism and, 157; protecting, 114, 185, 250; responsibility for, 186; sensual experience and, 77; sewage and, 72, 172, 174, 177; underfunding of, 52; undermining, 8
public space, 237, 248
public works, 37, 224
Public Works Administration, 95, 130, 191
public works departments, 92, 147, 160
"Pull Together Club, The," 148
pumping stations, 22, 38, 62
"pure ice" campaigns, 292n30
Pure Milk campaign, 48
purification, 59, 60, 184, 188, 193, 224
purifiers, advertisements for, 28
"Putting the Screws to Him" (cartoon), described, 121

Quality Hill, 135
Quantity Hill, 135, 136

344 | INDEX

Quindaro, 21, 36, 38, 39, 55; criticism of, 41, 60; expansion of, 59, 65, 66, 68, 108; reliance on, 62; settling basins at, 37, 53; waterworks in, 22, 40, 64, 68, 108, 202, 251, 267n157
Quindaro Bend, 38, 39
Quindaro Bottoms, 69
Quindaro waterworks, 22, 68, 107–108, 202

race, 6, 91, 148, 156; class and, 162; connections and, 133; infrastructure and, 163; power relation of, 132; prejudices of, 153; public rights and, 140–141
Race for the Rivers: Bringing People to the River, 247
race space, reform and, 142–145
racism, 5, 30, 55, 148, 151, 162
railroads, 3, 134, 139, 147; pollution and, 111; sacrifice zones for, 140
Ralph Sewer Law (1927), 92, 130
Randolph, Isham, 173
Ray, S. J., cartoon by, 95 (fig.)
Raytown, Missouri, 128, 226
real estate industry, 83, 93, 144, 146, 153
Red Cross, 158, 203
red-light zone, 138
Reed, James A., 19–20, 27
reform, 43, 187; campaigns, 42, 47; movement, 61, 162; progressive, 42, 45, 46; race space and, 142–145; social, 96, 168, 261n59; urban, 9; welfare, 159
Reform, 30
reformers, 9, 41, 61, 149, 155
refugees, 199, 206; African American, 24; help for, 23, 24; segregating, 24; shelter for, 23 (photo); smallpox and, 28; sympathy for, 24
regionalism, 105, 186; city services and, 127–131; environmental reality and, 186; planned, 127–131
regulation, 90, 142; government, 101–102
Relf, Claude, 218
rentals, 88, 89, 90, 135, 149
Report of the Board of Park and Boulevard Commissioners of Kansas City, Missouri, 83
Rhodes, Richard, 70

Rhodes, Stanley, 70
Richmond, Vicki, 14, 238, 242, 244, 246; cleanup by, 235; Healthy Rivers and, 23; photo of, 235
Riis, Jacob, 143
"Rime of the Ancient Mariner, The" (Coleridge), 28
Rising Sun, 30, 148; on race/public rights, 140–141
River Bluff Park, 238
Riverfront Heritage Trail, 244
Riverfront Park, 238, 240
River-Horse (Heat-Moon), 1
River Market, 248
River Miles (television program), 247
Rivers and Harbors Act (1899), 190
river wards, 133–142
Robinson, Jackie, 146
Rockhill District, 153
Rockwell, Norman, 205–206; painting by, 207 (fig.)
Roosevelt, Franklin D., 95
Roosevelt, Theodore, 180
Rosedale, Kansas, 110, 112, 113, 115, 116, 117, 128, 137, 138; annexation of, 281n88
Rothman, Hal, 231
Rothschild, Louis, 59, 62, 266n135
runoff, 71, 97, 221, 231, 249, 250

sacrifice zones, 139, 140, 283n21
Safe Drinking Water Act, 230
Saint Charles, Missouri, 247
Saint Joseph, Missouri, 185, 218, 228; waste from, 177, 178, 214, 219
Saint Louis, Missouri, 10, 218, 220, 221; solutions for, 292n27; treatment plants for, 229; waterworks, 173, 223
Saint Louis County Water Company, 220
Saint Louis World's Fair (1904), 174
Salmonella typhi, 176
saloons, 20, 133, 138
Salvation Army, 203
Sanborn Fire Insurance Company, maps, 136, 283n14, 285n48
sanitarians, 105, 169, 175, 184, 187, 189, 295n66

Sanitary and Ship Canal, 172, 291n19
sanitation, 6, 21, 34, 49, 68, 73, 91, 123, 176–178, 193, 216, 229; poor, 157; principles of, 122; standards for, 213
Santa Fe Railroad, 140
Schnarr, Steve, 247
Schneiders, Robert K., 194, 245, 300n39
Schultz, E. H., 180
Schuykill River, urban planning and, 84
sedimentation, 38, 53, 101, 184, 221, 267n153
segregation, 8, 91, 140–144, 146–147, 153, 159, 208, 236, 237, 253n4; class, 282n4; fighting, 158; health professionals and, 142; housing, 144; institutionalized, 141; law, 144; park, 151; racial, 198, 282n4; socioeconomic, 5; suburban, 209
septic tanks, 79, 90, 91
Serda, Daniel, 256n15
settlement houses, 47
settling basins, 37, 53, 54, 59, 62
sewage, 12, 35, 56, 70, 71, 74, 80, 87, 89, 106, 112, 122, 169, 232, 249; diluting, 73, 173, 175, 176, 189; disposal of, 72, 172, 177, 179, 185, 190; dumping, 181, 217, 225, 248; farming, 18; flooding and, 22; overflows, 119; pollution from, 53, 100, 185; public health and, 172, 177; treating, 113, 193; urban, 187–188
sewage plants. *See* treatment plants
Sewage Pollution of Interstate and International Waters with Special Reference to the Spread of Typhoid Fever: The Missouri River from Sioux City to Its Mouth, 183
sewerage, 5, 51, 81–89, 89–91, 96, 105, 110, 127, 129, 131, 152; agreement, 216; comprehensive, 78–79; cross-border, 111 (map); dilemmas with, 71–75, 77; emptying, 190; infrastructure and, 80; regional decisions on, 186
sewers, 12, 75, 78, 81, 89, 98–99, 100–101, 112, 117, 129, 131, 132, 133, 162, 191, 219, 279; access to, 92; as artifacts, 70–71; collapse of, 90; connections and, 163; construction of, 71, 74, 79; flushing, 77; importance of, 73–74;
municipal, 227; open, 79; public health and, 72; storm, 99; water service and, 74
Sheffield, Missouri, 30, 85, 86
Sheffield Car & Equipment, 86
showers, 48, 49; photo of, 48
Shuttle Creek, 127
Silent Spring (Carson), 211, 224–225
silt load, 221, 222
Sinclair, Upton, 143
Sinclair Oil, 107, 203, 219, 304n40
Sioux City, Iowa, 10, 177, 180, 182, 191, 194, 210, 219, 229, 235
Sloan, William, 195, 197
slum clearance, 208
slumlords, suburbanites and, 146–153
smallpox, 28
Smith, Elliot, 88
social classes, 83, 94, 131, 133
social history, 10, 236, 249
Social History of Missouri, A (painting), 96
social issues, 7–8, 13, 28–29, 46–47, 129, 157, 251
social outcomes, 145–146, 153
social units, 105; economic units and, 109
social uplift, 83, 150
social welfare, 142, 158
social workers, 41, 46, 47, 48; institutional barriers and, 158
socioeconomic conditions, 8, 13, 162, 248
Souls of Black Folk, The (Du Bois), 132
South Platte River, 213
Standard Oil, 148, 170
Standing Rock Sioux Tribe, 4, 298n121
Starkloff, Dr., 172
State of Missouri v. State of Illinois and the Sanitary District of Chicago, 169–170, 172–175
steamboats, 3, 12, 246
Steamboat Trace, 247
Stein, Murray, 14, 214, 217, 229, 230
Steinbeck, John, 10
stereotypes, 23, 140, 156, 157, 237
sterilization, 56–57, 263n95
Stewart-Peck Sand Company, 167
stockyards, 18, 75, 106, 200; photo of, 106
Stoeltzing, Amelia, 90

346 | INDEX

storm water, 70, 71, 72, 112, 249
Story of Kansas City, The (Cowick), 15
Stous, Dave, 246
Strong, Josiah, 266n140
Structural Steel, 199
Stubbs, W. R., 176
Sturdevant, Dan, 244
suburbanization, 128, 130, 146, 211
Summers, Clifford, 215
Sun, 145
Susan B. Anthony League, 46
swimming pools, 15, 48, 49, 156, 201
Swope, Thomas H., 54, 83, 263n88
Swope Park, 44, 84, 85, 86, 96, 97, 100; Blue River and, 83, 87, 88; housing development and, 151; segregation at, 151; visiting, 150–151
Swope Park Golf Club, 150
Swope Parkway, 149

taste problems, 35, 53, 213, 215, 216, 218, 221, 222, 224
taxation, 40, 79, 93, 100, 112, 116, 278n43
tenements, 22, 47, 136, 143, 144
Tennessee Valley Authority, 195
Ten-Year Plan, 160; bonds for, 95, 96, 97; cartoon about, 95 (fig.); civic improvement blueprint in, 99; New Deal and, 91–100, 130
There is No Wealth without Health trophy, 122, 123 (fig.)
Thomas, E. H., 28
Thornberry, John, 204, 299n30
Tomes, Nancy, 54
Topeka Owls, 198
topography, 3, 12, 74, 75, 78, 129
transportation, 19, 20, 38, 129, 147, 209; importance of, 210; railway, 37; river, 3–4
treatment, 56–57, 115, 124, 174, 181, 184, 193, 225–229, 267n153, 267n157; secondary, 228, 275n136; sewage, 72, 73, 81, 229, 231; softening, 60; typhoid fever and, 57
treatment plants, 12–13, 81, 100, 101, 116, 165, 191, 223, 228, 231; building, 188, 225, 226, 227, 282n99; photo of, 227

Troost Avenue, 136, 142
Truman, Bess, 198
Truman, Harry S., 129, 198, 204, 210
Truman, Margaret, 198
tuberculosis, 52, 56, 125, 263n90; African Americans and, 157, 159; bacilli of, 52; Mexicans and, 140
tunnels, 38, 40, 62, 63, 67, 112, 113, 116, 199, 200
turbidity, 220, 221, 222
Turkey Creek, 38, 62, 67, 80, 138, 201, 205; conditions in, 113, 118–119; described, 110; diverting, 112; drainage problem and, 110; flood and, 199, 200, 239; O.K. Creek and, 112–113, 115, 117, 118; One Hundred Yards War and, 110–119; pumping station, 21, 38, 62, 200; sewage in, 21, 35, 71, 73, 79, 114, 216; sewer dispute and, 106, 113, 118
Turkey Creek project, 106, 113, 115, 116–117; map of, 111; scandal over, 117–118
Turkey Creek Reservoir, 21, 37
Turkey Creek Tunnel, 67
Tuttle Creek Reservoir, 204, 205, 300n35
Twain, Mark, 38
typhoid fever, 6, 44, 52, 60, 157, 175, 184, 201, 232, 292n30; bacteria and, 47, 55, 192; battling, 53, 181, 186; carriers of, 55; concerns about, 187; deaths from, 28, 53, 159, 168 (fig.), 289n110; epidemics of, 28, 39; human-centric focus on, 189; rates of, 54, 56, 57, 85, 173, 174, 176, 295n82; shots for, 193; spread of, 55, 122, 172, 177; water treatment and, 57

Underground Railroad, 36
Union Station, 18, 74, 122, 145
University of Kansas, 124, 176
University of Kansas Medical School, 57, 121
Unruly River (Schneiders), 300n29
Unthank, T. C., 158
urban environment, 2, 5, 7, 45, 248; health and, 157
urban growth, 41, 134, 146, 248

urbanization, 35, 56, 161, 169, 175
Urban League, 141, 156
urban planning, 84, 248
US Army Corps of Engineers, 5, 10, 80, 96, 99, 111, 130, 173, 180, 193, 194, 195, 196, 203, 204, 213, 218, 222, 237, 240; Endangered Species Act and, 242; flooding and, 199; river work by, 192; sewage and, 190; Turkey Creek and, 116; water samples by, 192
US Bureau of Reclamation, 10, 195
US Constitution, 181, 214
US Fish and Wildlife Service (USFWS), 241, 242, 243
US Geological Survey, 101
US Navy, inoculations and, 201
US Public Health Service (USPHS), 27, 125, 178, 186, 191, 192, 212, 213, 214, 215, 217, 218, 219, 220, 221, 226, 228; monitoring by, 173; MRSC and, 183; report by, 182–183, 184
US Supreme Court, 40, 170, 173, 174, 208
US Surgeon General, 182
USFWS, 241, 242, 243
USPHS. *See* US Public Health Service

vaccines, 28, 29, 201
Van Brunt, John, 148
Van Valkenburgh, Arba, 64
vaults, 71, 90, 91; photo of, 137
Veatch, N. T., 65
venereal diseases, 122, 127
Veterans Administration, 190
vice district, 133, 138, 139
Volker, William, 62, 64, 261n59, 266n135

Waring, George E., Jr., 110, 152
Waring, R. P., 78
Washington, Booker T., 160
waste, 80, 92, 177, 217; industrial, 21, 22, 40, 70, 79, 186, 219, 229; inorganic, 223; management, 228; organic, 169, 223; public health and, 174; removal of, 91, 107, 172; social, 139; urban, 186
waste treatment plants. *See* treatment plants
water: access to, 47–48; bottled, 250; consumption of, 58, 79, 191, 196; distribution of, 5; flow of, 8, 77; interstate, 181; natural/ancient flow of, 77; privatization of, 250. *See also* drinking water
Water and Sewage Laboratory, 126, 295n66
Water and Sewage Law (1907), 123, 189
Water and Sewage Law (1927), 189
water companies, private, 35, 36, 130
Watermelon Hill, 151
water plants, 220; building, 33; first, 37; modern, 39; protecting, 198
Water Pollution Control Federation, 222, 229
water pressure, 41–50, 58
water quality, 12, 35–41, 50, 54, 125, 131, 184, 185, 218; agriculture and, 196; control over, 4, 115, 126, 179, 181, 186, 191, 233, 250; improving, 53, 59, 102, 115, 189, 232; issues, 127, 192, 215, 226; laws, 125, 229, 295n66; national policy for, 212; protecting, 53, 230; questions about, 192; studies on, 213; wildlife health and, 188–189
water quantity, 35–41
Watershed Councils, 247
watersheds, 56, 73, 105, 117; engineering, 118; interstate, 179; managing, 118; natural, 174; waste load of, 92
water supply, 55, 69, 129, 191, 258n16; consolidation of, 128; illnesses and, 55; inadequate, 41; protecting, 52; public, 224; risk for, 4; typhoid fever and, 55
waterways, 96, 101, 106, 115–116, 217, 250; engineering on, 98; floods on, 195; improving, 89, 102; natural, 174, 175; polluted, 53, 229; regulating, 191
Waterways Journal, 175
waterworks, 21, 47, 54, 173, 175, 184, 187, 191, 192, 211, 220, 221, 222, 223, 240, 250, 251; building, 49, 58–69; capacity, 128; first, 35, 36 (fig.), 244; flooding and, 68; fountain logo of, 7; hand-me-down, 39; increased consumption and, 71; intake of, 70; planning, 33; purchasing, 37; purification by, 53; side-by-side, 39–40

Wayne Miner housing, 209
wealth: demographics of, 24;
    development of, 196; health and, 11,
    42, 43, 169, 178–182; public, 71; social,
    168; touring, 106–110
welfare board, 48, 88, 143, 144, 148
Wendell, Mitchell, 229
West Bluffs, photo of, 154
West Bottoms, 23, 38, 113, 119, 122, 135,
    136, 248, 251; consumption in, 50;
    drainage in, 75; evacuation of, 22;
    flood in, 19 (photo), 199–200; pollution
    in, 111; population of, 256n15; sewer
    in, 75, 77, 129; wastewater treatment
    facility in, 12–13
Western College, 144, 152
Westport, Missouri, 79, 268n11
Westside, 144, 158

West Terrace Park, 201
wetlands, 5, 238, 241
Wheatley-Provident Hospital, 140
Whipple, George C., 178
White, William Allen, 94
Wixford, John, 174
Woman's City Club, 159
working classes, 6, 39, 85, 90, 109, 133,
    135; floods and, 20; homelessness
    and, 29
Works Progress Administration (WPA),
    95, 96, 97
Wyandotte County, Kansas, 40, 41, 110, 129

Yacht Club, 87, 88

zoning, 138, 152

www.ingramcontent.com/pod-product-compliance
Lightning Source LLC
Chambersburg PA
CBHW031753220426
43662CB00007B/390